Methods in Molecular Biotechnology

Methods in Molecular Biotechnology

Contributors

Tanya Chauhan, Kumar Rajiv et al.

www.aurisreference.com

Methods in Molecular Biotechnology

Contributors: Tanya Chauhan, Kumar Rajiv et al.

Published by Auris Reference Limited
www.aurisreference.com

United Kingdom

Copyright 2016
Printed in 2017 for Sale in the Indian Subcontinent

The information in this book has been obtained from highly regarded resources. The copyrights for individual articles remain with the authors, as indicated. All chapters are distributed under the terms of the Creative Commons Attribution License, which permit unrestricted use, distribution, and reproduction in any medium, provided the original author and source are credited.

Notice

Contributors, whose names have been given on the book cover, are not associated with the Publisher. The editors and the Publisher have attempted to trace the copyright holders of all material reproduced in this publication and apologise to copyright holders if permission has not been obtained. If any copyright holder has not been acknowledged, please write to us so we may rectify.

Reasonable efforts have been made to publish reliable data. The views articulated in the chapters are those of the individual contributors, and not necessarily those of the editors or the Publisher. Editors and/or the Publisher are not responsible for the accuracy of the information in the published chapters or consequences from their use. The Publisher accepts no responsibility for any damage or grievance to individual(s) or property arising out of the use of any material(s), instruction(s), methods or thoughts in the book.

Methods in Molecular Biotechnology

ISBN: 978-1-78154-965-0

British Library Cataloguing in Publication Data
A CIP record for this book is available from the British Library

Printed in the United Kingdom

Exclusively distributed by CBS Publishers & Distributors Pvt. Ltd.

Sales & Distribution Rights only for India, Pakistan, Bangladesh, Sri Lanka, Nepal and Bhutan. This book is not to be sold outside these territories.

Contents

List of Abbreviations vii
List of Contributors... ix
Preface.. xv

Chapter 1 Molecular Markers and their Applications in Fisheries and Aquaculture 1
Tanya Chauhan and Kumar Rajiv

Chapter 2 Assessing Spatial Genetic Structure From Molecular Marker Data Via Principal Component Analyses: A Case Study in a Prosopis Sp. Forest 27
Ingrid Teich, Aníbal Verga, and Mónica Balzarini

Chapter 3 Determining the Specific Status of The Iberian Sturgeons by Means Genetic Analyses of Old Specimens 49
Francisca Robles, Belén Cano-Roldán, Carmelo Ruiz Rejón, Luís Javier Martínez-González, María Jesús Álvarez-Cubero, José Antonio Lorente, José Antonio Riquelme Cantal, Pedro Aguayo de Hoyos, Javier Carrasco Rus, Miguel Cortés Sánchez, María Dolores Simón Vallejo, Manuel Ruiz Rejón, and Roberto de la Herrán

Chapter 4 The Efficacy of Molecular Markers Analysis with Integration of Sensory Methods in Detection of Aroma in Rice 67
H. Y. Yeap, G. Faruq, H. P. Zakaria, and J. A. Harikrishna

Chapter 5 Homogeneous Nature of Malaysian Marine Fish Epinephelus fuscoguttatus(Perciformes; Serranidae): Evidence Based on Molecular Markers, Morphology and Fourier Transform Infrared Analysis 79
A'wani Aziz Nurdalila, Hamidun Bunawan, Subbiah Vijay Kumar, Kenneth Francis Rodrigues, and Syarul Nataqain Baharum

Chapter 6 Synthesis, Molecular Docking and Biological Evaluation of Glycyrrhizin Analogs as Anticancer Agents Targeting EGFR 103
Yong-An Yang, Wen-Jian Tang, Xin Zhang, Ji-Wen Yuan, Xin-Hua Liu, and Hai-Liang Zhu

Chapter 7 Molecular Cloning of a Chitinase Gene from the Ovotestis of Kuroda's Sea Hare Aplysia kurodai 127
Gaku Matsunaga, Syuuji Karasuda, Ryo Nishino, Hideto Fukushima, and Masahiro Matsumiya

Chapter 8	Molecular Biology and Genetic Engineering 141	
Chapter 9	Molecular Cloning, Expression and Characterization of a Novel Geneβ-N-Acetylglucosaminidase From Bombyx Mori 241	
	Cheng Chang, Xiaoyong Liu, and Keping Chen	
Chapter 10	Teraherz Vibrational Spectroscopy of E. Coli and Molecular Constituents: Computational Modeling and Experiment 251	
	Tatiana Globus, Igor Sizov, and Boris Gelmont	
Chapter 11	Influence of Initial Molecular Substance on the Diffusion Flux Across Cell Membranes 275	
	Bum Joon Jung and Dae-Han Ki	
	Citations 293	
	Index 295	

List of Abbreviations

AFL	Anal Fin Length
AI	Artificial Insemination
ARS	Autonomously Replicating Sequence
ASA	Allele Specific Amplification
ATR	Attenuated Total Reflectance
BAC	Bacterial Artificial Chromosomes
BW	Body Weight
CBD	Chitin Binding Domain
CBD	Convention on Biological Diversity
CFL	Caudal Fin Length
DFL	Dorsal Fin Length
EAP	External Antisense Primer
EGFR	Epidermal Growth Factor Receptor
EM	Electro-Magnetic
ESP	External Sence Primer
EST	Expressed Sequence Tags
ET	Embryo Transfer
FL	Fork Length
FT	Fourier Transform
FTIR	Fourier-Transform Infra-Red
GFP	Green Fluorescent Protein
GH	Glycoside Hydrolase
GMO	Genetically Modified Organism
HL	Head Length
HRP	Horse Radish Peroxidase
IFAP	Internal Fragrant Antisence Primer
INSP	Internal Non-fragrant Sence Primer
MAS	Marker-Assisted Selection
MCS	Multiple Cloning Site
MD	Molecular Dynamics
ME	Molecular Evolution
ML	Mouth Length
MP	Maximum Parsimony
NJ	Neighbor-Joining
NT	Nuclear Transfer
ORF	Open Reading Frame
PC	Principal Components
PCA	Principal Component Analysis
PCR	Polymerase Chain Reaction
PFL	Pectoral Fin Length
PH	Partial Hepatectomy

PMF	Peptide Mass Fingerprinting
QTL	Quantitative Trait Loci
RFLP	Restriction Fragment Length Polymorphism
RNA	Ribonucleic Acid
RTPCR	Reverse Transcription Polymerase Chain Reaction
SCNT	Somatic Cell Nuclear Transfer
SGS	Spatial Genetic Structure
SL	Standard Length
SMG	Selectable Marker Genes
SNP	Single Nucleotide Polymorphism
SPC	Spatial Principal Component
SSR	Simple Sequence Repeat
TKR	Tyrosine Kinase Receptor
TL	Total Length
TMV	Tobacco Mosaic Virus
YAC	Yeast Artificial Chromosomes

List of Contributors

Tanya Chauhan
National Institute of Criminology and Forensic Science, Rohini, New Delhi, Delhi, India

Kumar Rajiv
Department of Chemistry, (SC), University of Delhi, Delhi, India.

Ingrid Teich
Statistics and Biometry, Faculty of Agricultural Sciences, National University of Córdoba-CONICET, Córdoba, Argentina

Aníbal Verga
Centro de Investigaciones Agropecuarias, Instituto Nacional de Tecnología Agropecuaria, Córdoba, Argentina

Mónica Balzarini
Statistics and Biometry, Faculty of Agricultural Sciences, National University of Córdoba-CONICET, Córdoba, Argentina

Francisca Robles
Departamento de Genética, Facultad de Ciencias, Universidad de Granada, Granada, Spain

Belén Cano-Roldán
Departamento de Genética, Facultad de Ciencias, Universidad de Granada, Granada, Spain

Carmelo Ruiz Rejón
Departamento de Genética, Facultad de Ciencias, Universidad de Granada, Granada, Spain

Luís Javier Martínez-González
Centro Pfizer, Universidad de Granada, Junta de Andalucía de Genómica e Investigación Oncológica, Centro de Investigación Biomédica, Av. del Conocimiento s/n, Armilla, Granada, Spain

María Jesús Álvarez-Cubero
Centro Pfizer, Universidad de Granada, Junta de Andalucía de Genómica e Investigación Oncológica, Centro de Investigación Biomédica, Av. del Conocimiento s/n, Armilla, Granada, Spain

José Antonio Lorente
Centro Pfizer, Universidad de Granada, Junta de Andalucía de Genómica e Investigación Oncológica, Centro de Investigación Biomédica, Av. del Conocimiento s/n, Armilla, Granada, Spain

José Antonio Riquelme Cantal
Consejería de Cultura, Junta de Andalucía, Sevilla, Spain

Pedro Aguayo de Hoyos
Departamento de Prehistoria y Arqueología. Facultad de Filosofía y Letras, Universidad de Granada, Granada, Spain

Javier Carrasco Rus
Departamento de Prehistoria y Arqueología. Facultad de Filosofía y Letras, Universidad de Granada, Granada, Spain

Miguel Cortés Sánchez
Faculdade de Ciências Humanas e Sociais, Universidade do Algarve, Faro, Portugal

María Dolores Simón Vallejo
Fundación Cueva de Nerja, Malaga, Spain.

Manuel Ruiz Rejón
Departamento de Genética, Facultad de Ciencias, Universidad de Granada, Granada, Spain

Roberto de la Herrán
Departamento de Genética, Facultad de Ciencias, Universidad de Granada, Granada, Spain

H. Y. Yeap
Institute of Biological Sciences, Faculty of Science, University of Malaya, 50603 Kuala Lumpur, Malaysia

G. Faruq
Institute of Biological Sciences, Faculty of Science, University of Malaya, 50603 Kuala Lumpur, Malaysia

H. P. Zakaria
Institute of Biological Sciences, Faculty of Science, University of Malaya, 50603 Kuala Lumpur, Malaysia

J. A. Harikrishna
Institute of Biological Sciences, Faculty of Science, University of Malaya, 50603 Kuala Lumpur, Malaysia
Centre for Research in Biotechnology for Agriculture, University of Malaya, 50603 Kuala Lumpur, Malaysia

A'wani Aziz Nurdalila
Institute of Systems Biology, Universiti Kebangsaan Malaysia, UKM Bangi, 43600 Selangor, Malaysia

Hamidun Bunawan
Institute of Systems Biology, Universiti Kebangsaan Malaysia, UKM Bangi, 43600 Selangor, Malaysia

Subbiah Vijay Kumar
Biotechnology Research Institute, Universiti Malaysia Sabah, Jalan UMS, 88400 Kota Kinabalu Sabah, Malaysia

Kenneth Francis Rodrigues
Biotechnology Research Institute, Universiti Malaysia Sabah, Jalan UMS, 88400 Kota Kinabalu Sabah, Malaysia

Syarul Nataqain Baharum
Institute of Systems Biology, Universiti Kebangsaan Malaysia, UKM Bangi, 43600 Selangor, Malaysia

Yong-An Yang
State Key Laboratory of Pharmaceutical Biotechnology, Nanjing University, Nanjing 210093, China

Wen-Jian Tang
School of Pharmacy, Anhui Medical University, Hefei 230032, China

Xin Zhang
State Key Laboratory of Pharmaceutical Biotechnology, Nanjing University, Nanjing 210093, China

Ji-Wen Yuan
State Key Laboratory of Pharmaceutical Biotechnology, Nanjing University, Nanjing 210093, China

Xin-Hua Liu
State Key Laboratory of Pharmaceutical Biotechnology, Nanjing University, Nanjing 210093, China
School of Pharmacy, Anhui Medical University, Hefei 230032, China

Hai-Liang Zhu
State Key Laboratory of Pharmaceutical Biotechnology, Nanjing University, Nanjing 210093, China

Gaku Matsunaga
Department of Marine Science and Resources, College of Bioresource Sciences, Nihon University, Kanagawa, Japan

Syuuji Karasuda
Department of Marine Science and Resources, College of Bioresource Sciences, Nihon University, Kanagawa, Japan

Ryo Nishino
Department of Marine Science and Resources, College of Bioresource Sciences, Nihon University, Kanagawa, Japan

Hideto Fukushima
Department of Marine Science and Resources, College of Bioresource Sciences, Nihon University, Kanagawa, Japan

Masahiro Matsumiya
Department of Marine Science and Resources, College of Bioresource Sciences, Nihon University, Kanagawa, Japan

Bum Joon Jung
Department of Biomedical Engineering, Rensselaer Polytechnic Institute, New York, USA
Proton Therapy Center, National Cancer Center, Goyang, South Korea

Dae-Han Ki
Proton Therapy Center, National Cancer Center, Goyang, South Korea

Cheng Chang
Institute of Life Sciences, Jiangsu University, Zhenjiang, China.

Xiaoyong Liu
Institute of Life Sciences, Jiangsu University, Zhenjiang, China.

Keping Chen
Institute of Life Sciences, Jiangsu University, Zhenjiang, China.

Tatiana Globus
Department of Electrical and Computer Engineering, University of Virginia, Charlottesville, USA
Vibratess, LLC, Charlottesville, USA

Igor Sizov
Department of Electrical and Computer Engineering, University of Virginia, Charlottesville, USA
Vibratess, LLC, Charlottesville, USA

Boris Gelmont
Department of Electrical and Computer Engineering, University of Virginia, Charlottesville, USA
Vibratess, LLC, Charlottesville, USA

Preface

The text *Methods in Molecular Biotechnology* explores contemporary techniques and applications of molecular biotechnology. Molecular biotechnology is the use of laboratory techniques to study and modify nucleic acids and proteins for applications in areas such as human and animal health, agriculture, and the environment. First chapter deals with molecular markers and their applications in fisheries and aquaculture. In second chapter, we clarify the use of principal component analysis (PCA) to tackle the study of spatial genetic patterns from molecular marker data. The aim of third chapter is to determine the specific status of the Iberian sturgeons by means genetic analysis of old specimens. In fourth chapter, we evaluate the efficacy of molecular markers and integration of sensory methods with molecular markers for the detection of aroma in different rice genotypes. In fifth chapter, we use the 16s rRNA, cytochrome oxidase subunit I (COI) and III (COIII) genes as molecular markers to study the Malaysian marine fishes *Epinephelus fuscoguttatus and Epinephelus hexagonatus. Epinephelus fuscoguttatus*. Synthesis, molecular docking, and biological evaluation of glycyrrhizin analogs as anticancer agents targeting EGFR have been focused in sixth chapter. In seventh chapter, we report that we successfully cloned and sequenced a chitinase gene from the ovotestis of Kuroda's sea hare Aplysia kurodai. Eighth chapter focuses on molecular biology and genetic engineering. Ninth chapter deals with molecular cloning, expression and characterization of a novel gene β-N-acetylglucosaminidase from Bombyx mori. In tenth chapter, we present the results of research of E. coli cells and cellular components, DNA and protein thioredoxin, using highly resolved sub-Terahertz (THz) vibrational spectroscopy. The influence of initial molecular substance on the diffusion flux across cell membranes has been investigated in last chapter.

Chapter 1

MOLECULAR MARKERS AND THEIR APPLICATIONS IN FISHERIES AND AQUACULTURE

Tanya Chauhan[1] and Kumar Rajiv[2]

[1]National Institute of Criminology and Forensic Science, Rohini, New Delhi, Delhi, India

[2]Department of Chemistry, (SC), University of Delhi, Delhi, India.

ABSTRACT

Genetic variation in a species enhances the capability of organism to adapt to changing environment and is necessary for survival of the species. Genetic variation arises between individuals leading to differentiation at the level of population, species and higher order taxonomic groups. The genetic diversity data has varied application in research on evolution, conservation and management of natural resources and genetic improvement programs, etc. Development of Molecular genetic markers has powerful ability to detect genetic studies of individuals, populations or species. These molecular markers combined with new statistical developments have revolutionized the analytical power, necessary to explore the genetic diversity. Molecular markers and their statistical analysis revolutionized the analytical power, necessary to explore the genetic diversity. Various molecular markers, protein or DNA (mt-DNA or nuclear DNA such as microsatellites, SNP or RAPD) are now being used in fisheries and aquaculture. These markers provide various scientific observations which have importance in aquaculture practice recently such as: 1) Species Identification 2) Genetic variation and population structure study in natural populations 3) Comparison between wild and hatchery populations 4) Assessment of demographic bottleneck in natural population 5) Propagation assisted rehabilitation programmes. In this review article, we have concentrated on the basics of molecular genetics, overview of commonly used markers and their application along with their limitations (major classes of markers) in fisheries and aquaculture studies.

INTRODUCTION

All organisms are subject to mutations because of normal cellular operations or interactions with the environment, leading to genetic variation (polymorphism). Genetic variation in a species enhances the capability of organism to adapt to changing environment and is necessary for survival of the species [1]. In conjunction with other evolutionary forces like selection and genetic drift, genetic variation arises between individuals leading to differentiation at the level of population, species and higher order taxonomic groups. Molecular genetic markers are powerful tools to detect genetic uniqueness of individuals, populations or species [2,3]. These markers have revolutionized the analytical power, necessary to explore the genetic diversity [4]. The conclusion from genetic diversity data has varied application in research on evolution, conservation and management of natural resources and genetic improvement programmes, etc [5-10]

In addition to protein markers, application of DNA markers is finding wide acceptance in population genetics. With DNA markers, it is theoretically possible to observe and exploit genetic variation in the entire genome. Both genomic and mitochondrial DNA is used for varied applications. The commonly used technique are allozyme analysis, types of restriction fragment length polymorphism (RFLP), randomly amplified polymorphic DNA (RAPD), amplified fragment length polymorphism (AFLP), microsatellite typing, single nucleotide polymorphism (SNP), and expressed sequence tag (EST) markers, etc.

Molecular markers can be classified into type I and type II markers. Type I markers are associated with genes of known function, while type II markers are associated with anonymous genomic regions [11]. Under this classification, allozyme markers are type I markers because the protein they encode has known function. RAPD markers are type II markers because RAPD bands are amplified from anonymous genomic regions via the polymerase chain reaction (PCR). Microsatellite markers are also type II markers unless they are associated with genes of known function. The significance of type I markers is becoming extremely important for aquaculture genetics. Type I markers serve as a bridge for comparison and transfer of genomic information from a map-rich species into a relatively map-poor species. In general, type II markers such as RAPDs, microsatellites, and AFLPs are considered non-coding and therefore selectively neutral. Such markers have found widespread use in population genetic studies to characterize genetic divergence within and among the populations or species [12]

ALLOZYME MARKERS

Analysis of allozyme loci remained one of the most popular approaches in examining population genetics and stock structure questions in fishes [13]. The technique is rapid, relatively inexpensive and provides an independent estimate of level of variation within a population without an extensive morphological and quantitative survey [14]. Isohyets are structurally different molecular forms of an enzyme system with qualitatively the same catalytic function encoded by one or more loci [15]. Isohyets, which are encoded by different alleles of the same gene locus, are designated as "allozymes" or "alloenzymes" [16]. Amino acid differences in the polypeptide chain of the different allelic forms of an enzyme reflect changes in the underlying DNA sequence. Depending on the nature of the amino acid changes, the resulting protein products may migrate at different rates (due to charge and size differences) when run through a gel subjected to an electrical field. Differences in the relative frequencies of alleles are used to quantify genetic variation and distinguish among genetic units at the levels of populations, species, and higher taxonomic designations. Disadvantages associated with allozymes include occasional heterozygote deficiencies due to null (enzymatically inactive) alleles and sensitive to the amount as well as quality of tissue samples. In addition, some changes in DNA sequence are masked at the protein level, reducing the level of detectable variation. Some changes in nucleotide sequence do not change the encoded polypeptide (silent substitutions), and some polypeptide changes do not alter the mobility of the protein in an electrophoretic gel (synonymous substitutions). At present 75 isozyme systems representing several hundred genetic loci are known [17]. With the strength as codominant marker, ease of use, and low cost, the allozyme markers are popular in population structure and phylogenetic studies, though has limited role in aquaculture genetics.

MITOCHONDRIAL DNA MARKERS

Mitochondrial DNA (mtDNA) analysis is being increasingly used in recent population and phylogenetic surveys of organisms. Studies of vertebrate species generally have shown that sequence divergence accumulates more rapidly in mitochondrial than in nuclear DNA [18]. This has been attributed to a faster mutation rate in mtDNA that may result from a lack of repair mechanisms during replication [19] and smaller effective population size due to the strict maternal inheritance of the haploid mitochondrial genome [20]. Due to its rapid rate of evolution, mtDNA analysis has proven useful in clarifying relationships among closely related species. Different parts of the mitochondrial genome are known to evolve at different rates [21]. Almost the entire mtDNA molecule is transcribed except for the approximately 1-kb control region (D-loop), where replication and transcription of the molecule is initiated. In general, non-coding segments like the D-loop exhibit elevated levels of variation relative to coding sequences such as the cytochrome b gene [22], presumably due to reduced functional constraints and relaxed selection pressure. The 16S rRNA gene in the mitochondrial genome is one of the slowest evolving genes [21] whereas rapidly evolving regions are control regions [23,24]. Due to non-Mendelian mode of inheritance, the mtDNA molecule is considered as a single locus [2]. In addition, because mtDNA is maternally inherited, the phylogenies and population structures derived from mtDNA data may not reflect complete picture of the nuclear genome if gender-biased migration or selection [20] or introgression [25] exists.

Analyses of mtDNA markers have been used extensively to investigate stock structure in a variety of vertebrates including fishes [26-30], birds [31-34], mammals [35] and reptiles [36-39].

RANDOM AMPLIFIED POLYMORPHIC DNA (RAPD) MARKERS

RAPD markers are the amplified products of less functional part of the genome that do not strongly respond to selection on the phenotypic level. Such DNA regions may accumulate more nucleotide mutations with potential to assess inter-population genetic differentiation [40]. The amplification of genomic DNA by PCR with arbitrary nucleotide sequence primers, RAPD can detect high levels of DNA polymorphisms [41,42]. The technique detects coding as well as non-coding DNA sequences, and many of the most informative polymorphic sequences are those derived from repetitive (non-coding) DNA sequences in the genome [43]. Because 90% of the vertebrate nuclear genome is non-coding, it is presumed that most of the amplified loci will be selectively neutral. RAPD loci are inherited as Mendelian markers in a dominant fashion and scored as present/absent. RAPDs have all the advantages of a PCR-based marker, with the added benefit that primers are commercially available and do not require prior knowledge of the target DNA sequence or genome organization. Other advantages of RAPDs include the ease with which a large number of loci and individuals can be screened simultaneously. Shortcomings of this type of marker include the difficulty of demonstrating Mendelian inheritance of the loci and the inability to distinguish between homozygotes and heterozygotes. Analysis follows the assumption that populations under study follow Hardy-Weinberg expectations. In addition, the presence of paralogous PCR product (different DNA regions which have the same lengths and thus appear to be a single locus), low reproducibility due to the low annealing temperature used in the PCR amplification, have limited the application of this marker in fisheries science [44].

SINGLE NUCLEOTIDE POLYMORPHISM (SNP)

Single nucleotide polymorphism (SNP) describes polymorphisms caused by point mutations that give rise to different alleles containing alternative bases at a given nucleotide position within a locus. SNPs are becoming a focal point in molecular marker development since they represent the most abundant polymorphism in any organism's genome (coding and non-coding regions), adaptable to automation, and reveal hidden polymorphism not detected with other markers and methods [9, 10]. Theoretically, a SNP within a locus can produce as many as two alleles, each containing one of two possible base pairs at the SNP site. Therefore, SNPs have been regarded as bi-allelic. SNP markers are inherited as co-dominant markers. Several approaches have been used for SNP discovery including SSCP analysis [45], heteroduplex analysis, and direct DNA sequencing. DNA sequencing has been the most accurate and most used approach for SNP discovery. SNPs are not without their limitations, however, might provide marginal additional, or even less, utility in some applications (e.g. relatedness) [9].

MICROSATELLITE MARKERS

Microsatellites consist of multiple copies of tandemly arranged simple sequence repeats (SSRs) that range in size from 1 to 6 base pairs [e.g., ACA or GATA; 46,47]. Abundant in all species studied to date, microsatellite motifs have been estimated to occur as often as once every 10 kb in fishes [48]. Microsatellites tend to be evenly distributed in the genome on all chromosomes and all regions of the chromosome. However, data from whole genome sequencing has somewhat contradicted this statement. They have been found inside gene coding regions [49], introns, and in the non-gene sequences. Most microsatellite loci are relatively small, ranging from a few to a few hundred repeats. Regardless of specific mechanisms, changes in numbers of repeat units can result in a large number of alleles at each microsatellite locus in a population. Microsatellites have been inherited in a Mendelian fashion as codominant markers. Microsatellites were found to be informative in several species, which showed almost no variation at other markers [50]. However, use of microsatellite markers involves a large amount of up-front investment and effort. Each microsatellite locus has to be identified and its flanking region sequenced to design of PCR primers. Due to polymerase slippage during replication, small size differences between alleles of a given microsatellite locus (as little as 2 bp in a locus comprised of di-nucleotide repeats) are possible. Microsatellites recently have become an extremely popular marker type in a wide variety of genetic investigations.

NEW DEVELOPING MARKERS IN FISHERIES AND AQUACULTURE

Various type of DNA markers have been developed, including Allozymes, microsatellites, RAPDs, mt-DNA and SNPs. These markers in fish populations have revealed high levels of genetic variation distributed throughout the fish genome. A recent initiative has been made to accelerate efforts of DNA marker development, genome mapping and species identification. Major progress has been made toward Expressed Sequence Tags (EST) and DNA barcode development in several aquaculture species.

EXPRESSED SEQUENCE TAGS (ESTS)

Expressed sequence tags (ESTs) are single-pass sequences generated from random sequencing of cDNA clones [51]. The EST is use to identify genes and analyze their expression by means of expression profiling. It helps for rapid and valuable analysis of genes expressed in specific tissue types, under specific physiological conditions, or during specific developmental stages. ESTs offer the development of cDNA microarrays that allow analysis of differentially expressed genes to be determined in a systematic way [52], in addition to their great value in genome mapping [53]. For genome mapping, ESTs are most useful for linkage mapping and physical mapping in animal genomics such as those of cattle and swine, where radiation hybrid panels are available for mapping non-polymorphic DNA markers [54]. A radiation panel is composed of lines of hybrid cells, with each hybrid cell containing small fragments of irradiated chromosomes of the species of interest. Typically, the cells from species of interest are radiated to break chromosomes into small fragments. The radiated cells are unable to survive by themselves. However, the radiated cells can be fused with recipient cells to form hybrid cells retaining a short segment of the radiated chromosome. Characterization of the chromosomal break points within many hybrid cell lines would allow linkage and physical mapping of markers and genes. In spite of its popularity in mammalian genome mapping [55, 56], radiation hybrid panels are not yet available for any aquaculture species. Development of radiation hybrid panels from aquaculture species is not expected in the near future, given the fact that physical mapping using BAC libraries can provide even higher resolution and the fact that BAC libraries are already available from several aquaculture species. Therefore, ESTs are useful for mapping in aquaculture species only if polymorphic ESTs are identified [57]. The value of EST resources and applications of bioinformatics in aquaculture genetics/genomics is inevitable, and it is expected that various EST databases will serve as rich sources of genomic information not only for aquaculture geneticists, but also for aquaculture physiologists, immunologists and biotechnologists.

DNA BARCODING

The principle of conservation biology is the preservation and management of biodiversity. The two major problems to such an endeavor are the difficulty of developing an assessment of this diversity for prioritization of hotspots of species richness [58] and the identification of lineages particularly worthy, or in need, of preservation [59-64]. Understudied taxa are greatly susceptible to extinction [65], suggesting there is a conservation penalty for our ignorance. Even there are millions of unidentified and unknown species [66]. DNA barcodes, segments of approximately 600 base pairs of the mitochondrial gene cytochrome oxidase I (COI), have been proposed as a fast, efficient, and inexpensive technique to catalogue all biodiversity [67-70]. Barcoding is the use of universal polymerase chain reaction (PCR) primers to amplify and sequence an approximately 600-basepair fragment of the COI gene. That portion of sequence is then compared using distance-based algorithms with an existing database of "known" sequences from specimens previously identified by taxonomists. DNA barcodes from a small portion of the mitochondrial genome might seem like an effective and rapid way to assess at least some, perhaps minimal, level of biodiversity. And for groups that are already relatively well known, especially birds and mammals, molecular studies based on barcode sized sequences have revealed cryptic DNA lineages and may be helpful [70].

APPLICATION OF MOLECULAR MARKERS SPECIES IDENTIFICATION

The inter-specific genetic divergence established through species specific diagnostic molecular markers provides precise knowledge on phylogenetic relationships and also resolve taxonomic ambiguities [71-74]. These markers can be used to detect hybrid and introgressed or backcrossed individuals [75], distinguish early life history stage of morphologically close species [76] both in hatchery and in natural populations.

Species-specific allozyme markers have been identified in many fishes [Tilapia: 72,77,78; Sciaenid: 73; Anguilla sp: 79; Mugilidae: 80] Specific diagnostic allozyme loci were used for different species: apache trout (Oncorhynchus apache), cutthroat (Oncorhynchus clarki) and rainbow trout (Oncorhynchus mykiss) [81] and Gambusia affinis and G. holbrooki [82]. Allozyme markers have also been used for individual classification in cyprinid species Zacco pachycephalus and Z. platypus [83], in cyprinodontid species V. letourneuxi and V. hispanica [84], in mullets Mullus barbatus and M. surmuletus [85] and hake species Merluccius australis and M. hubbsi [86].

Species-specific diagnostic RAPD fingerprints were generated in several fish species and their taxonomic relationship has been analyzed. The RAPD-PCR technique was employed to identify three endemic morphologically similar Spanish species of Barbus: Barbus bocagei, B. graellsii and B. sclateri that have similar morphologies [87]. RAPD markers were characterized to identify five species of family Cyprinidae: Chondrostoma lemmingii, Leuciscus pyrenaicus, Barbus bocagei, Barbus comizo, all endemic in the Iberian Peninsula, and introduced Alburnus alburnus [88], for studying genetic relationship and diversities in four species of Indian Major carps (family Cyprinidae): rohu (Labeo rohita), kalbasu (L. calbasu), catla (Catla catla) and mrigal (Cirrhinus mrigala) [89], for identification of three eel species, A. japonica, A. australis and A. bicolor [90] and to estimate the population structure and phylogenetic relationships among the eight species of the genus Barbus [88].

Large variation in mtDNA sequences among species can be utilized to produce species-specific markers. Since the structures of mitochondrial RNA genes (tRNA and rRNA) and the functional molecule of the 16S rRNA are highly conserved among the animal taxa that are related even distantly [21], change of even few nucleotides in such a gene between closely related taxa might indicate a substantial degree of genetic divergence [2]. Mt-DNA sequences have been used as useful marker for species-specific identification in many fishes [Tuna: 91; Billfish: 92 Snappers: 29, 93; Myctophidae: 94; Grey mullets: 95]. Comparable levels of divergence based on 12S rRNA and 16S rRNA sequences have been reported for several recently diverged fish species [genus Sternoptyx: 96; Cyclothone sp: 97]. Sequence variation in the control region (D-loop) of the mitochondrial DNA (mtDNA) was examined to assess the genetic distinctiveness of the short-jaw cisco, Coregonus zenithicus [98] and revealed high similarity of C. zenithicus and the related species C. artedi, C. hoyi, C. kiyi, and C. clupeaformis Identification of Astyanax altiparanae (Teleostei, Characidae) in the Iguacu River, Brazil, was done on the basis of mitochondrial DNA and RAPD markers [99]. Two species, Acipenser baeri, and A. stellatus, was studied using mitochondrial DNA (D-loop, cytochrome b (cyt-b) and ND5/6 genes) sequencing to determine whether traditionally defined subspecies correspond to taxonomic entities and conservation management units [100].

GENETIC VARIATION AND POPULATION STRUCTURE STUDY IN NATURAL POPULATIONS

Molecular markers provide direct assessment of pattern and distribution of genetic variation [5] thus helping in answering, "if the population is single unit or composed of subunits". Several evolutionary forces affect the amount and distribution of genetic variation among populations and thereby population differentiation [101]. Geographic distance and physical barriers enhance reproductive isolation by limiting the migration and increase genetic differentiation between populations [102]. Impact of migration and gene flow on genetic differentiation also depends upon effective size of receiving population and number of migrants. Increased computational power and mathematical models have enhanced the scope of conclusions that can be drawn out of genotype data generated through molecular markers. Some of the possibilities are assignment of migrants [103], determination of genetic bottlenecks [104], effective breeding population estimates [105] besides genetic variation and differentiation estimations [106-108]. These markers have been extensively employed across various taxonomic groups [mosquito: 109; turtle: 39; amphibians: 7; panda: 110; five vertebrate classes including fish, amphibian, reptiles, birds and mammals: 6]. Experiments on fish populations have significantly contributed towards development of science of population genetics, models and analytical softwares.

Population genetic structure has been investigated using allozyme markers in many fish species [Oncorhynchus gorbuscha: 111; Tenualosa ilisha: 112 and Lal et al., 113; Pagrus auratus: 114].

Fifteen random primers were used to analyze the genome DNA of Jian carp (Cyprinus carpio var jian) by the RAPD technique [115]. Study on cold tolerant traits for common carp Cyprinus carpio was conducted by Chang et al. [116] and nine RAPD-PCR markers associated with cold tolerance of common carp were identified. The genetic diversity has been studied using RAPD markers in Carassius auratus [117], Epinephelus merra population [118] and Solea solea [119].

Genetic variation have been assessed with Allozyme and RAPD markers on Mullus surmuletus L., [120] and three species of Pimelodidae catfish [121].

Population structure has been examined using microsatellite markers of sockeye salmon [122], Chinook salmon [123] and Arctic charr populations [124]. Genetic variation have been assessed using microsatellite genetic markers to identify the population structure of brook charr, Salvelinus fontinalis [125] and 14 populations of northern pike (Esox lucius) in the North Central United States and in six populations from Quebec, Alaska, Siberia, and Finland [126].

Based on five microsatellite loci, the genetic structure of endangered fish species Anaecypris hispanica was studied in eight distinct populations in the Portuguese Guadiana drainage to determine levels of genetic variation within and among populations and suggested implications for conservation of the species [127].

Combination of allozyme and microsatellites was used to investigate genetic divergence in Salmo trutta [128] and Salmo salar [129].

Alarcon et al. [130] represents population genetic analysis of gilthead sea bream (Sparus aurata) and Kanda [131], Kanda and Allendorf [132] examine population genetic structure of bull trout Salvelinus confluentus using a combination of allozyme, microsatellite and mtDNA variation.

Genetic variability of Salmo trutta [133] and Sparus aurata [130] was evaluated on the basis of Allozyme, Microsatellites and RAPD markers.

Patterns of population subdivision and the relationship between gene flow and geographical distance in the tropical estuarine fish Lates calcarifer (Centropomidae) were investigated using mtDNA control region sequences [134].

Allozymes and mtDNA sequences were assessed to evaluate the genetic variability in small marine fish Pomatoschistus microps [135], brown trout [136] and Macquaria novemaculeata [137].

COMPARISON OF GENETIC VARIATION BETWEEN WILD AND HATCHERY POPULATIONS

Molecular markers also find application in aquaculture to assess loss of genetic variation in hatcheries through, comparison of variation estimates between hatchery stocks and wild counterparts. The information is useful obtained in monitoring farmed stocks against inbreeding loss and to plan genetic up gradation programmes. A major aspect such studies address is concerned with the assessment of farm escapes into the natural population and introgression of wild genome.

Brook trout Salvelinus fontinalis from ustocked waters, naturalized lakes, and hatcheries in New York and Pennsylvania were analyzed electrophoretically for allozyme expression [138]. All wild-unstocked samples were highly differentiated populations and significantly different from each other and from hatchery samples.

Genetic diversity was investigated using microsatellites between farmed and wild populations of Atlantic salmon [139]. Farmed salmon showed less genetic variability than natural source population in terms of allelic diversity.

Variation in allozymes and three microsatellite loci was assessed in populations of wild and cultured stocks of Sparus aurata [140] and Sparius auratus [130]. The microsatellite heterozygosity values were high in wild, but lower in the cultured samples.

ASSESSMENT OF DEMOGRAPHIC BOTTLENECK IN NATURAL POPULATION

Demographic bottlenecks occur when populations experience severe, temporary reduction in size. Because bottlenecks may influence the distribution of genetic variation within and among populations, the genetic effects of reductions in population size have been studied extensively by evolutionary biologist [141,142].

It may often be necessary to perform genetic analyses of temporal replicates to estimate the significance of spatial variation independently from that of temporal variation in order to ensure the reliability of estimates of a defined population structure. Such estimates provide understanding about changes in genetic variation, effective population size and other historical bottlenecks and can be extrapolated to define evolutionary trends of species. Today various models are available that can resolve bottlenecks or effective population size changes through use of heterozygosity excess, linkage disequilibrium etc. However, estimates through temporal changes are considered more accurate. Analysis of temporal changes is limited due to lack of historical data as well as samples. Therefore, such studies are limited and mostly use archived samples, wherever available. In vertebrates, a limited number of studies have specifically assessed the temporal changes in genetic variation for more than one generation.

Microsatellite DNA markers have been used to assess bottlenecks in many fish species. A microsatellite analysis of DNA was performed, from archived scales to compare the population structure among four sympatric landlocked populations of Atlantic salmon [143], Atlantic salmon [144], European hake [145] and steelhead from [146].

Larson et al. [147] recommended close monitoring of negative effects on sea otter population based on the conclusion from mtDNA, D-loop, microsatellite variability comparison between prefur trade and present population. Prefur trade DNA samples were obtained from excavated bones.

PROPOGATION ASSISTED REHABILITATION PROGRAMMES

Habitat alterations and over harvesting have contributed to the decline or disappearance of numerous natural populations. In addition, reinforcement programs of wild populations based on releases of hatchery reared fish of non-native origin compromise the conservation of remnant native trout resources. Effect of these programmes through releases in natural populations has been studies in many fishes through molecular markers.

Beaudou et al. [148] found through allozyme polymorphism that brown trout (Salmo trutta L.) in the Abatesco river basin on the eastern coast of Corsica restoration was mainly due to the populations of the tributaries, which had been less disturbed by the spate. This study has shown that the wild population was primarily restored by the surviving individuals, particularly those from the tributaries that escaped the spate.

To assess the levels of gene introgression from cultured to wild brown trout populations, four officially stocked locations and four non-stocked locations were sampled for one to three consecutive years and compared to the hatchery strain used for stocking. Allozyme analysis for 25 loci included providing allelic markers distinguishing hatchery stocks and native populations [133]. Different levels of hybridization and introgression with hatchery individuals were detected in stocked drainages as well as in protected locations.

The foregoing review incorporates the wide spectrum of information that the molecular markers provide. The literature indicates that different markers have been employed depending upon the question to be answered. The importance of the research on molecular markers improved due to enhanced computational power, large data available that has enabled researchers to derive various mathematical estimators. Such innovations provide insight concerning the population bottleneck, migration patterns besides the genetic structure in natural populations.

REFERENCES

1. Fisher, R.A. (1930) The Genetical Theory of Natural Selection. Oxford University Press, UK.
2. Avise, J.C. (1994) Molecular Markers, Natural History and Evolution. Chapman and Hall, New York, London.
3. Linda, K.P. and Paul, M. (1995) Developments in molecular genetic techniques in fisheries. In: G.R. Carvalho and T.J. Pitcher, Eds., Molecular Genetics in Fisheries, Chapman and hall, London, 1-28.

4. Hillis, D.M., Mable, B.K. and Moritz, C. (1996) Applications of molecular systematics: The state of the field and a look to the future. In: Hillis, D.M., Moritz, C. and Mable, B.K. Eds., Molecular systematics, Sinauer Associates, Massachusetts, 515-543.
5. Ferguson, A., Taggart, J.B., Prodohl, P.A., McMeel, O., Thompson, C., Stone, C., McGinnity, P. and Hynes, R.A. (1995) The application of molecular markers to the study and conservation of fish populations with special reference to Salmo. Journal of Fish Biology, 47(A), 103-126.
6. Neff, B.D. and Gross, M.R. (2001) Microsatellite evolution in vertebrates: Inference from AC dinucleotide repeats. Evolution, 55(9), 1717-1733.
7. Jehle, R. and Arntzen, J.W. (2002) Microsatellite markers in amphibian conservation genetics. Herpetological Journal, 12, 1-9.
8. Wasko, A.P., Martins, C., Oliveira, C. and Foresti, F. (2003) Non-destructive genetic sampling in fish. An improved method for DNA extraction from fish fins and scales. Hereditas, 138(3), 161-165.
9. Morin, P.A., Luikart, G., Wayne, R.K. and the SNP working group, SNPs in ecology, evolution and conservation. Trends in Ecology and Evolution, 19(4), 208-216.
10. Liu, Z.J. and Cordes, J.F. (2004) DNA marker technologies and their applications in aquaculture genetics. Aquaculture, 238, 1-37.
11. O'Brien, S.J. (1991) Molecular genome mapping: lessons and prospects. Current Opinion in Genetic Development, 1(1), 105-111.
12. Brown, B. and Epifanio, J. (2003) Nuclear DNA. In: Hallermann, E.M. Ed., Population Genetics: Principles and Applications for Fisheries Scientists, American Fisheries Society, Bethesda, 458-472.
13. Suneetha, B.K. (2000) Interspecific and inter specific genetic variation in selected mesopelagic fishes with emphasis on microgeographic variation and species characterization. Dr. Scient. Dissertation, Department of Fisheries and Marine Biology, University of Bergen, Bergen, Norway.
14. Menezes, M.R., Naik, S. and Martins, M. (1993) Genetic characterization in four sciaenid species from the Arabian Sea. Journal of Fish Biology, 43(1), 61-67.
15. Markert, C.L. and Moller, F. (1959) Multiple forms of enzymes: Tissue, ontogenetic and species-specific patterns. Proceedings of the Naionall Academy of Science (USA), 45(5), 753-763.
16. Starck, M.G. (1998) Isozymes in Molecular tools for screening biodiversity. In: Angela, K., Peter, G.I. and David, S.I. Eds., Chapmann and Hall, London, 75-80.

17. Murphy, R.W., Sites, J.J.W., Buth, D.G. and Haufler, C.H. (1996) Proteins I: Isozyme electrophoresis. In: Hillis, D.M., Moritz, C. and Mable, B.K. Eds., Molecular Systematics, Sinauer Associates, Sunderland, 51-132.
18. Brown, W.M. (1985) The mitochondrial genome of animals. In: MacIntyre, R.J. Ed., Molecular Evolutionary Genetics, Plenum, New York, 95-130.
19. Wilson, A.C., Cann, R.L., Carr, S.M., George, M., Gyllensten, U.B., Helm-Bychowski, K.M., Higuchi, R.G., Palumbi, S.R. and Prager, E.M. (1985) Mitochondrial DNA and two perspectives on evolutionary genetics. Biological Journal of Linnean Society, 26(4), 375-400.
20. Birky, C.W., Fuerst, P. and Maruyama, T. (1989) Organelle gene diversity under migration, mutation, and drift: equilibrium expectations, approach to equilibrium, effect of heteroplasmic cells, and comparison to nuclear genes. Genetics, 121(3), 613-627.
21. Meyer, A. (1993) Evolution of mitochondrial DNA in fishes. In: Mochachka, P.W. and Mommsen, T.P. Eds., Biochemistry and molecular biology of fishes, Elsevier Press Amsterdam, New York, 1-38.
22. Brown, J.R., Bechenbach, A.T. and Smith, M.J. (1993) Intraspecific DNA sequence variation of the mitochondrial control region of white sturgeon (Acipenser transmontanus). Molecular Biology Evolution, 10(2), 326-341.
23. Chow, S., Okamoto, H., Uozumi, Y., Takeuchi, Y. and Takeyama, H. (1997) Genetic stock structure of the swordfish (Xiphias gladius) inferred by PCR-RFLP analysis of the mitochondrial DNA control region. Marine Biology, 127(3), 359-367.
24. Gold, J.R. Sun, F. and Richardson, L.R. (1997) Population structure of red snapper from the Gulf of Mexico as inferred from analysis of mitochondrial DNA. Transaction of American Fisheries Society, 126(3), 386-396.
25. Chow, S. and Kishino, H. (1995) Phylogenetic relationships between tuna species of the genus Thunnus (Scombriidae: Teleosrei): Inconsistent implications from morphology, nuclear and mitochondrial genomes. Journal of Molecular Evolution, 41, 741-748.
26. Avise, J.C., Helfman, G.S., Saunders, N.C. and Hales, L.S. (1986) Mitochondrial DNA differentiation in North Atlantic eels: Population genetic consequences of an unusual life history pattern. Proceeding of the National Academy Science (USA), 83(12), 4350-4354.
27. Graves, J.E., McDowell, J.R. and Jones, M.L. (1992) A genetic analysis of weakfish Cynoscion regalis stock structure along the mid-Atlantic

coast. Fisheries Bulletin, 90, 469-475.

28. Gold, J.R., Richardson, L.R., Furman, C. and King, T.L. (1993) Mitochondrial DNA differentiation and population structure in red drum (Sciaenops ocellatus) from the Gulf of Mexico and Atlantic Ocean. Marine Biology, 116(2), 175-185.

29. Chow, S., Clarke, M.E. and Walsh, P.J. (1993) PCR-RFLP analysis of thirteen western Atlantic snappers (subfamily Lutjaninae): A simple method for species and stock identification. Fisheries Bulletin (US), 91, 619-627.

30. Heist, E.J. and Gold, J.R. (1999) Microsatellite DNA variation in sandbar sharks (Carcharhinus plumbeus) from the Gulf of Mexico and mid-Atlantic bight. Copeia, 1, 182-186.

31. Baker, A.J. and Marshall, H.D. (1997) Molecular evolution of the mitochondrial genome. In: Mindell, D.P. Ed., Avian Molecular Evolution and Systematics, Academic Press, San Diego, 51-82.

32. Greenberg, R., Cordero, P.J., Droege, S. and Fleischer, R.C. (1998) Morphological adaptation with no mitochondrial DNA differentiation in the coastal plain swamp sparrow. Journal of American Ornithologists' Union, 115(3), 706-712.

33. Mila, B., Girman, D.J., Kimura, M. and Smith, T.B. (2000) Genetic evidence for the effect of a postglacial population expansion on the phylogeography of a North American songbird. Proceedings of the Royal Society London Series B, 267(1447), 1033-1040

34. Zink, R.M., Barrowclough, G.F., Atwood, J.L. and Blackwell-Rago, R.C. (2000) Genetics, taxonomy, and conservation of the threatened California gnatcatcher. Conservation Biology, 14(5), 1394-1405.

35. Menotti-Raymond, M. and O'Brien, S.J. (1993) Dating the genetic bottleneck of the African cheetah. Proceedings of the National Academy of Science (USA.), 90(8), 3172-3176.

36. Avise, J.C., Walker, D. and Johns, G.C. (1998) Speciation durations and Pleistocene effects on vertebrate phylogeography. Proceedings of the Royal Society of London Series B, 265(1407), 1707-1712.

37. Serb, J.M., Phillips, C.A. and Iverson, J.B. (2001) Molecular phylogeny and biogeography of Kinosternon flavenscens based on complete mitochondrial control region sequences. Molecular Phylogenetics and Evolution, 18(1), 149-162.

38. Riberon, A., Sotiriou, E., Miaud, C., Andreone, F. and Taberlet, P. (2002) Lack of genetic diversity in Salamandra lanzai revealed by cytochrome b gene sequences. Copeia, 2002, 229-232.

39. Shanker, K., Ramadevi, J., Choudhaury, B.C., Singh, L. and Aggarawal, R.K. (2004) Phylogeny of olive ridley turtles (Lepidochelys olivacea) on the east coast of India: implications for conservation theory. Moelcular Ecology, 13(7), 1899-1909.

40. Mamuris, Z., Sfougaris, A.I., Stamatis, C. and Suchentrunk, F. (2002) Assessment of genetic structure of Greek Brown Hare (Lepus europeaus) populations based on variation in Random Amplified Polymorphic DNA (RAPD). Biochemical Genetics, 40(9-10), 323- 338.

41. Williams, J.G.K., Kubelik, A.R., Livak, K.J., Rafalski, J.A. and Tingey, S.V. (1990) DNA polymorphisms amplified by arbitrary primers are useful as genetic markers. Nucleic Acids Research, 18(22), 6531-6535.

42. Welsh, J. and McClelland, M. (1990) Fingerprinting genomes using PCR with arbitrary primers. Nucleic Acids Research, 18(24), 7213-7218.

43. Haymer, D.S. (1994) Random amplified polymorphic DNAs and microsatellites: What are they, and can they tell us anything we don't already know? Annals of Entomological Society of American, 87, 717- 722.

44. Wirgin, I.I. and Waldman, J.R. (1994) What DNA can do for you? Fisheries, 19, 16-27.

45. Hecker, K.H., Taylor, P.D. and Gjerde, D.T. (1999) Mutation detection by denaturing DNA chromatography using fluorescently labeled polymerase chain reaction products. Analytical Biochemistry, 272(2), 156-164.

46. Tautz, D. (1989) Hypervariability of simple sequences as a general source for polymorphic DNA markers. Nucleic Acids Research, 17(16), 6463-6471.

47. Litt, M. and Luty, J.A. (1989) A hypervariable microsatellite revealed by in-vitro amplification of dinucleotide repeat within the cardiac muscle actin gene. American of Journal of Human Genetics, 44(3), 397-401.

48. Wright, J.M. (1993) DNA fingerprinting in fishes. In: W. Hochachka, P. and Mommsen, T. Eds., Biochemistry and Molecular Biology of Fishes, Elsevier, Amsterdam, 58-91.

49. Liu, Z.J., Li, P., Kocabas, A., Ju, Z., Karsi, A., Cao, D. and Patterson, A. (2001) Microsatellite-containing genes from the channel catfish brain: evidence of trinucleotide repeat expansion in the coding region of nucleotide excision repair gene RAD23B. Biochemical Biophysical Research Communication, 289(2), 317-324.

50. Taylor, A.C., Sherwin, W.B. and Wayne, R.K. (1994) Genetic variation of microsatellite loci in a bottlenecked species: The northern hairy-nosed

wombat Lasiorhinus krefftii. Molecular Ecology, 3(4), 277-290.

51. Adams, M.D., Kelley, J.M., Gocayne, J.D., Dubnick, M., Polymeropoulos, M.H., Xiao, H., Merril, C.R., Wu, A., Olde, B., Moreno, R.F., Kerlavage, A.R., McCombie, W.R. and Venter, J.C. (1991) Complementary DNA sequencing: Expressed sequence tags and human genome project. Science, 252(5013), 1651-1656.

52. Wang, K., Gan, L., Jeffry, E., Gayle, M., Gown, A.M., Skelly, M., Nelson, P.S., Ng, W.V., Schummer, M., Hood, L. and Mulligan, J. (1999) Monotoring gene expression profile changes in ovarian carcinomas using cDNA microarray. Gene, 229(1-2), 101-108.

53. Boguski, M.S. and Schuler, G.D. (1995) Establishing a human transcript map. Nature Genetics, 10(4), 369-371.

54. Cox, D.R., Burmeister, M., Price, E., Kim, S. and Myers, R.M. (1990) Radiation hybrid mapping: a somatic cell genetic method for constructing high-resolution map of mammalian chromosomes. Science, 250(4978), 245-250.

55. Korwin-Kossakowska, A., Reed, K.M., Pelak, C., Krause, E., Morrison, L. and Alexander, L.J. (2002) Radiation hybrid mapping of 118 new porcine microsatellites. Animal Genetics, 33(3), 224-227.

56. McCoard, S.A., Fahrenkrug, S.C., Alexander, L.J., Freking, B.A., Rohrer, G.A., Wise, T.H. and Ford, J.J. (2002) An integrated comparative map of the porcine X chromosome. Animal Genetics, 33(3), 178-185.

57. Liu, Z.J., Karsi, A. and Dunham, R.A. (1999) Development of polymorphic EST markers suitable for geneticlinkage mapping of catfish. Marine Biotechnology, 1(5), 437-447.

58. Dobson, A.P., Rodriguez, J.P., Roberts, W.M. and Wilcove, D.S. (1997) Geographic distribution of endangered species in the United States. Science, 275(5299), 550- 555.

59. Daugherty, C.H., Cree, A., Hay, J.M. and Thompson, M.B. (1990) Neglected taxonomy and continuing extinctions of tuatara (Sphenodon). Nature, 347(6289), 177- 179.

60. Faith, D.P. (1994) Genetic diversity and taxonomic priorities for conservation. Biological Conservation, 68(1), 69-74.

61. Crozier, R.H. (1997) Preserving the information content of species: Genetic diversity, phylogeny, and conservation worth. Annual Review of Ecology and Systematics, 28(1), 243-268.

62. Haig, S.M. (1998) Molecular contributions to conservation. Ecology, 79(2), 413-425.

63. Soltis, P.S. and Gitzendanner, M.A. (1999) Molecular systematics and the conservation of rare species. Conservation Biology, 13(3), 471-483.
64. Moritz, C. (2002) Strategies to protect diversity and the evolutionary processes that sustain it. Systematic Biology, 51(2), 238-254.
65. McKinney, M.L. (1999) High rates of extinction and threat in poorly studied taxa. Conservation Biology, 13(6), 1273-1281.
66. Novotny, V., Basset, Y., Miller, S.E., Weiblen, G.D., Bremer, B., Cizek, L. and Drozd, P. (2002) Low host specificity of herbivorous insects in a tropical forest. Nature, 416, 841-844.
67. Hebert, P.D.N., Ratnasingham, S. and deWaard, J.R. (2003) Barcoding animal life: Cytochrome c oxidase subunit 1 divergences among closely related species. Royal Society London, 270 (Suppl 1), S96-S99.
68. Stoeckle, M. (2003) Taxonomy, DNA, and the bar code of life. BioScience, 53(9), 2-3.
69. Stoeckle, M., Janzen, D., Hallwachs, W., Hanken, J. and Baker, J. (2003) Taxonomy, DNA, and the barcode of life," Draft conference report. Rubinoff DNA Barcodes and Conservation Barcode Conference, The Rockefeller University, New York, http://phe.rockefeller.edu/Barcode Confeence/docs/B2summary.doc
70. Hebert, P.D.N., Stoeckle, M.Y., Zemlak, T.S. and Francis, C.M., (2004) Identification of birds through DNA barcodes. PLoS Biology, 2(10), 312-316.
71. Rocha-Olivares, A., Moser, H.G. and Stannard, J. (2000) Molecular identification and description of pelagic young of the rockfishes Sebastes constellatus and Sebastes ensifer. Fisheries Bulletin, 98, 353-363.
72. Backer, J., Bentzen, P. and Moran, P. (2002) Molecular markers distinguish coastal cutthroat trout from coastal rainbow trout/steelhead and their hybrids. Transaction of American Fisheries Society, 131(3), 404-417.
73. Asensio, L., Gonzalez, I., Fernandez, A., Rodriguez, M.A., Lobo, E., Hernandez, P.E., Garcia, T. and Martin, R. (2002) Application of random amplified polymorphic DNA (RAPD) analysis for identification of grouper (Epinephelus guaza), wreckfish (Polyprion americanus), and nile perch (Lates niloticus) fillets. Journal of Food Product, 65(2), 432-435.
74. Rasmussen, C., Ostberg, C.O., Clifton, D.R., Holloway, J.L. and Rodriguez, R.J. (2003) Identification of a genetic marker that discriminates ocean-type and stream-type Chinook salmon in the Columbia River Basin. Transaction American Fisheries Society, 132(1), 131-142.

75. Compton, D. and Utter, F.M. (1985) Natural hybridization between steelhead trout (Salmo gairdneri) and coastal cutthroat trout (Salmo clarki clarki) in two Puget sound streams. Canadian Journal of Fisheries and Aquatic Science, 42, 110-119.

76. Olivar, M.P., Moser, H.G. and Beckley, L.E. (1999) Lantern fish larvae from the Agulhas current (SW Indian Ocean). Science of Marine, 63, 101-120.

77. McAndrew, B.J. and Majumdar, K.C. (1983) Tilapia stock identification using electrophoretic markers. Aquaculture, 30(1-4), 249-261.

78. Bartly, D.M. and Gall, G.A.E. (1991) Genetic identification of native cutthroat trout (Onchorynchus clarki) and introgressive hybridization with introduced rainbow trout (O. mykiss) in streams associated with the Alvord basin, Oregon and Nevada. Copeia, 3, 854-859.

79. Lee, S.C., Tosi, S.C.M., Cheng, H.L. and Chang, J.T. (1997) Identification of Anguilla Japonica and A. marmorata elvers by allozyme electrophoresis. Journal of Fish Biology, 51(1), 208-210.

80. Rossi, A.R., Capula, M., Crosetti, D., Sola, L. and Campton, D.E. (1998) Allozyme variation in global populations of striped mullet, Mugil cephalus (Pisces: Mugilidae). Marine Biololgy, 131(2), 203-212.

81. Carmichael, G.J., Hanson, J.N., Schmidt, M.E. and Morizot, D.C. (1993) Introgression among apache, cutthroat and rainbow trout in Ariozona. Transaction of American Fisheries Society, 122, 121-130

82. Wooten, M.C. and Lydeard, C. (1990) Allozyme variation in a natural contact zone between Gambusia affinis and Gambusia holbrooki. Biochemical Systematics and Ecology, 18(2-3), 169-173.

83. Wang-Hurng, Y., Lee, S.C. and Yu, M.J. (1997) Genetic evidence to clarify the systematic status of the genra Zacco and Candidia (Cypriniformes: Cprinidae). Zoological Studies, 36(3), 170-177.

84. Perdices, A., McChordom, A. and Doadrio, I. (1996) Allozyme variation and relationships of the endangered cyprinodontid genus Valencia and its implications for conservation. Journal of Fish Biology, 49(6), 1112-1127.

85. Mamuris, Z., Apostolidis, A.P. and Triantaphyllidis, C. (1998) Genetic protein variation in red mullet (Mullus barbatus) and striped red mullet (M. surmuletus) populations from the Mediterranean sea. Marine Biology, 130(3), 353-360.

86. Roldán, M.I. and Pla, C. (2001) Species identification of two sympatric hakes by allozyme markers. Science of Marine, 65, 81-84.

87. Callejas, C. and Ochando, M.D. (1998) Identification of Spanish barbel

species using the RAPD technique. Journal of Fish Biology, 53(1), 208-215.

88. Callejas, C. and Ochando, M.D. (2002) Phylogenetic relationships among Spanish Barbus species (Pisces, Cyprinidae) shown by RAPD markers. Heredity, 89(1), 36- 43.

89. Barman, H.K., Barat, A., Yadav, B.M., Banerjee, S., Meher, P.K., Reddy, P.V.G.K. and Jana, R.K. (2003) Genetic variation between four species of Indian major carps as revealed by random amplified polymorphic DNA assay. Aquaculture, 217(1-4), 115-123.

90. Takagi, M. and Taniguchi, N. (1995) Random amplified polymorphic DNA (RAPD) for identification of three species of Anguilla, A. japonica, A. australis and A. bicor. Fisheries Science, 61, 884-885.

91. Chow, S. and Inoue, S. (1993) Intra-and interspecific restriction fragment length polymorphism in mitochondrial genes of Thunnus tuna species. Bulletin of National Research Institute of Far Seas Fisheries, 30, 229-248.

92. Finnerty, J.R. and Block, B.A. (1992) Direct sequencing of mitochondrial DNA detects highly divergent haplotypes in blue marlin (Makaira nigricans). Molecular Marine Biology and Biotechnology, 1(3), 206-214.

93. Hare, J.A., Crown, R.K., Zehr, J.P., Juanes, F. and Day, K.H. (1998) A correction to: biological and oceanographic insights from larval labrid (Pisces: Labridae) identification using mtDNA sequences. Marine Biology, 130(4), 589-592.

94. Suneetha, B.K. and Dahle, G. (2000) Analysis of mitochondrial DNA sequences from Benthosema glaciale and two other myctophids (Pisces: Myctophiformes): Intraand interspecific genetic variation. Interspecific and inter specific genetic variation in selected mesopelagic fishes with emphasis on microgeographic variation and species characterization, University of Bergen, Bergen, Norway.

95. Murgia, R., Tola, G., Archer, S.N., Vallerga, S. and Hirano, J. (2002) Genetic identification of grey mullet species (Mugilidae) by analysis of mitochondrial DNA sequence: application to identify the origin of processed ovary products (bottarga). Marine Biotechnology, 4(2), 119-126.

96. Miya, M. and Nishida, M. (1998) Molecular phylogeny and evolution of the deep-sea fish genus Sternoptyx. Molecular Phylogenetics and Evolution, 10(1), 11-22.

97. Miya, M.and Nishida, M. (1996) Molecular phylogenetic perspective on the evolution of the deep-sea fish genus Cyclothone (Stomiiformes:

Gonostomatidae). Ichthyolgical Research, 43(4), 375-398.
98. Reed, K.M., Dorschner, M.O., Todd, T.N. and Phillips, R.B. (1998) Sequence analysis of the mitochondrial DNA control region of ciscoes (genus Coregonus): Taxonomic implications for the Great Lake species flock. Molecular Ecology, 7(9), 1091-1096.
99. Prioli, S.M.A.P., Prioli, A.J. and Julio, H.F.J. (2002) Identification of Astyanax altiparanae (Teleostei, Characidae) in the Iguacu River, Brazil, based on mitochondrial DNA and RAPD markers. Genetics and Molecular Biology, 25(4), 421-430.
100. Doukakis, P., Birstein, V.J., Ruban, G.I. and Desalle, R. (1999) Molecular genetic analysis among subspecies of two Eurasian sturgeon species, Acipenser baerii and A. stellatus. Molecular Ecology, 8(12 Suppl 1), 117-127.
101. Felsenstein, J. (1985) Confidence limits on phylogenies: An approach using the bootstrap. Evolution, 39, 783-791.
102. Ryman, N., (2002) Population genetic structure. NOAA Technical Memoranda, Northwest Fisheries Science Centre Publication Page,http://www.nwfsc.noaa.gov/publications/tecmemos/index.cfm
103. Piry, S., Alapetite, A., Cornuet, J.M., Paetkau, D., Baudoiin, L. and Estoup, A. (2004) Geneclass2: A software for genetic assignment and first generation migrant detection. Journal of Heredity, 95(6), 536-539.
104. Luikart, G. and Cornuet, J.M. (1998) Empirical evaluation of a test for identifying recently bottlenecked populations from allele frequency data. Conservation Biology, 12(1), 228-237.
105. Luikart, G. and Cornuet, J.M. (1999) Estimating the effective number of breeders from heterozygote excess in progeny. Genetics, 151(3), 1211-1216.
106. Weir, B.S. and Cockerham, C.C. (1984) Estimating F-statistics for the analysis of population structure. Evolution, 38(6), 1358-1370.
107. Nei, M. (1983) Genetic Polymorphism and the role of mutation in evolution. In: Nei, M. and Kohen, R.K. Eds., Evolution of Gene and Proteins, Sinaver Associates, Sunderlans, 165-190.
108. Raymond, M. and Rouseet, F. (1995) An exact test for population differentiation. Evolution, 49, 1280-1283.
109. Fonseca, D.M., Dennis, A., Pointe, L. and Fleischer, C. (2000) Bottlenecks and multiple introductions: Population genetics of the vector of avian malaria in Hawaii. Molecular Ecology, 9(11), 1803-1814.
110. Zhang, Y.P., Wang, X.X., Ryder, O.A., Li, M.P., Zhang, Y. Yong, H.M.

and Wang, P.Y. (2002) Genetic diversity and conservation of endangered animal species. Pure Applied Chemestry, 74(4), 575-584.

111. Efremov, V.V., (2002) Allozyme Variation in Pink Salmon Oncorhynchus gorbuscha from Sakhalin Island. Journal of Ichthyology, 42, 339-347.

112. Salini, J.P., Milton, D.A., Rahaman, M.J. and Hussein, M.G. (2004) Allozyme and morphological variation throughout the geographic range of the tropical shad, hilsa Tenualosa ilisha. Fisheries Research, 66(1), 53-69.

113. Lal, K.K., Kumar, D., Srivastava, S.K., Mukherjee, A., Mohindra, V., Prakash, S., Sinha, M. and Ponniah, A.G. (2004) Genetic variation in Hilsa shad (Tenualosa ilisha) population in River Ganges. Indian Journal of Fisheries, 51, 33-42.

114. Meggs, L.B., Austin, C.M. and Coutin, P.C. (2003) Low allozyme variation in snapper, Pagrus auratus, in Victoria, Australia. Fisheries Management and Ecology, 10(3), 155-162.

115. Dong, Z., Zhu, J., Yuan, X. and Wang, J. (2002) RAPD analysis of the genome DNA of Jian carp. Journal of Zhanjiang Ocean University, 22, 3-6.

116. Chang, Y., Sun, Y. and Liang, A. (2003) Study on cold tolerant traits for common carp Cyprinus carpio. Journal of Shanghi Fisheries University, 12, 102-105.

117. Feng, J., Zhang, X., Zhou, X., Chen, J. and Wang, L. (2003) RAPD markers and genetic diversity of Carassius auratus in the Qihe River. Transaction of Oceanology and Limnology, 4, 90-94.

118. Zheng, L. and Liu, C. (2002) Studies on random amplified polymorphic DNA (RAPD) of Epinephelus merra Bloch. Journal of Zhanjiang Ocean University, 22, 14- 18.

119. Exadactylos, A., Geffen, A.J., Panagiotaki, P. and Thorpe, J.P. (2003) Population structure of Dover sole Solea solea: RAPD and allozyme data indicate divergence in European stocks. Marine Ecollogy Progress Series, 246, 253-264.

120. Mamuris, Z., Stamatis, C. and Triantaphyllidis, C. (1999) Intraspecific genetic variation of striped red mullet (Mullus surmuletus L.) in the Mediterranean Sea assessed by allozyme and random amplified polymorphic DNA (RAPD) analysis. Heredity, 83(pt1), 30-38.

121. Almeida, F.S., Fungaro, M.H.P. and Sodré, L.M.K. RAPD and isoenzyme analysis of genetic variability in three allied species of catfishes (Siluriformes: Pimelodidae) from the Tibagi river, Brazil. Journal of

Zoology, 253, 113-120.

122. Nelson, R.J., Wood, C.C., Cooper, G., Smith, C. and Koop, B. (2003) Population structure of sockeye salmon of the central coast of British Columbia: Implications for recovery planning. North American Journal of Fisheries Management, 23, 703-720.

123. Beacham, T.D., Supernault, K.J., Wetklo, M., Deagle, B., Labaree, K., Irvine, J.R., Candy, J.R., Miller, K.M., Nelson, R.J. and Withler, R.E. (2003) The geographic basis for population structure in Fraser River chinook salmon (Oncorhynchus tshawytscha). Fisheries Bulletin, 101, 229-242.

124. Brunner, P.C., Douglas, M.R. and Bernatchez, L. (1998) Microsatellite and mitochondrial DNA assessment of population structure and stocking effects in Arctic charr Salvelinus alpinus (Teleostei: Salmonidae) from central Alpine lakes. Molecular Ecology, 7, 209-223.

125. Adams, B.K. and Hutchings, J.A. (2003) Microgeographic population structure of brook charr: A comparison of microsatellite and mark-recapture. Journal of Fish Biology, 62(3), 517-533.

126. Senanan, W. and Kapuscinski, A.R. (2000) Genetic relationships among populations of northern pike (Esox lucius). Canadian Journal Fisheries Aquatic Science, 57, 391-404.

127. Salgueiro, P., Carvalho, G., Collares-Pereira, M.J. and Coelho, M.M. (2003) Microsatellite analysis of genetic population structure of the endangered cyprinid Anaecypris hispanica in Portugal: Implications for conservation. Biological Conservation, 109(1), 47-56.

128. Palm, S., Dannewitz, J., Jaervi, T., Petersson, E., Prestegaard, T. and Ryman, N. (2003) Lack of molecular genetic divergence between sea-ranched and wild sea trout (Salmo trutta). Molecular Ecology, 12(8), 2057- 2071.

129. Elliott, N.G. and Reilly, A. (2003) Likelihood of bottleneck event in the history of the Australian population of Atlantic salmon (Salmo salar L.). Aquaculture, 215(1-4), 31-44.

130. Alarcon, J.A., Magoulas, A., Georgakopoulos, T., Zouros, E. and Alvarez, M.C. (2004) Genetic comparison of wild and cultivated European populations of the Gilthead Sea bream (Sparus aurata). Aquaculture, 230(1-4), 65-80.

131. Kanda, N. (1998) Genetics and conservation of Bull trout: Comparison of population genetic structure among different genetic markers and hybridization with brook trout. Dissertation Abstracts International Part B: Science and Engineering, University of Montana, Missoula.

132. Kanda, N. and Allendorf, F.W. (2001) Genetic population structure of Bull trout from the Flathead River basin as shown by microsatellite and mitochondrial DNA marker. Transaction of American Fisheries Society, 130, 92-106.

133. Cagigas, M.E., Vazquez, E., Blanco, G. and Sanchez, J.A. (1999) Genetic effects of introduced hatchery stocks on indigenous brown trout (Salmo trutta L.) populations in Spain. Ecological of Freshwater Fish, 8(3), 141-150.

134. Chenoweth, S.F., Hughes, J.M., Keenan, C.P. and Shane, L. (1998) Concordance between dispersal and mitochondrial gene flow: Isolation by distance in a tropical teleost, Lates calcarifer (Australian barramundi). Heredity, 80, 1897-1907.

135. Gysels, E.S., Hellemans, B., Pampoulie, C. and Volckaert, F.A. (2004) Phylogeography of the common goby, Pomatoschistus microps, with particular emphasis on the colonization of the Mediterranean and the North Sea. Molecular Ecololgy, 13(2), 403-417.

136. Marzano, F.N., Corradi, N., Papa, R., Tagliavini, J. and Gandolfi, G. (2003) Molecular evidence for introgression and loss of genetic variability in Salmo (trutta) macrostigma as a result of massive restocking of Apennine populations (Northern and Central Italy). Environmental Biology of Fishes, 68(4), 349-356.

137. Jerry, D.R. (1997) Population genetic structure of the catadromous Australian bass from through out its range. Journal of Fish Biology, 51(5), 909-920.

138. Perkins, D.L. and Krueger, C.C. (1993) Heritage brook trout in northeastern USA: Genetic variability within and among populations. Transaction of American Fisheries Society, 122, 1515-1532.

139. Norris, A.T., Bradley, D.G. and Cunningham, E.P. (1999) Microsatellite genetic variation between and within farmed and wild Atlantic salmon (Salmo salar) populations. Aquaculture, 180(3-4), 247-264.

140. Palma, J., Alarcon, J.A., Alvarez, C., Zouros, E., Magoulas, A. and Andrade, J.P. (2001) Developmental stability and genetic heterozygosity in wild and cultured stocks of gilthead sea bream. Journal of Marine Biological Association of United Kingdom, 81(2), 283-288.

141. Wright, S. (1931) Evolution in Mendelian populations. Genetics, 16(2), 97-159.

142. Nei, M., Maruyama, T. and Chakraborty, R. (1975) The bottleneck effect and genetic variability in populations. Evolution, 29(1), 1-10.

143. Tessier, N. and Bernatchez, L. (1999) Stability of population structure and genetic diversity across generations assessed by microsatellites among sympatric populations of landlocked Atlantic salmon (Salmo salar L.). Molecular Ecology, 8, 169-179.

144. Nielsen, E.E., Hansen, M.M. and Loeschcke, V. (1999) Genetic variation in time and space: Microsatellite analysis of extinct and extant populations of Atlantic salmon. Evolution, 53, 261-268.

145. Lundy, J.C., Rico, C. and Hewitt, M.G. (2000) Temporal and spatial genetic variation in spawning grounds of European hake (Merluccius merluccius) in the Bay of Biscay. Molecular Ecology, 9(12), 2067-2079.

146. Heath, D.D., Bryden, C.A., Shrimpton, J.M., Iwama, G.K., Kelly, J. and Heath, J.W. (2002) Relationships between heterozygosity, allelic distance (d^2), and reproductive traits in Chinook salmon, Oncorhynchus tshawytscha. Canadian Journal of Fisheries and Aquatic Science, 59, 77-84.

147. Larson, S., Jameson, R., Etnier, M., Flemings, M. and Bentzen, P. (2002) Loss of genetic diversity in sea otters (Enhydra lutris) associated with the fur trade of the 18th and 19th centuries. Molecular Ecology, 11(10), 1899-1903.

148. Beaudou, D., Baril, D., Roche, Â.B. and Le. Baron, M. (1995) Recolonization in a devastated Corsican river: Respective contribution of wild and domestic brown trout. Bulletin Francais de la Peche et de la Pisciculture, 337, 259-266.

Chapter 2

ASSESSING SPATIAL GENETIC STRUCTURE FROM MOLECULAR MARKER DATA VIA PRINCIPAL COMPONENT ANALYSES: A CASE STUDY IN A PROSOPIS SP. FOREST

Ingrid Teich[1], Aníbal Verga[2], and Mónica Balzarini[1]

[1]Statistics and Biometry, Faculty of Agricultural Sciences, National University of Córdoba-CONICET, Córdoba, Argentina

[2]Centro de Investigaciones Agropecuarias, Instituto Nacional de Tecnología Agropecuaria, Córdoba, Argentina

ABSTRACT

Advances in genotyping technology, such as molecular markers, have noticeably improved our capacity to characterize genomes at multiple loci. Concomitantly, the methodological framework to analyze genetic data has expanded, and keeping abreast with the latest statistical developments to analyze molecular marker data in the context of spatial genetics has become a difficult task. Most methods in spatial statistics are devoted to univariate data whereas the nature of molecular marker data is highly dimensional. Multivariate methods are aimed at finding proximities between entities characterized by multiple variables by summarizing information in few synthetic variables. In particular, Principal Component analysis (PCA) has been used to study genetic structure of geo-referenced allele frequency profiles, incorporating spatial information with a posteriori analysis. Conversely, the recently developed spatially restricted PCA (sPCA) explicitly includes spatial data in the optimization criterion. In this work, we compared the results of the application of PCA and sPCA in the study of the spatial genetic structure at fine scale of a Prosopis flexuosa and P. chilensis hybrid swarm. Data consisted in the genetic characterization of 87 trees sampled in Córdoba, Argentina and genotyped at six microsatellites, which yielded 72 alleles. As expected, principal components explained more variance than sPCA components, but were less spatially autocorrelated. The maps obtained by the interpolation of sPC1 values allowed a better visualization of a patchy spatial pattern of genetic variability than the PC1 synthetic map. We also proposed a PC-sPC scatter plot of allele loadings to better understand the allele contributions to spatial genetic variability.

INTRODUCTION

Advances in molecular biology have led to the introduction of many new types of molecular markers, which provide cheap and high-throughput methods to characterize genomes at multiple loci. However, the amount of available information in biological studies has increased dramatically not only at the molecular level but also at other levels of organization and nowadays molecular marker data are often complemented with other covariates, like spatial and temporal coordinates. The joint analysis of genetic and spatial data can lead to a better understanding of evolutionary and ecological processes, such as drift, population expansions, bottlenecks, and selection and mutation regimes [1,2] and it is rapidly becoming the norm in population genetics. Spatial genetic structure (SGS) in natural populations, i.e. the nonrandom spatial distribution of genotypes, is expected to occur frequently at fine spatial scales within continuous plant populations [3]. SGS can result from different processes, including selection pressures or historical events. At a fine spatial scale, however, the most prevalent cause is the formation of local pedigree structures as a result of limited gene dispersal. In this context, genetic similarity is expected to be higher among neighbors (positive autocorrelation) than among more distant individuals, and the theory of isolation by distance [4,5], predicts the expected pattern of SGS at drift-dispersal equilibrium. Many empirical studies have investigated fine-scale SGS within plant populations from molecular marker data, often using spatial autocorrelation coefficients [6].

Regarding the statistical framework to study genetic variability from genetic data with known sampling site positions, spatial statistical genetics has become a rapidly evolving field. When implementing a spatially explicit approach to analyze georeferenced molecular marker data, it is important to consider that different statistical methods provide different types of information. The statistical dependence between geographic and genetic distances is usually carried out using the Mantel test, a permutational procedure to test the statistical significance of the correlation between matrices [7]. A common approach to quantify autocorrelation is the Moran index I [8], which has been extensively used and in genetic studies has been frequently applied to test the spatial structure of single alleles. Many methods for the analysis of SGS have been developed for single-locus, diploid genotypic data such as the one provided by isozymes [9]. However, genetic data are highly multidimensional and it is currently obtained from multiple loci with molecular markers. To deal with multivariate molecular data, dimension reduction techniques have proven to be useful [10-13].

Principal Components Analysis (PCA) [14] is one dimension reduction technique that can be applied to summarize molecular marker profiles into a few uncorrelated components. It finds an orthogonal basis for the data in such a way that the first axis of the new spanned space is along the direction of greatest variation of the original data, providing a set of eigenvectors and their corresponding eigenvalues. Eigenvectors contain the weight coefficients to build the linear combinations, which indicate the relative importance of variables to explain variability among the biological entities (e.g. trees) on each axis. Once the synthetic variables (principal components) of interest have been chosen, they can be used to give scatter plots of observations with optimal properties to study the underlying variability among entities. One advantage of the use of synthetic variables is that they collapse the multidimensional genetic characterization of individuals, allowing the construction of synthetic maps of genetic variability. For mapping purposes, individual scores on the principal components can be interpolated, by the prediction of the variable (PC) in spatial points. This technique allows visualizing the spatial pattern of genetic variability [15-18]. Plotting the values of the resulting synthetic variables (components) onto a geographic map as a way to explore the spatial structure of genetic variance, has been pioneered by Cavalli-Sforza [10] for the reconstruction of the early history of human populations. The power of PCA with large spatial genomic data sets became evident in Novembre et al. [19], who observed a very high correlation between the positions in a PCA plot and human geographic origin, showing that Single Nucleotide Polymorphisms (SNPs) were spatially structured. However, PCA was not properly designed to investigate spatial patterns and consequently spatial information was used as the posteriori analysis. The first principal components explain variance among observations rather than autocorrelation and therefore PCA may fail to detect spatial structuring if this is not associated with the most pronounced genetic differentiation.

For a more complete characterization of spatial structures in genomic data, the analysis of the principal components has to focus on the part of the multidimensional variance that is spatially structured. This can be accomplished using the spatial information within the optimization criterion used to find the synthetic variables. This issue was previously tackled in the context of ecological data by Thioulouse et al. [20], who built on the work of Wartenberg [21] to test the statistical significance of spatial structures in the context of multivariate analyses. The main concept was to introduce the neighboring relationship between sampling units in the analysis. Jombart et al. [22] developed a spatial Principal Component Analysis (sPCA) suitable for genetic allelic frequency data which relied on a modification of PCA such that not only the variance of the synthetic variables, but also their spatial autocorrelation, was optimized. The spatial information is stored inside a spatial weighting matrix which contains positive terms corresponding to some measurement (often binary) of spatial proximity among entities. Such terms can be derived from a connection network, or a neighboring graph, which is created by connecting the neighboring observations on a map [23]. For example, the Delaunay neighboring graph [24] is suited to evenly distributed observations, but may also connect unrelated peripheral observations, whereas the Gabriel neighboring graph [25] is a subset of the Delaunay graph without peripheral connections. In sPCA this spatial weighing, matrix is used to compute the spatial autocorrelation using the Moran's index statistic. The optimization criterion defined in the sPCA allows us to take into account both the spatial structure and the variability of the data. The eigenvalues provided by the sPCA are highly positive when the synthetic variables have a large variance and exhibit positive autocorrelation; and conversely, sPCA eigenvalues are largely negative when the spatial principal components have a high variance and display negative autocorrelation.

In this work, we attempt to clarify the use of PCA to tackle the study of spatial genetic patterns from molecular marker data. To achieve this, we compare the results of the application of PCA and sPCA on microsatellite data in the study of the SGS of a tree species. We also propose a PC-sPC scatter plot of allele loadings to better understand the allele contributions to spatial genetic variability. The value of the simultaneous use of both types of principal component analysis is demonstrated for a hybrid swarm between Prosopis chilensis and P. flexuosa, two arboreal species with economic and ecological importance in Argentina.

METHODS

Data

The data [26] contains the genetic characterization of geo-referenced trees (observations) of a hybrid swarm between Prosopis chilensis and Prosopis flexuosa. The study was carried out in a 4700 m^2 plot included in a continuous forest located in the Natural Reserve Chancaní, in Córdoba, Argentina (Lat. 31°23'S, Long. 65°27'W). In the study plot, a total of 87 flowering Prosopis trees (adult population) were identified as P. flexuosa, P. chilensis or hybrid using a taxonomic key based on quantitative characters [27]. The position of each tree in the plot was measured in the field using polar coordinates (distances and angles) and then converted to Cartesian coordinates. Genetic structure and variation was characterized in the adult population using six polymorphic microsatellites (SSR) originally developed for P. chilensis [28]. The total number of alleles found over all individuals was 72 and the number of alleles per locus ranged from 3 to 16. Allelic frequencies were calculated and centered by subtracting the mean allele frequency from all observations. Therefore, all analyses were performed on an 87 × 72 data matrix, corresponding to the 87 trees and 72 alleles.

Univariate Analysis

To better interpret the multivariate output, we evaluated the variance and the spatial autocorrelation of each allele independently. To estimate autocorrelation, we first built two spatial weighting matrices, one using Gabriel neighboring graph and the other using Delaunay's triangulation, with the choose CN function of adegenet [29] library in R software [30]. The number of neighbors for each individual obtained with both methods were compared through their frequency distribution. Each allele's spatial autocorrelation was estimated with the Moran Index and both spatial weighing matrices, using the function moran.test of spdep library [31] in R software. Finally, we plotted the Moran Index of each allele against its corresponding variance. On this plot we identified the four alleles with higher variances and the four alleles with more autocorrelation. We focused our analysis on the spatial structures with positive autocorrelation.

Application of PCA and sPCA

Both PCA and sPCA were performed on the 87 × 72 allele frequency data matrix using R software. For PCA the dudi.pca function in ade4 library [32] was applied and sPCA was run with the spca function in the adegenet library using Gabriel's Graph connection network. A number of components which explain a relevant amount of genetic variance were analyzed. To select the number of components we considered not only the variance they explained but also its distribution among eigenvalues in a screeplot. Additionally, biplots for both analyses were obtained with InfoStat software [33]. In the biplots the individuals were identified as P. chilensis, P. flexuosa or hybrid.

Comparison Criteria

The results obtained with both PCA and sPCA were compared by three criteria. First we contrasted the variance and spatial autocorrelation explained by the Principal Components (PC) and the spatial Principal Components (sPC). For this purpose we calculated the autocorrelation of the PCs and sPCs with the Moran Index using the spatial weighing matrix obtained by Gabriel Graph with the moran.test function of spdep library of R software. The Moran Index of each PC and each sPC against their corresponding variance were plotted. On this plot we identified the PCs and sPCs with the highest variances and autocorrelation. Secondly, we compared the maps of genetic variability built with the first synthetic variables yielded by both methods. To achieve this we plotted the PC1 and sPC1 scores of each tree positioned by its spatial coordinates. In this map the different sizes of the used symbols (squares) represent different absolute values of the synthetic variables: trees with large black squares are well differentiated from trees with large white squares and observations represented with small squares are less differentiated among them. This type of map was performed using the s.value function in ade 4. We also generated a surface using a local interpolation of principal component scores (function s.image in library ade 4), using grey levels and contour lines. The closer the contour lines are from each other, the steepest the genetic differentiation is. Finally, we compared the allele's contribution to the PCA and sPCA axes. To achieve this we proposed to build a PC-sPC scatter plot of the allele loadings of both synthetic variables and identify those alleles with high inertia in one axis (e.g. PC) and low inertia in the other (e.g. sPC).

RESULTS AND DISCUSSION

Univariate Analysis

The connection networks used to calculate the Moran Index of each allele (CN) are shown in Figure 1. With the Delaunay CN the total number of connections was 492, with an average number of 5.6 neighbors per individual, higher than with the Gabriel's CN, which rendered an average of 3.5 links per individual and a total of 310 connections. The most connected individual in the Delaunay CN had 9 neighbors and all individuals had 3 or more neighbors, whereas with the Gabriel CN, two individuals had the maximum number of links, which was 7 and 17 individuals had less than 3 neighbors. As shown in Figure 1 the most frequent number of links was 5 for Delaunay CN and 4 with Gabriel CN. The frequency distributions suggest that a higher number of neighbors per individual will be used to estimate spatial autocorrelation with Delaunay Triangulation than with Gabriel Graph.

Among the 72 alleles, only four explained more than 6% of total variance each (Figure 2). These were L1.3, L5.10, L1.2 and L4.03, which together explained 25% of total variance. Among these alleles, L4.03 showed the highest spatial autocorrelation, with a Moran's I of 0.081. In general, spatial autocorrelation was higher when calculated with Gabriel CN. Nevertheless, the four alleles with the highest positive autocorrelation are the same using both connection networks. The allele with highest Moran Index was L5.04 with a Moran I of 0.25 or 0.1 if calculated with Gabriel CN or Delaunay CN, respectively. Alleles L6.05, L5.05 and L3.04 also showed relatively high Moran Indexes. The alleles with more spatial autocorrelation did not account for high proportions of the total variance (5.6%).

As Figure 1 shows, many peripherical connections are included with Delaunay CN, connecting individuals which may not be actual neighbors in space. When there is information regarding the actual connectivity among the biological entities, such information should be used to choose or build a connection network. For example, in some data sets it might be better to adapt the connection network manually in order to exclude contacts across geographical barriers or to include long-range contacts which for biological reasons might have genetic exchange. When this information is not available, an algorithm has to be used to build it [23]. The R software provides many tools to perform this task though they are spread through different packages. For our case study we preferred to use the Gabriel CN.

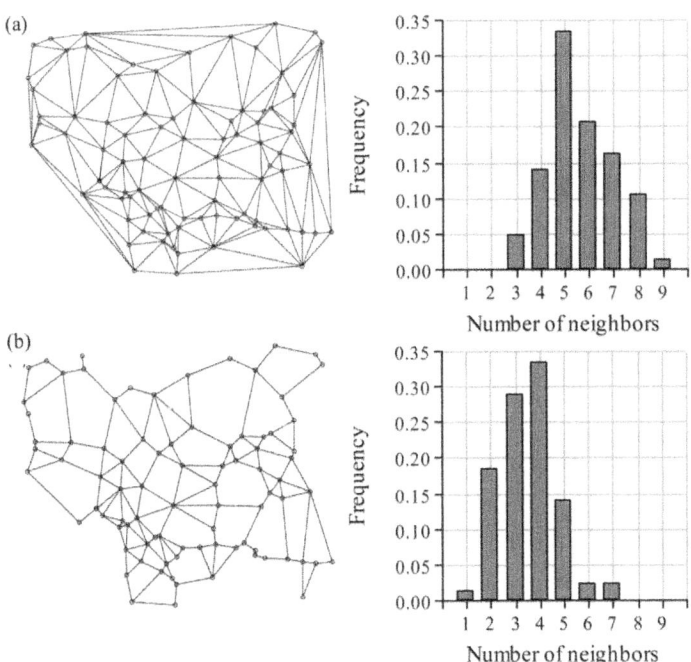

Figure 1. Connection networks for spatial analyses calculated using (a) Delaunay triangulation and (b) Gabriel's Graph. Bar plots indicate the frequency of individuals with each neighborhood size expressed as number of neighbors.

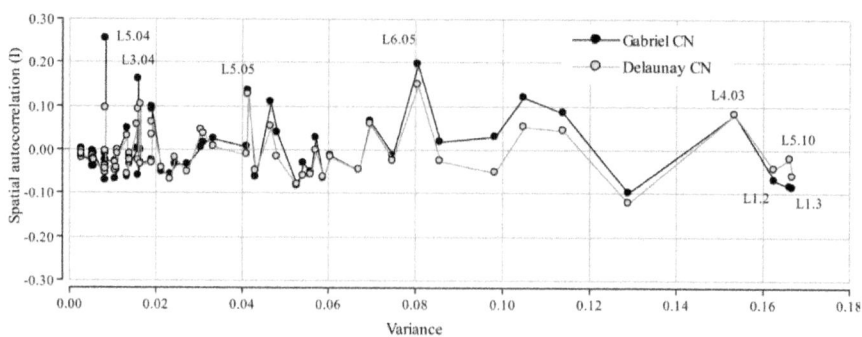

Figure 2. Spatial Moran's index of each allele plotted against the corresponding variances. Results correspond to the two connection networks: Delaunay triangulation and Gabriel's Graph.

Analysis of the Genetic Variability with PCA

The first synthetic variable of PCA (PC 1) had a variance of 0.42 and PC 2 had a variance of 0.36, accounting for 16.1% and 13.7% of the total genetic variability, respectively. With highly dimensional data, such as the provided by molecular markers, which are not necessarily linked among each other, PC1 and PC2 should not be expected to explain a high percentage of the total variance. For example, Novembre et al. [19] analyzed population SGS measured by 500,568 SNP loci on the space generated by a PC1 and a PC2 which explained 0.30% and 0.15%, respectively. However, the SGS of the populations was evident in the synthetic space. In our case study set, in which 72 allele's frequencies were analyzed, a first plane explaining 29.8% of total genetic variance was regarded as sufficient to explain the main pattern of alleles (co)-variability. In addition, the screeplot (Figure 3(b)) shows a sharp decay between PC2 and PC3, indicating that most of the variance in the data can be explained with the first two synthetic variables. The screeplot is a complementary tool to the axis variances and both should be used to decide the number of synthetic variables to be analyzed. Jombart et al. [11] cites two contrasting studies illustrating the need of using both indicators. In one study [34], the first two PC explaining a high percentage (80%) of total genetic variability of yak (Poephagus grunniens) populations were not as much informative in terms of genetic differentiation as in another study [35] in which they explained 10% of total variability, providing insights about the phylogeny of different maize subspecies. In our study, the analysis was performed using the first two PCs and the difference between species was clearly visible in the biplot.

As showed in the PCA biplot (Figure 3(a)) PC 1 separates P. flexuosa individuals from the hybrids and P. chilensis individuals. These trees show higher allelic frequencies of allele 3 in locus 4 (L4.03) and lower frequencies of allele 5 in locus 10 (L5.10). Alleles 3 and 2 of locus 1 (L1.3 and L1.2) are the alleles with more contribution in PC2 variability, which is not associated with a between species variance. The group with higher within genetic variability was P. flexuosa and these individuals were the most separated on the PC2 axis. As expected, the four alleles identified with higher variances in the univariate analysis (Figure 2) are the four alleles with more contribution in the first two PCA synthetic variables. In the biplot, the length of the arrows representing the alleles is proportional to the amount of genetic variability explained by the allele. Allele frequencies were centered but not scaled, maintaining the inherent variance of the alleles. This approach allows identifying the alleles that contribute most to the total genetic variability even in high dimensional data sets. Centering of allele frequencies is common but their scaling is discussed [36]. Scaling allele frequencies could mask differences in the genetic variability contained by informative and non-informative markers, ultimately hiding structures in the data [11]. Many studies that apply PCA on genetic data represent either the entities in the variables (alleles) space or the alleles in the space spanned by the observations. Here we used the biplot representation of the allele frequencies data which is useful because both the alleles and the trees can be visualized in the same plot. Different types of biplots have been used to graphically represent genetic variability from molecular markers profiles [12,37].

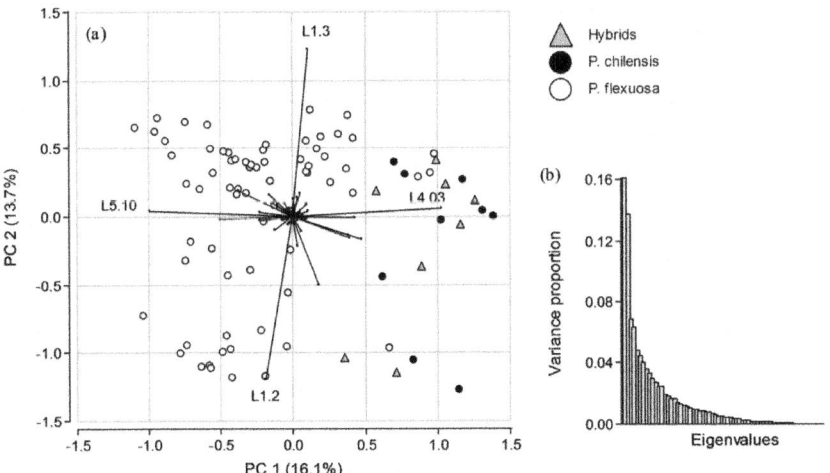

Figure 3. Results obtained by principal component analysis. (a) Biplot of first and second axis of PCA, individuals are colored according to classification in P. chilensis,

P. flexuosa and hybrid and segment lines represent the alleles; (b) Screeplot of PCA eigenvalues.

Analysis of Genetic Variability with sPCA

The first synthetic variable (sPC 1) had a variance of 0.26 accounting for 10% of the total inertia. The first two principal components of sPCA explained 14% of the data structure. Similar levels of total inertia explained by the first two spatial principal components (sPC) were accounted for other genetic studies [22,38]. Analogous to the classical biplot used to represent PCA results, we built a symmetrical biplot from the two sPCs. The sPCA biplot (Figure 4(b)) shows that sPC 1 also allows to separate P. flexuosa individuals from the hybrids and P. chilensis. The sPCA screeplot shows a sharp decay between the first and the second eigenvalue, indicating that the analysis of sPC1 variability may be enough to explain SGS in this Prosopis hybrid swarm. Although sPCA was also applied on centered and not scaled allele frequencies, the lengths of the vectors representing the alleles are similarly distributed. sPCA is related to multivariate spatial correlation [21] but it allows alleles to have different variances. Scaling allele frequency data in sPCA has the same negative effect discussed above for PCA.

The allele with the highest contribution in sPC1 is L4.03 and allele L6.05 is the second allele with a relatively high loading in sPC1. This allele is one of the four alleles with high spatial autocorrelation and from these four, the one with most variance (Figure 2). When sPCA is performed, negative eigenvalues, which account for negative autocorrelation structure, arise. In our study case the highest negative eigenvalue explained less percentage of total variance than the first positive eigenvalue. In addition there is no evidence of a sharp decay between two negative eigenvalues in the sPCA screeplot (Figure 4(b)). For this reason we only analyzed the SGS related to positive autocorrelation.

Comparison of PCA and sPCA Results

PCA eigenvalues were larger in magnitude and much less spatially autocorrelated (Figure 5). The first two PCs, which account for 16% and 14% of total genetic variance, had no spatial autocorrelation, with low and not statistically significant Moran's Indexes ($I_1^{PCA} = 0.05$, $I_2^{PCA} = -0.07$, $p > 0.05$) (Table 1). On the contrary, the variance of the first two sPCs was much lower (10% and 3.8% of total variance) but they had higher and significant Moran Indexes ($I_1^{sPCA} = 0.29$, $I_2^{sPCA} = -0.39$, $p < 0.05$).

Table 1. Variance and spatial autocorrelation of the first 2 PCA and sPCA eigenvalues.

Analysis	Eigenvalue	Variance	Proportion of Total Variance	Moran Index
PCA	1	0.42	0.161	0.05
	2	0.36	0.138	−0.07
sPCA	1	0.26	0.100	0.29
	2	0.10	0.038	0.39

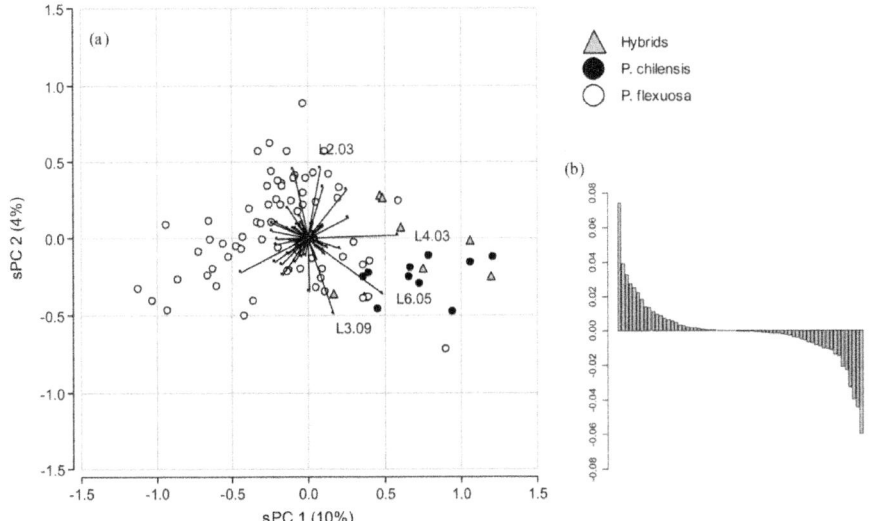

Figure 4. Results obtained by spatial principal component analysis (sPCA). (a) Biplot of first and second axis of sPCA, individuals are colored according to classification in P. chilensis, P. flexuosa and hybrids; (b) Screeplot of sPCA eigenvalues.

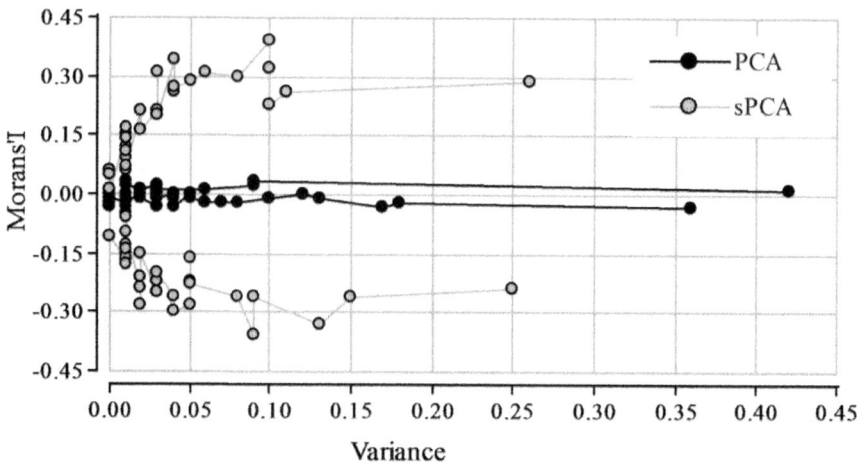

Figure 5. Spatial Moran's index of sPCA and PCA eigenvalues plotted against the corresponding variances.

The spatial principal component with most positive spatial auto correlation was sPC 2 ($I_2^{sPCA} = 0.39$) and the PC with most positive spatial autocorrelation was PC 9 ($I_9^{PCA} = 0.22$). These results show that the first principal components are associated with alleles that explain variance instead of spatial correlation. PCA axes, which spatial autocorrelation was very low, might fail to identify relevant spatial patterns in this Prosopis hybrid zone. On the contrary, sPCA detects additional spatially structured components. As discussed by Jombart et al. [22], the variance associated to the first axis in sPCA was lower than the variance of PC1, however, it captures a spatial pattern associated to the spatially structured genetic differentiation. The relative value of sPCA over PCA depends on the nature of the structure underlying the data. When the spatially genetic variability is not associated to the alleles with higher variability among entities, the relative sPCA value increases. In our study case the alleles with most spatial autocorrelation were not those with highest variances. Therefore sPCs provide new information to the study of SGS. In other cases, when the most spatially structured alleles also have the higher variances, the first sPCs correspond to the first coordinates of the unrestricted PCA [38].

As both principal component analyses suggest, the spatial pattern of genetic variability in this hybrid swarm shows at least two patches of genotypes in space (Figure 6). One patch is constituted by individuals with high positive scores (black) on the principal components and the other with high negative scores (white). In both types of maps the spatial structure is clearer with sPC1 scores than with PC1 scores. In the interpolated maps, contour lines are closer together in the sPC1 map, indicating that the magnitude of the gradient is larger. Therefore, the sPC1 allows a better visualization of a patchy spatial pattern of genetic variability in our study case. As higher scores of sPC1 are associated to hybrids and P. chilensis, the darker patch is associated to them, whereas the lighter patch is associated to P. flexuosa.

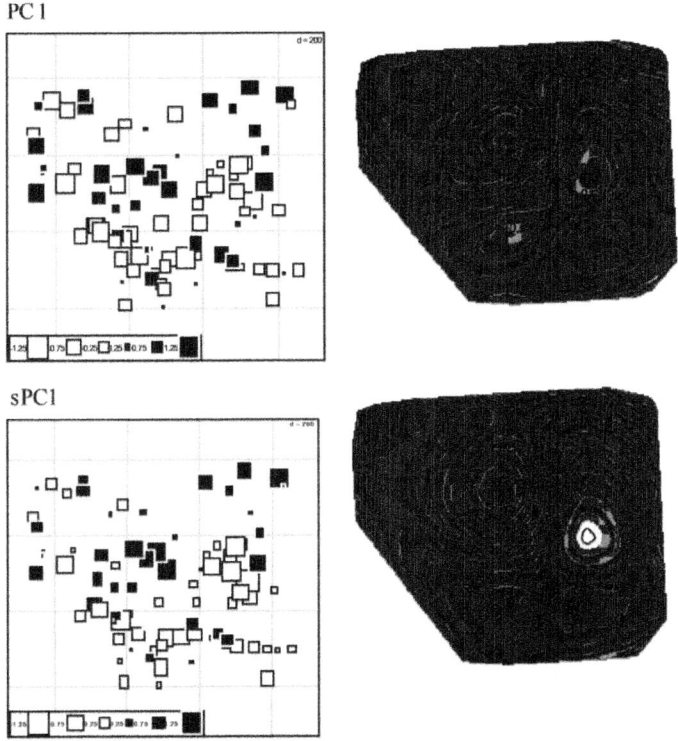

Figure 6. Spatial analysis of PCA and sPCA results. Scores of the first principal components obtained with PCA (above) and sPCA (below). Left: Each square corresponds to the score of an individual and it is positioned by its spatial coordinates. Right: Map of the scores, values obtained by the interpolation of the principal components.

Our results are in concordance with the findings of Bessega et al. [39], who studied the genetic structure of P. alba, a very similar species. They conclude that pollen and seed dispersion is limited, estimating the average pollen dispersal distance to be between 5.36 and 30.92 m. Their findings explain the strong genetic structure of the P. alba population, which was studied through its mating system, but not through the spatial distribution of the genotypes.

To visually identify those alleles that contribute most to the structures captured by the both the PC1 and the sPC1, as well as those alleles that either have a lot of inertia in one axis but not on the other, we built a scatter plot of the allele loadings in PC1 and sPC1 (Figure 7). Statistical validation of identified markers was carried out through comparison of variance and Moran Index between both types of principal components (Table 2). In the PC-sPC scatter plot, three types of areas are identified by different colors. These areas were defined using the mean ± a standard deviation of the allele loadings in a synthetic variable. For all synthetic variables, loadings have a mean of 0 and standard deviation of 0.12 because of the normalization of eigenvectors. The white square in the middle of the PC1-sPC1 scatter plot corresponds to loadings that range between −0.12 and 0.12 in both the sPC1 and the PC1; the alleles that belong to this area do not have much inertia in either the PC1 or the sPC1.

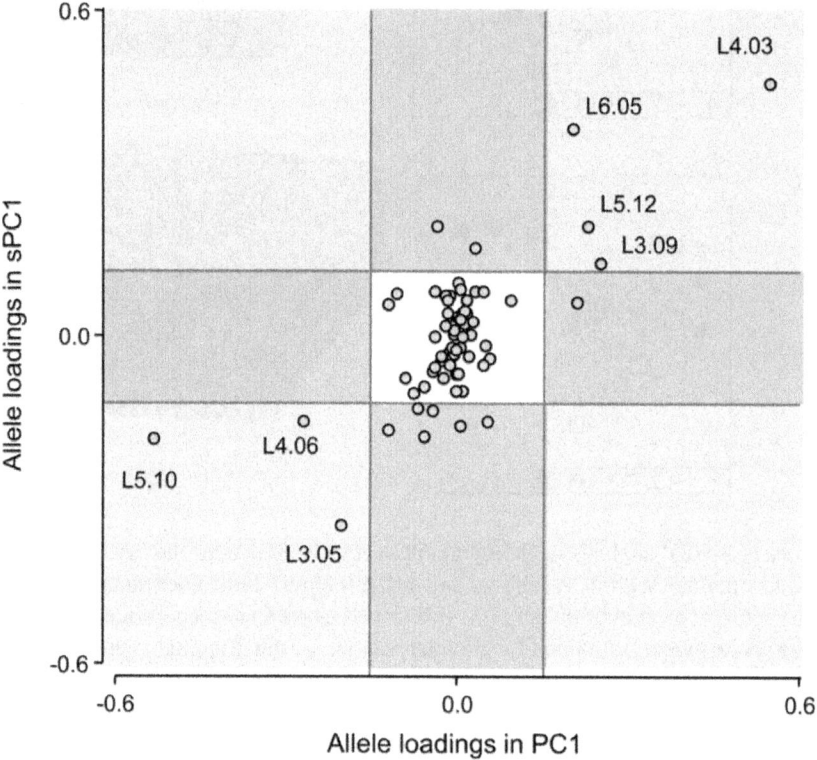

Figure 7. PC-sPC scatter plot: allele loadings of the first sPCA axis vs. PCA axis.

Table 2. Variance, Moran Index and loadings in the first PC and sPC of identified alleles.

Allele	Variance	Moran Index	PC1	sPC1
L5.10	0.17	−0.08	−0.53	−0.19
L4.06	0.10	0.03	−0.27	−0.16
L3.05	0.12	0.08	−0.20	−0.35
L5.12	0.07	0.06	0.05	0.20
L6.05	0.08	0.20	0.20	0.38
L3.09	0.10	0.12	0.25	0.13
L4.03	0.15	0.08	0.55	0.47

Therefore, these alleles do not contribute much to the SGS. On the contrary, the four light grey squared areas correspond to alleles that have high inertia in both PC1 and sPC1. These alleles explain variability between species as shown in the biplots (Figures 3 and 4) and their variances were relatively high (Table 2). The four remaining darker areas correspond to alleles with a high contribution in one synthetic variable and a low inertia in the other. The horizontal dark rectangles correspond to alleles with high loadings in PC1 and low loadings in sPC1. Only one allele falls in this category (L5.09). This allele is important in terms of between species variability but is associated to a type of variability that is not spatially structured. The vertical dark grey rectangles correspond to areas of high loadings in sPC1 and low loadings in PC1. In our study eight alleles were found in these areas, corresponding to alleles that do not contribute much to the main axis of genetic differentiation between species but that their variability is spatially structured. However, the interpretation is not simple in these cases, as it is important to consider that high loadings in a sPC are associated to alleles with a relatively high product between their variances and their spatial autocorrelation.

The sPCA biplot (Figure 4(a)) shows that allele's loadings have a more uniform distribution than in PCA. This fact is probably associated with the lower variability of the product between allele's variance and Moran Index than the variability of allele's variances, which are the optimization criteria of sPCA and PCA, respectively. The cut-off values of 0.12 and −0.12 which were used as selection criterion for groups of contributing allele markers will have effects on the outcome of the biological results. In other words different cutoffs can render different biological results. Therefore the whole process was performed with several others cut-offs on the first 2 PCs. This analysis showed that the cut-off based on one standard deviation of allele loadings highlighted markers which have either high variance and/or high autocorrelation (data not shown).

Our results show that to effectively understand the relative contribution of alleles to spatial genetic structure, the joint application of both principal component analyses is useful. However, the results shown before were obtained by applying the combination of PCA and sPCA on all available markers. To explore the results of both PCA and sPCA when performed on the selected subset of markers we applied both methods on the 16 alleles outside the white square of the PC1-sPC1 scatter plot. As expected, the results show that the main pattern of species differentiation was no different from the overall effects present in the whole dataset. This is another way to validate the interpretation of the allele contributions in the PC-sPC scatter plot.

Both techniques, PCA and sPCA, have been applied in studies of the SGS of animals, such as the Scandinavian brown bear (Ursus arctus) and domestic ruminants in Europe [22,38]. As compared to most animal species, adults from plant species do not move and plants' propagules, i.e. pollen and seeds, often show moderate to strong spatial restriction in their dispersal leading to strong SGS. In particular, the study of these structures in forests provides vital information for their conservation and management. This is of utmost importance in Argentina, where 70% of forest cover has been lost and forest emergency has recently been declared (National Law 26.331). Among other native tree species, the genus Prosopis constitute a very important natural resource for dry zones that need strong conservation actions [40]. Although the genus Prosopis presents no difficulty of identification, individual species are in some cases difficult to determine due to the occurrence of many natural hybrid combinations within the genus [41-43]. Because frequent events of interspecific natural hybridization with fertile hybrid production in areas of sympatry occur, isolation mechanisms between Prosopis species seem to be weak or incomplete [26,44]. Natural interspecific hybridization has been recognized as playing an important role in plant evolution and hybrid zones are viewed as active sites of evolutionary change that constitute sources of new recombinant types [45-47]. Hybrid zones are characterized by a continuous variation in morphological and genetic traits and the loss of differentiation of pure species. Therefore, cryptic and continuous patterns of spatial genetic variability are expected even at small spatial scales, which might be difficult to identify and characterize. In this case, recovery plans and management of forests can particularly benefit from the joint use of both type of principal component analysis of spatial molecular marker data. They provide a useful insight into the problem of selecting founding populations and particularly, in selecting individuals within populations, where sometimes the spatial genetic structure is overlooked. Spatial analysis techniques provide a suitable framework to integrate the knowledge derived from genetic, demographic and ecological approaches to species conservation, allowing the formulation of management strategies that take into account different considerations.

CONCLUSION

After the application of PCA and sPCA and visual inspection of the allele contribution to both types of synthetic variables, interesting markers to investigate genetic spatial structure can be selected. The combination of PCA and sPCA, as demonstrated here, is a valuable tool in forests molecular marker data analysis because more information is available on the allele contributions to the spatial genetic structure. The PC-sPC scatter plot can be used to split and visualize the different components of genetic variability yielded by molecular markers. Considering the spatial genetic structure of the studied Prosopis sp. hybrid swarm, two groups of tree genotypes (corresponding to different Prosopis species) were distinguished at a small spatial scale. The patchy spatial pattern observed could be explained by the existence of a patchy spatial structure of available safe sites for the establishment of the different genotypes and by limited gene dispersal.

ACKNOWLEDGEMENTS

We thank Martin Mottura for providing the Prosopis sp. data set. I.T. is a recipient of a postdoctoral fellowship of the National Council of Technical and Scientific Research (CONICET), respectively.

REFERENCES

1. Keitt, T.H., Bjørnstad, O.N., Dixon, P.M. and CitronPousty, S. (2002) Accounting for spatial pattern when modeling organism-environment interactions. Ecography, 25, 616-625. http://dx.doi.org/10.1034/j.1600-0587.2002.250509.x

2. Vucetich, J. and Waite, T. (2003) Spatial patterns of demography and genetic processes across the species' range: Null hypotheses for landscape conservation genetics. Conservation Genetics, 4, 639-645. http://dx.doi.org/10.1023/A:1025671831349

3. Vekemans, X. and Hardy, O.J. (2004) New insights from fine-scale spatial genetic structure analyses in plant populations. Molecular Ecology, 13,

921-935.http://dx.doi.org/10.1046/j.1365-294X.2004.02076.x
4. Wright, S. (1943) Isolation by distance. Genetics, 28, 114-138.
5. Wright, S. (1946) Isolation by distance under diverse systems of mating. Genetics, 31, 39-59.
6. Epperson, B.K. (1993) Spatial and space-time correlations in systems of subpopulations with genetic drift and migration. Genetics, 133, 711-727.
7. Mantel, N. (1967) The detection of disease clustering and a generalized regression approach. Cancer Research, 27, 209-220.
8. Moran, P.A.P. (1950) Notes on continuous stochastic phenomena. Biometrika, 37, 17-23.
9. Smouse, P.E. and Peakall, R. (1999) Spatial autocorrelation analysis of individual multiallele and multilocus genetic structure. Heredity (Edinb), 82, 561-573.http://dx.doi.org/10.1038/sj.hdy.6885180
10. Cavalli-Sforza, L.L. (1966) Population structure and human evolution. Proceedings of the Royal Society of London. Series B, Biological Sciences, 164, 362-379.http://dx.doi.org/10.1098/rspb.1966.0038
11. Jombart, T., Pontier, D. and Dufour, A.B. (2009) Genetic markers in the playground of multivariate analysis. Heredity (Edinb), 102, 330-341.
12. Balzarini, M., Teich, I., Bruno, C. and Peña, A. (2011) Making genetic biodiversity measurable: A review of statistical multivariate methods to study variability at gene level. Revista de la Facultad de Ciencias Agrarias de la Universidad Nacional de Cuyo, 43, 261-275.
13. Wang, C., Zöllner, S. and Rosenberg, N.A. (2012) A quantitative comparison of the similarity between genes and geography in worldwide human populations. PLoS Genetics, 8, e1002886. http://dx.doi.org/10.1371/journal.pgen.1002886
14. Hotelling, H. (1933) Analysis of a complex of statistical variables into principal components. Journal of Educational Psychology, 24, 417-441. http://dx.doi.org/10.1037/h0071325
15. Manel, S., Joost, S., Epperson, B.K., Holderegger, R., Storfer, A., Rosenberg, M.S., et al. (2010) Perspectives on the use of landscape genetics to detect genetic adaptive variation in the field. Molecular Ecology, 19, 3760- 3772. http://dx.doi.org/10.1111/j.1365-294X.2010.04717.x
16. Manel, S., Schwartz, M.K., Luikart, G. and Taberlet, P. (2003) Landscape genetics: Combining landscape ecology and population genetics. Trends in Ecology & Evolution, 18, 189-197. http://dx.doi.org/10.1016/S0169-5347(03)00008-9

17. Manel, S. and Segelbacher, G. (2009) Perspectives and challenges in landscape genetics. Molecular Ecology, 19, 1821-1822. http://dx.doi.org/10.1111/j.1365-294X.2009.04151.x

18. Storfer, A., Murphy, M.A., Evans, J.S., Goldberg, C.S., Robinson, S., Spear, S.F., et al. (2007) Putting the "landscape" in landscape genetics. Heredity (Edinb), 98, 128- 142.http://dx.doi.org/10.1038/sj.hdy.6800917

19. Novembre, J., Johnson, T., Bryc, K., Kutalik, Z., Boyko, A.R., Auton, A., et al. (2008) Genes mirror geography within Europe. Nature, 456, 98-101.http://dx.doi.org/10.1038/nature07331

20. Thioulouse, J., Chessel, D. and Champely, S. (1995) Multivariate analysis of spatial patterns: A unified approach to local and global structures. Environmental and Ecological Statistics, 2, 1-14. http://dx.doi.org/10.1007/BF00452928

21. Wartenberg, D. (1985) Multivariate spatial correlation: A method for exploratory geographical analysis. Geographical Analysis, 17, 263-283. http://dx.doi.org/10.1111/j.1538-4632.1985.tb00849.x

22. Jombart, T., Devillard, S., Dufour, A.B. and Pontier, D. (2008) Revealing cryptic spatial patterns in genetic variability by a new multivariate method. Heredity (Edinb), 101, 92-103. http://dx.doi.org/10.1038/hdy.2008.34

23. Legendre, P. and Legendre, L. (1998) Numerical ecology. Elsevier Science B.V., Amsterdam.

24. Upton, G.J.G. and Fingleton, B. (1985) Spatial data analysis by example. Wiley, Chichester/New York.

25. Gabriel, K.R. and Sokal, R.R. (1969) A new statistical approach to geographic variation analysis. Systematic Biology, 18, 259-278.

26. Mottura, M.C. (2006) Development of microsatellites in Prosopis spp. and their application to study the reproduction system. Library of Lower Saxony State and GeorgAugust University of Göttingen, Göttingen.

27. Verga, A. (2000) Clave para la identificación de híbridos entre Prosopis chilensis y P. flexuosa sobre la base de carcateres cuantitativos. Multequina, 9, 17-22.

28. Mottura, M.C., Finkeldey, R., Verga, A.R. and Gailing, O. (2005) Development and characterization of microsatellite markers for Prosopis chilensis and Prosopis flexuosa and cross-species amplification. Molecular Ecology Notes, 5, 487-489.http://dx.doi.org/10.1111/j.1471-8286.2005.00965.x

29. Jombart, T. (2008) Adegenet: A R package for the multivariate analysis of genetic markers. Bioinformatics, 24, 1403-1405. http://dx.doi.org/10.1093/bioinformatics/btn129

30. R Development Core Team, R. (2011) R: A language and environment for statistical computing. Vienna, Austria.
31. Bivand, R., Altman, M., Anselin, L., Assunção, R., Berke, O., Bernat, A., et al. (2011) Spdep: Spatial dependence: weighting schemes, statistics and models. R package version 0.5-31.
32. Dray, S. and Dufour, A.B. (2007) The ade4 package: Implementing the duality diagram for ecologists. Journal of Statistical Software, 22, 1-20.
33. Di Rienzo, J.A., Casanoves, F., Balzarini, M.G., Gonzalez, L., Tablada, M. and Robledo, C.W. (2011) InfoStat.
34. Xuebin, Q., Jianlin, H., Lkhagva, B., Chekarova, I., Badamdorj, D., Rege, J.E.O., et al. (2005) Genetic diversity and differentiation of Mongolian and Russian yak populations. Journal of Animal Breeding and Genetics, 122, 117-126.http://dx.doi.org/10.1111/j.1439-0388.2004.00497.x
35. Matsuoka, Y., Vigouroux, Y., Goodman, M.M., Sanchez G., J., Buckler, E. and Doebley, J. (2002) A single domestication for maize shown by multilocus microsatellite genotyping. Proceedings of the National Academy of Sciences, 99, 6080-6084.
36. Weir, B.S. (1996) Genetic data analysis II: Methods for discrete population genetic data. Sinauer Associates, Sunderland.
37. Demey, J.R., Vicente-Villardón, J.L., Galindo-Villardón, M.P. and Zambrano, A.Y. (2008) Identifying molecular markers associated with classification of genotypes by External Logistic Biplots. Bioinformatics, 24, 2832-2838.http://dx.doi.org/10.1093/bioinformatics/btn552
38. Laloë, D., Moazami-Goudarzi, K., Lenstra, J.A., Marsan, P.A., Azor, P., Baumung, R., et al. (2010) Spatial trends of genetic variation of domestic ruminants in Europe. Diversity, 2, 932-945. http://dx.doi.org/10.3390/d2060932
39. Bessega, C., Pometti, C.L., Ewens, M., Saidman, B.O. and Vilardi, J.C. (2012) Strategies for conservation for disturbed Prosopis alba (Leguminosae, Mimosoidae) forests based on mating system and pollen dispersal parameters. Tree Genetics & Genomes, 8, 277-288.http://dx.doi.org/10.1007/s11295-011-0439-6
40. Pasiecznik, N.M., Felker, P., Harris, P.J.C., Harsh, L.N., Cruz, G., Tewari, J.C., et al. (2001) The Prosopis juliflora-Prosopis pallida complex: A monograph. HDRA, Coventry.
41. Palacios, R. (1998) Taxonomía numérica (Descriptores). Prosopis en la Argentina. Facultad de Ciencias Agrarias, Universidad Nacional de Córdoba, Argentina, 91-96.

42. Saidman, B.O., Bessega, C.F., Ferreira, L.I., Julio, N. and Vilardi, J. (2000) The use of genetic markers to assess population structure and relationships among species of the genus Prosopis (Leguminosae). Boletín de la Sociedad Argentina de Botánica, 35, 315-324.

43. Ferreyra, L., Vilardi, J., Verga, A., López, V. and Saidman, B. (2013) Genetic and morphometric markers are able to differentiate three morphotypes belonging to Section Algarobia of genus Prosopis (Leguminosae, Mimosoideae). Plant Systematics and Evolution, 299, 1157- 1173. http://dx.doi.org/10.1007/s00606-013-0786-x

44. Verga, A.R. (1995) Genetische untersuchungen an Prosopis chilensis und P. flexuosa (Mimosaceae) im trockenen Chaco Argentiniens. Göttingen Research Notes in Forest Genetics. Abteilung für Forstgenetik und Forstpflanzenzüchtung der Universität Göttingen, 19, 1-96.

45. Harrison, R.G. (1990) Hybrid zones: Windows on evolutionary process. Oxford Surveys in Evolutionary Biology, 7, 59.

46. Barton, N.H. (2001) The role of hybridization in evolution. Molecular Ecology, 10, 551-568. http://dx.doi.org/10.1046/j.1365-294x.2001.01216.x

47. Soltis, P.S. and Soltis, D.E. (2009) The role of hybridization in plant speciation. Annual Review of Plant Biology, 60, 561-588. http://dx.doi.org/10.1146/annurev.arplant.043008.092039

Chapter 3

DETERMINING THE SPECIFIC STATUS OF THE IBERIAN STURGEONS BY MEANS GENETIC ANALYSES OF OLD SPECIMENS

Francisca Robles[1], Belén Cano-Roldán[1], Carmelo Ruiz Rejón[1], Luís Javier Martínez-González[2], María Jesús Álvarez-Cubero[2], José Antonio Lorente[2], José Antonio Riquelme Cantal[3], Pedro Aguayo de Hoyos[4], Javier Carrasco Rus[4], Miguel Cortés Sánchez[5], María Dolores Simón Vallejo[6], Manuel Ruiz Rejón[1], and Roberto de la Herrán[1]

[1]Departamento de Genética, Facultad de Ciencias, Universidad de Granada, Granada, Spain

[2]Centro Pfizer, Universidad de Granada, Junta de Andalucía de Genómica e Investigación Oncológica, Centro de Investigación Biomédica, Av. del Conocimiento s/n, Armilla, Granada, Spain

[3]Consejería de Cultura, Junta de Andalucía, Sevilla, Spain

[4]Departamento de Prehistoria y Arqueología. Facultad de Filosofía y Letras, Universidad de Granada, Granada, Spain

[5]Faculdade de Ciências Humanas e Sociais, Universidade do Algarve, Faro, Portugal

[6]Fundación Cueva de Nerja, Malaga, Spain.

ABSTRACT

To clarify the species status of sturgeon from rivers of the Iberian Peninsula, eight molecular markers (4 nuclear and 4 mitochondrial) have been analysed in different specimens from historical museum samples and prehistoric samples from archaeological sites. These analyses indicate that one of these specimens (UGP captured in the Guadalquivir River in the 19th century) is A. sturio, based on all the eight molecular markers, four of them used from the first time in this study. In previous analyses based on 5 genetic markers, our group assigned two specimens captured in this river in the 1970-80s (EBD8173 and EBD8401) to the species A. naccarii, suggesting the presence of this species in the Iberian Peninsula. In this work, this conclusion is drawn after successfully obtaining a mitochondrial marker in a very old scute from a prehistoric site (Acinipo, about 1500 BC, from the Guadalquivir River basin). On the other hand, in the specimen EBD8174 captured in the Guadalquivir in 1975, we have obtained two new mitochondrial markers confirming that it can be considered A. sturio for all the mitochondrial markers, but nuclear ones identify it as A. naccarii. Finally, two very old samples (Nerja E-VI and Nerja N/62-63) were not successfully characterized by any molecular markers. Some aspects and consequences of our results are discussed, such as the origin of the "mosaic" specimen EBD8174 and, above all, the native status of A. naccarii in historic and prehistoric times in the southern Iberian Peninsula.

INTRODUCTION

The identification of sturgeon species inhabiting a certain geographical region has interest not only from the basic scientific standpoint but also for the conservation and recovery of this group of ancient fish so important from the evolutionary as well as the economic perspective [1]. Thus, the specific status of the Iberian Peninsula sturgeons is a debatable matter because, bearing in mind that they are currently almost extinct, it becomes necessary to analyze old museum specimens and even archaeological remains. In this sense, during the second half of the 20th century, it was traditionally considered that in the seas and southern rivers of Western Europe and, more concretely, in the southern Iberian Peninsula, there was only one sturgeon species, Acipenser sturio (Linnaeus 1758). However, from end of last century, the idea arose that until recently at least two species could have coexisted. In fact, based on morphologic and mainly genetic studies (including mitochondrial and nuclear markers) of old museum specimens of sturgeons from this region, it has been shown [2-4] that, in addition to specimens belonging to A. sturio, it is possible to find specimens belonging to another species, A. naccarii (Bonaparte 1836). This situation had been previously proposed by different authors who historically, although forgotten, cited A. naccarii in the Iberian Peninsula [5-13]. All these results would indicate that, in recent historical times, this latter species (A. naccarii), until now considered only endemic to the Adriatic region, would also have lived in rivers of the Iberian Peninsula.

However, these results have been questioned partly by other studies, which have not provided data to indicate the presence of the species A. naccarii in this region [14-16]. Finally, recently Ludwig et al. [17] studying the mitochondrial region control in five scutes of sturgeons from archaeological locations of historical times in the Iberian Peninsula, have recently found only mitochondrial haplotypes of A. sturio. Therefore, it becomes necessary to continue delving into the analysis of this issue.

In this work, our group, which has contributed to opening this new vision of the distribution of sturgeons in Southern Europe (i.e. the coexistence of A. naccarii with A. sturio), analyses and discusses the attempts to obtain eight molecular markers (mitochondrial and nuclear) in seven old specimens of historic and prehistoric times in southern Spain. These molecular markers are compared in several sturgeon species. Thus, the results previously reported by our group have been corroborated in four historical specimens. In addition, we have tried to clarify the specific status of three new sturgeon samples from archaeological sites. Emphasis is placed mainly on the positive results for one of these samples, in a scute of a very old specimen dating from 1500 BC, which again verifies the presence of the species A. naccarii in this region.

MATERIALS AND METHODS

Samples

In this work, DNA was extracted from seven old sturgeon specimens from the southern Iberian Peninsula. Four of the specimens analyzed had been captured in the Guadalquivir River, EBD8173, EBD8401, EBD8174 and UGP. Three of them (labelled EBD), captured in the 1970-80s, are conserved in the Biological Station of Doñana (Spain). The samples EBD8173 and EBD8401 are preserved in ethanol, whereas the EBD8174 is a dry skin. The fourth sample (labelled UGP) is a skin conserved in the Department of Biology Animal of the University of Granada and was also captured in the Guadalquivir River (19th century).

Additionally, three prehistoric samples are analyzed for first time. One of them corresponds to a scute from 1500 BC which was found in the Acinipo archaeological deposit (Ronda, Malaga, Spain) (**Figure 1**). The archaeological deposit of Ronda la Vieja (called Acinipo, the name of the Roman city built on this site; [18]) is located in the depression of Ronda, 20 km from the city.

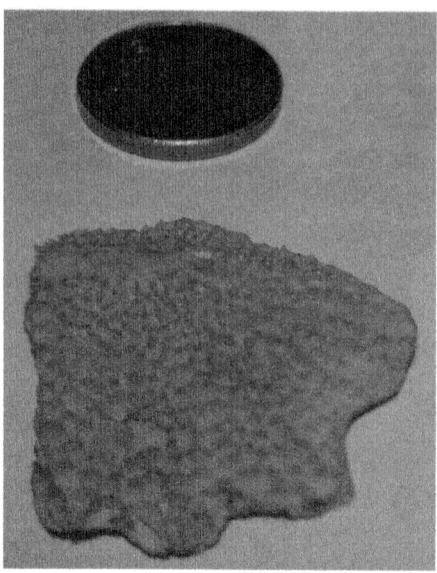

Figure 1. Scute dated in 1500 BC found in the archaeological deposit of Acinipo (Ronda, Malaga, Spain).

The site is situated on a large limestone plateau, which provides a strategic view of the surrounding territory and provides communication with other areas, including the countryside of the Guadalquivir River. The bony sample of sturgeon analyzed corresponds to the archaeological phase Acinipo III [19], prior to the Phoenician colonization around the second half of the II millennium B.C. Although it is difficult to assign its origin to the Guadalquivir River, the dates and the zone where it has been found would indicate its origin from this river. Finally, an attempt was made to extract DNA and amplify the different molecular markers from two very old scutes of sturgeons found at another prehistoric deposit (Cave of Nerja, Malaga, Spain). The Cave of Nerja has a long ichthyoarchaeological record of the excavations made basically in the room of the Vestíbulo [20] on the stratum VII (about 12,000 years old). This level is correlated with Magdalenian occupation in the cave [21].

DNA Extraction

The extraction and purification of DNA was carried out using ancient DNA techniques and according to the protocol described in Martínez-Espín et al. [22]. The first step consisted of cleaning the tissue samples in a polymethacrylate (PMMA) box. A miniature Dremel drill was used to eliminate any polluting agents adhering to the surface. Then, the tissue samples were pulverized in liquid nitrogen using a Freezer Mill. After pulverization, the powdered sample was transferred to a sterile 15 ml conical polypropylene tube.

To improve DNA recovery, in older samples (the scute from Acinipo and the two scutes from Nerja), we made some changes in the protocols. For these three samples, a protocol was adapted for demineralization of skeletal remains frequently used in mommies and historical identification [23,24]. To minimize the possibility of contamination by contemporary DNA of extraneous sources, these samples were extracted in the minimal-humanremains laboratory, where an animal sample had never before been extracted. Here, possible contamination was eliminated from the old samples. Only one specimen was cleaned and processed at the same time and a negative control was included with the analysis of each specimen. After adding demineralization buffer, the samples were incubated on an orbital shaker at 56°C for 20-30h. The tubes were angled during agitation to ensure thorough mixing. At the beginning of the extraction, we first added 50 µl of proteinase K (20 mg/ml) and 25 µl again 18 h later. The extracts were purified using sterile water washes in Microcon YM-30 Millipore centrifugal filter units; in the other samples, Microcon YM-100 was used. As a final point, the concentrator was discarded, and 200 µl of the purified DNA were obtained. In this case, many inhibitors were also obtained owing to the fact that tissue is adsorbed into a mineral matrix, after the death of the animal. The following step was the purification with the GENECLEAN® (BIO 101) for Ancient DNA Kit (using the recommended protocol). To guarantee the absence of inhibitor, the Quantifiler® kit for 7500 Real-Time PCR (Applied Biosystems) was used. The Internal Positive Control detectors indicated the absence of PCR inhibitor in all samples.

Amplification, Cloning, and Sequencing of Molecular Markers

For each specimen an effort was made to characterize the following genetic markers: 1) four nuclear markers corresponding to two satellite-DNA families: the family HindIII [25] and the family PstI [26]; non-transcribed sequences of 5S ribosomal gene (NTS) [27] and 230 base pairs from nuclear DNA flanking the microsatellite Aox-23 [28]. 2) four mitochondrial markers corresponding to two fragments of the cytochrome b gene of 212 bp and 265 bp, respectively [29,30], one fragment of 210 bp corresponding to the mitochondrial region control, d-loop, [30], and one fragment of the 12S ribosomal gene of 139 bp [16]. In each case, the PCR reactions were carried out with the amplification conditions described in each of the references.

Each marker was cloned using the vector TOPO TA (TOPO TA Cloning® kit PCR® 2.1) and were used to transform the cells DH5α of E. coli, according to the supplier recommendations (Invitrogen Carlsbad, CA, USA).

Recombinant plasmids were sequenced on both strands using Big Dye⁰ Terminator Cycle Sequencing Kit (Applied Biosystems) and T7 and M13 primers in an ABI Prims⁰ 3100-Avant Genetic Analyzer DNA Sequencer (Applied Biosystems).

Sequence Analysis

Multiples alignments of sequences obtained from the samples and reference sequences from GenBank database were performed using ClustalX software [31]. Phylogenetic and molecular evolutionary analyses were conducted using MEGA version 4 [32]. Sequence divergences were calculated according to the Jukes-Cantor method and distance trees produced by UPGMA [33] and the neighbor-joining method [34].

RESULTS AND DISCUSSION

The (**Table 1**) presents a summary for all the seven sturgeon specimens from Iberian Peninsula analysed for different molecular markers. This table includes the data obtained in this new study, completed with the data from previous analyses made by us.

The specimen UGP (**Table 1**) had previously been analyzed for four markers (HindIII and PstI satellite DNA family, 212-bp cytochrome b and 12S mitochondrial gene) and catalogued as A. sturio [4]. Considering nuclear markers, Garrido-Ramos et al. [4] analyzed this specimen for the HindIII satellite DNA family, showing the lack of this repetitive sequence in its genome (its absence is characteristic of the species A. sturio; [35]) In the same study, these researchers showed that the sequences corresponding to PstI satellite DNA family analyzed for this specimen UGP were grouped, in a phylogenetic tree, together with the sequences of A. sturio.

Now, nine clones have been sequenced for non-transcribed sequences of the 5S ribosomal nuclear genes (NTS), and their sequences were aligned with NTS ribosomal genes from other sturgeon species (**Figure 2**). Characteristic positions for A. sturio and A. oxyrinchus are present in the sequences isolated from UGP. Thus, in a phylogenetic tree based in genetics distances, all sequences belonging to this sample were grouped together with the NTS sequences of A. sturio (**Figure 2**).

Additionally, a new nuclear marker Aox23 locus [28] was amplified in this specimen. The sequence found, when compared with sequences of A. sturio and A. oxyrinchus taken from GenBank, proved similar to those of A. sturio (data not shown).

Table 1. Summary of sturgeon specimens analysed.

Specimen	Provenance (year of catch)	Sampling location (preservation)	Traditional Classification	HindIII	PstI	NTS	Aox23	212-bp Cyt b	265-bp Cyt b	d-loop	12S mitochondrial gene	Molecular status
UGP	Guadalquivir river (nineteenth century)	Museum of the Animal Biology Department, Facultad de Ciencias, Univ of Granada, Spain (stuffed)	A. sturio	np	6 FN256417 to FN256422	9* FN256408 to FN256416	9* FN256399 to FN256407	FN256388	* FN256392	* FN256381	FN256367	A. sturio
EBD8173	Guadalquivir river, Alcalá del Río, Seville, Spain (1974)	Doñana Biological Station, Seville, Spain (ethanol)	A. sturio	Z50744	4 AJ543450, AJ543451, AJ543458, AJ543459	2 AJ543472, AJ543473		AJ543488	-	-	AJ543480	A. naccarii
EBD8401	Guadalquivir river, Coria del Río, Seville, Spain (1981)	Doñana Biological Station, Seville, Spain (ethanol)	A. sturio	2 AJ543464, AJ543465	6 AJ543452 to AJ543457	6 AJ543466 to AJ543471		AJ543485	-	-	AJ543482	A. naccarii
EBD8174	Guadalquivir river, Alcalá del Río, Seville, Spain (1975)	Doñana Biological Station, Seville, Spain (stuffed)	A. sturio	AJ543463	3 AJ543460 to AJ543462	5 AJ543474 to AJ543478		AJ543486	* FN256395	* FN256386	AJ543479	A. naccarii (nuclear) A. sturio (mitochondrial)
Acinipo	Archaeological Deposit of Ronda la Vieja (Ronda, Malaga, Spain)	Municipal Archaeological Museum of Ronda, Malaga, Spain	?	-	-	-	-	-	-	-	* FN256368	A. naccarii
Nerja E-VI 1963	Cave of Nerja (Malaga, Spain)	Provincial Archaeological Museum of Malaga, Spain	?									?
Nerja N/62-63 J6/VII Capa 25	Cave of Nerja (Malaga, Spain)	Provincial Archaeological Museum of Malaga, Spain	?									?

List of sturgeon specimens analysed, their current specific status and the results for the markers analysed in each specimen and the number of units sequenced (with their accession number) for each nuclear repetitive marker or the number of mitochondrial clones sequenced for each mitochondrial marker. np: not present; na: not amplified; the asterisk (*) shows the sequences found in this study; question mark (?) indicates unknown Traditional Classification and/or Molecular status.

With respect to mitochondrial markers, GarridoRamos et al. [4] analysed in this specimen the fragments of the mitochondrial DNA 212-bp cytochrome b and 12S gene and considered UGP as A. sturio. In the present work, two new mitochondrial markers have been amplified for this sample (265-bp cytochrome b and d-loop). It was found that all diagnostic positions for these markers correspond to the species A. sturio (Figures 3(a) and (b)).

Therefore, the results of eight nuclear and mitochondrial markers confirm the classification of this sample (UGP) as A. sturio. This affirmation is not surprising if we bear in mind, as mentioned in the Introduction, that the species A. sturio has been broadly described in most of rivers of the Iberian Peninsula.

On the other hand, previous molecular analyses carried out by our group in three samples from the Biological Station of Doñana (**Table 1**), identified two of them, EBD8173 and EBD8401, as A. naccarii, based both on the mitochondrial and on the nuclear markers [2-4]. Thus, the samples EBD8173 and EBD8401 have the HindIII satellite DNA family in their genome. This satellite DNA, as commented above, is absent in the A. sturio genome [2].

The presence of this repetitive sequence means that these two samples cannot be assigned to A. sturio, the only species that had previously been considered to live in the rivers of the Iberian Peninsula. Additional results using the markers PstI satellite DNA, non-transcribed sequences of 5S ribosomal gene (**Figure 2**), 212-bp cytochrome b and 12S mitochondrial gene (Figures 3(a) and 4) confirmed that EBD8173 and EBD8401 belong to A. naccarii [3,4].

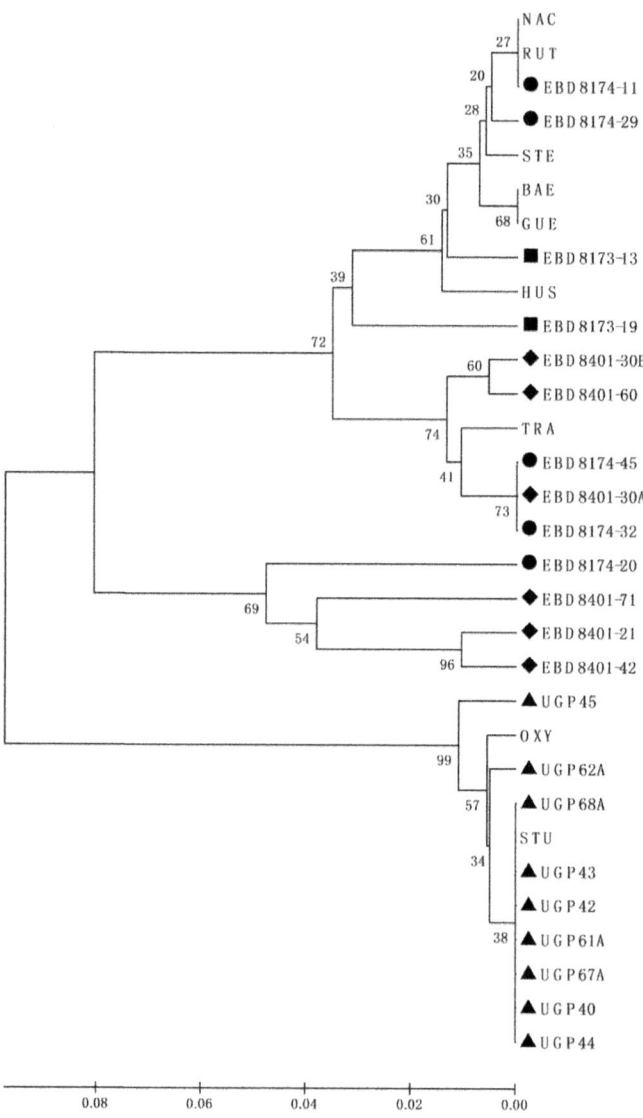

Figure 2. UPGMA tree based on NTS sequences of 5S ribosomal nuclear genes.

However, the sample EBD8174 (**Table 1**) is a special specimen from the genetic perspective. For all the nuclear markers analyzed to date, this sample EBD8174 cannot be assigned to A. sturio but to A. naccarii. The presence of HindIII satellite DNA family and the fact that all the sequences corresponding to the nuclear markers (the HindIII itself and satellite PstI and NTS) are not grouped with A. sturio but with A. naccarii are indicative of this fact [3,4]. However, previous mitochondrial DNA studies using 212-bp cytochrome b and 12S mitochondrial gene DNA markers [3,15,16], conclude that, in this specimen, mitochondrial DNA markers are similar to A. sturio.

UPGMA tree based on NTS sequences and Juckes Cantor distances calculated in MEGA 4. The tree shows the close relationships between the 9 NTS sequences from UGP (▲) specimen and NTS sequences of A. sturio (STU AJ550044) and A. oxyrinchus (OXY AJ555397), and between the 13 NTS sequences from the three EBD specimens -EBD8173 (■), EBD8174 (●) and EBD8401 (♦)- and NTS sequences of A. naccarii (NAC AJ550039) and other sturgeon species as A. transmontanus (TRA AJ555360), A. baerii (BAE AJ555351), A. gueldenstaedtii (GUE AJ555353), A. stellatus (STE AJ555385), Huso huso (HUS AJ555358) and A. ruthenus (RUT AJ555393). The code name species and the accession number are show into parenthesis. Numbers indicate bootstrap support for each node (10000 replicates).

Thus, the results of nuclear and mitochondrial DNA are contradictory in this specimen because, the nuclear DNA markers indicate its assignment to A. naccarii but mitochondrial DNA markers show identities to A. sturio. To confirm this situation, we analyzed two new mitochondrial markers, 265-bp cytochrome b and d-loop (Figures 3(a) and (b)). And the results coincided with previous ones, demonstrating that this sample corresponds to A. sturio for all mitochondrial markers. In fact, in the mitochondrial sequences analysed in this study (265-bp cytochrome b and d-loop), we found positions fixed with those of the species A. sturio Thus, the specimen EBD8174 could be considered a "mosaic" sturgeon: having nuclear characteristics of A. naccarii but mitochondrial markers of A. sturio. Hybridization or introgression processes between A. sturio and A. naccarii could explain this phenomenon. In sturgeons, genetic evidence of hybridisation phenomena between sympatric sturgeon species has been shown for example in Arefjev [36], and more recently between A. ruthenus and A. baerii in the Danube River [37].

Also, similar introgression processes have been described previously in the Adriatic region (A. gueldenstaedtii introgressed into the A. naccarii) [38] and in the population of the Baltic Sea of A. sturio (A. oxyrinchus introgressed into the A. sturio; [39]).

Finally, we have tried to clarify the specific status of three samples from archaeological sites (**Table 1**). Two of these samples (about 12,000 years old found at an older prehistoric settlement, the Cave of Nerja) were not successfully analysed. Unfortunately, none of the markers used could be characterized for these samples. However, we succeeded in amplifying a fragment of the 12S mitochondrial gene from the prehistoric scute (Ronda, Malaga, of about 3500 years of antiquity) found at the archaeological site of Acinipo (**Table 1**). These results are tentative because the first samples were very old and it was difficult to extract enough quality DNA to amplify the molecular markers. However, in previous studies some samples with similar antiquity at Acinipo, have been used successfully in species identification [17,40].

The 12S mitochondrial gene obtained from the Acinipo sample was compared with other 12S sequences from different species of sturgeons in the GeneBank database. The diagnostic positions for this marker did not coincide with A. sturio, ruling out its assignment to this species (**Figure 4**). In fact, all diagnostic sites coincided with A. naccarii, although they are not exclusive of this species, sharing them with other sturgeon species such as A. gueldenstaedtii, A. baerii, A. persicus and A. nudiventris with a distribution far away from the Iberian Peninsula. Thus, in a phylogenetic tree, based on genetic distances, the sequence from the Acinipo scute is grouped with the sequences from A. naccarii (**Figure 5**).

CONCLUSIONS

The nuclear and mitochondrial markers show that the specimens EBD8173 and EBD8401 belong to the species A. naccarii, and the sample UGP to A. sturio. The specimen EBD8174, using mitochondrial markers can be catalogued as A. sturio, or as A. naccarii according to nuclear markers. Hybridization or introgression processes between A. sturio and A. naccarii, could explain this phenomenon, common in sturgeons in these species. On the other hand, we were able to analyse the 12S mitochondrial marker for the ACINIPO sample (3500 years old) demonstrating that it belongs to species A. naccarii. These analyses provide insights into the existence of specimens belonging to A. naccarii in the southern Iberian Peninsula in historic (EBDs samples) and prehistoric (ACINIPO) times. Thus, our analyses confirm old references mentioning the presence of A. naccarii in the Iberian Peninsula [5-13]. Therefore, although A. naccarii is currently considered endemic of the Adriatic Sea, in the past it could have had a broader distribution area, extending to the Iberian Peninsula, including the Gua Neighbour-joining tree based on 12S mitochondrial gene sequences and Jukes-Cantor distances calculated in MEGA 4.

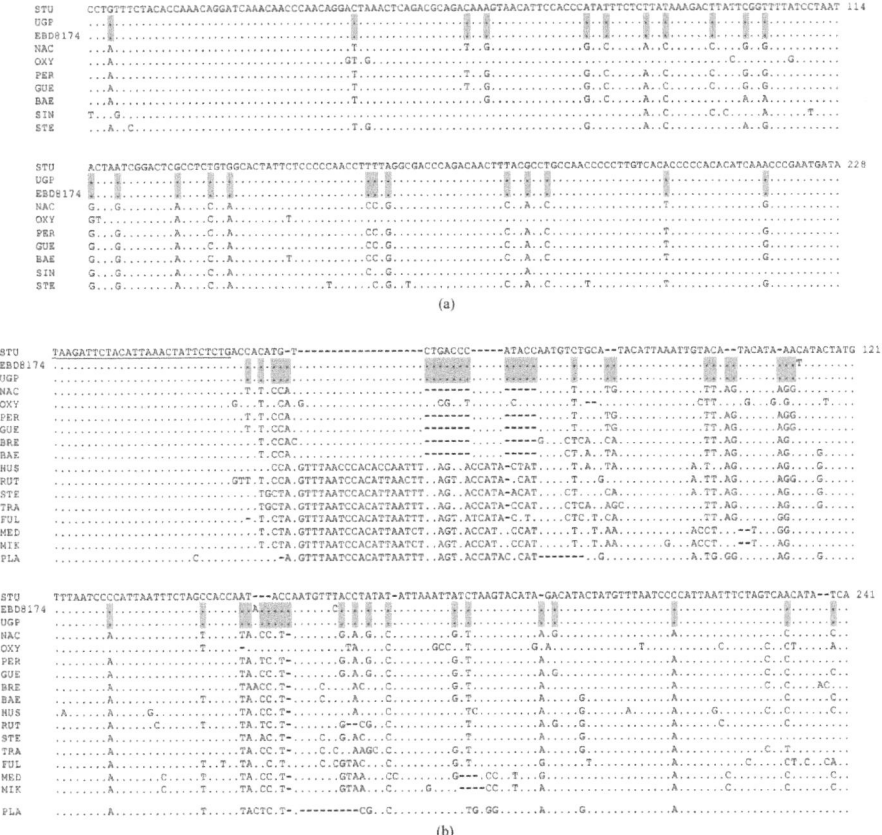

Figure 3. (a) Alignment of sequences of a 265-bp cytochrome b fragment. Multiple alignment of the sequences of a 265-bp cytochrome-b fragment from UGP and EBD8174, respectively. They are compared with the same mitochondrial DNA region from A. sturio (STU AJ245839), A. naccarii (NAC AJ245834), A. oxyrinchus (OXY AJ245838), A. persicus (PER AJ245835), A. gueldenstaedtii (GUE AJ245827), A. bacrii (BAE AJ245825), A. sinensis (SIN AJ252186), A. stellatus (STE AY846686). The grey boxes show the diagnostic sites used in the analysis. The primer sequence is not used in the alignment; (b) Alignment of partial d-loop sequences. Multiple alignment of the sequences of the d-loop from EBD8174 and UGP, respectively. These are compared with sequences of the same mitochondrial DNA region from 15 different species of sturgeon: A. sturio (STU AJ428274), A. naccarii (NAC AJ275199), A. oxyrinchus (OXY AJ249670), A. persicus (PER AJ275205), A. gueldenstaedtii (GUE AJ249668), A. brevirostrum (BRE AJ275194), A. baerii (BAE AJ249660), H. huso (HUS AJ249675), A. ruthenus (RUT AJ249671), A. stellatus (STE AJ249672), A. transmontanus (TRA AJ249674), A. fulvescens (FUL AJ249661), A. medirostris (MED AJ275188), A. mikadoi (MIK AJ275189) and S. platorynchus (PLA AJ249676).

The grey boxes show the diagnostic sites used in the analysis. The partial tRNAPro sequences are underlined. The alignment does not show the primer sequence.

```
STU      GGAAAGAAATGGGCTACATTTTCTGACACAGAAAACACACGAATAATACTGTGAAACCAGTGATTGAAGGTGGATTTAGCAGTAAAAAGAAAATAGAAA 99
UGP      ................................................................................................
EBD8174  ................................................................................................
EBD1873  ............................T...........C.................G....................................
EBD8401  ............................T...........C.................G....................................
Acinipo  ............................T...........C.................G....................................
NAC      ............................T...........C.................G....................................
OXY      .....................T......T...........C......................................................G.
GUE      ............................T...........C.................G....................................
BAE      ............................T...........C.................G....................................
PER      ............................T...........C.................G....................................
NUD      ............................T...........C.................G....................................
HUS      ............................T...........C.................G....................G...............
STE      ............................T...........C.................G....................G...............
RUT      ............................T...........C......................................................
FUL      .........................................C......................................................
BRE      ............................T...........C......................................................
TRA      ............................T...........C......................................................
SIN      ............................T...........C......................................................
SCH      ............................T...........C......................................................
DAU      ............................T...........C......................................................
MIK      ............................T...........C......................................................
MED      ............................T...........C......................................................
PLA      ............................T...........C......................................................
ALB      ............................T...........C......................................................

SUS      ............................T...........C......................................................
```

Figure 4. Alignment of sequences of 12S mitochondrial gene from Acinipo. Multiple alignment of sequences of 12S mitochondrial gene from Acinipo. These are compared with the same mitochondrial-DNA region from the three EBD and UGP specimens, A. sturio (STU AJ549115), A. naccarii ((NAC AJ549114) A. oxyrinchus (OXY AF402894), A. gueldenstaedtii (GUE FJ392605), A. baerii (BAE AY544135), A. persicus (PER AY544139), A. nudiventris (NUD AY544138), H. huso (HUS AY544146), A. stellatus (STE AY544144), A. ruthenus (RUT AY544140), A. fulvescens (FUL AF402885), A. brevirostrum (BRE AF402886), A. transmontanus (TRA AF402893), A. sinensis (SIN AY544143), A. schrenckii (SCH AY544142), H. dauricus (DAU AY544147), A. mikadoi (MIK AY544141), A. medirostris (MED AF125598), S. platorynchus (PLA AF402901), S. albus (ALB AY430247) and S. suttkusii (SUS AF402900). The grey boxes show the diagnostic sites used in the analysis. The alignment does not show the primer sequence.

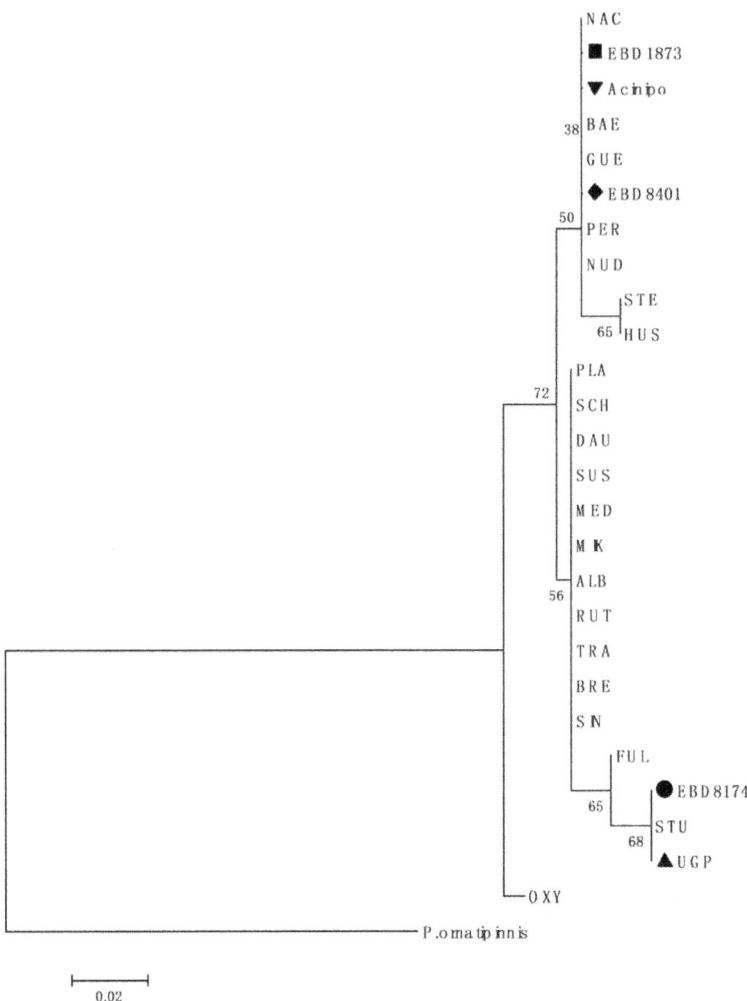

Figure 5. Neighbour-joining tree based on 12S mitochondrial gene sequences.

The tree shows the close relationships between sequences from Acinipo (▼), EBD8173 and EBD8104 with A. naccarii and between sequences from UGP and EBD8174 with A. sturio. Numbers indicate the bootstrap support for each node (10000 replicates). Polypterus. ornatinnis (Bichir NC001778) is used as outgroup. dalquivir River. Similarly, A. sturio was distributed not so long ago throughout Europe whereas, at the present, only one population exists, in the Gironde-GaronneDordogne River, France [41-44]. Furthermore, to propose a broad distribution area for A. naccarii is consistent with the general observation that most sturgeon species inhabited vast areas of continents and river basins [45]. Thus, observations based on molecular analyses, as we present in this paper, or the finding of an "American" species in Europe (i.e. the movement of A. oxyrinchus into Europe during the Little Middle Ages [46,47]), require more studies in order to establish a more complete vision of the distribution of different sturgeon species in Western Europe.

ACKNOWLEDGEMENTS

This research has been financed by grants of the Junta de Andalucía, Consejería de Innovación, Ciencia y Empresa (Proyecto de Investigación de Excelencia P07-CVI-03296) (F. Robles is a postdoctoral grant holder in this Project). Nerja Cave was analysed in the Project "Revisión, estudio y contextualización cronoestratigráfica de los restos arqueológicos procedentes de las antiguas excavaciones del Patronato de la Cueva de Nerja", authorized by Consejería de Cultura de la Junta de Andalucía to one of the authors (MCS). We also thank our colleague D. Nesbitt for revising our English text.

REFERENCES

1. Carmona, R., Domezain, A., García-Gallego, M., Hermando, J.A., Rodríguez, F. and Ruiz-Rejón, M. (Ed.) (2009) Biology, Conservation and Sustainable Development of Sturgeons, Fish & Fisheries, Series 29, Springer.

2. Garrido-Ramos, M.A., Soriguer, C., de la Herrán, R., Jamilena, M., Ruiz-Rejón, C., Domezain, A., Hernando, J.A. and Ruiz-Rejón, M. (1997) Morphometric and genetic analysis as proof of the existence of two sturgeon species in the Guadalquivir river. Marine Biology, 129(1), 33-39.

3. De la Herrán, R., Martínez-Espín, E., Lorente, J.A., Ruiz-Rejón, C., Garrido-Ramos, M.A. and Ruiz-Rejón, M. (2004) Genetic identification of western Mediterranean sturgeons and its implication for conservation. Conservation Genetics, 5(4), 545-551.

4. Garrido-Ramos, M.A., Robles, F., de la Herrán, R., Martínez-Espín, E., Lorente, J.A., Ruiz-Rejón, C. and Ruiz-Rejón, M. (2009) Analysis of mitochondrial and nuclear DNA markers in old museum sturgeons yield insights about the species existing in Western Europe: A. sturio, A. naccarii and A. oxyrinchus. In: Carmona, et al. Ed., Biology, Conservation and Sustainable Development of Sturgeons, Fish & Fisheries, Series 29, Springer, 25- 49.

5. Capello, F.B. (1869) Catalogo dos peixes do Portugal que existem do Museu de Lisboa. Jornal de Sciencias Mathematicas, Physicas, e Naturaes, 1(2), 131-193.

6. Capello, F.B. (1880) Catalogo dos peixes do Portugal. Mémoires de l'Académie Royale des Sciences, Lisboa, 6, 1-78.

7. Osório, B. (1894) D'algunas especies a juntar ao Catalogo dos peixes do Portugal de Capello. Jornal de Sciencias Mathematicas, Physicas, e Naturaes, 3(11), 186-188.

8. Nobre, A. (1931) Peixes das aguas doces de Portugal. Boletim Ministério da Agricultura, 13(2), 73-112.

9. Nobre, A. (1935) Fauna Marinha de Portugal. Vertebrados, Porto, 579.

10. Gonçalves, B.C. (1942) Colecçao ocenográfica de D. Carlos I. Catálogo dos Peixes. Travaux Station de Biologie Marine de Lisbonne, 46, 1-108.

11. Helling, H. (1943) Novo catalogo dos Peixes do Portugal em colecçao no Museu de Zoologia da Universidade de Coimbra. Memórias e Estudos do Museu Zoológico da Universidade de Coimbra, 149, 1-110.

12. Albuquerque, R.M. (1956) Peixes do Portugal e ilhas adjacentes. Chaves para a sua determinação. Portugaliae Acta Biológica, 5B, 195.

13. Bauchot, M.L. (1987) Poissons osseux. In: Fischer, W., Bauchot, M.L. and Schneider, M., Eds., Fiches FAO d'identification pour les besoins de la pêche (rev.1). Méditerranée et mer Noire. Zone de pêche 37, Commission des Communautés Européennes and FAO, Rome, 2, 891-1421.

14. Rincón, P.A. (2000) Putative morphometric evidence of the presence of Acipenser naccarii Bonaparte, 1836 in Iberian rivers, or why ontogenetic allometry needs adequate treatment. Boletín Instituto Español de Oceanografía, 16(1-4), 217-229.

15. Almodóvar, A., Machordom, A. and Suárez, J. (2000) Preliminary results from characterization of the Iberian Peninsula sturgeon based on analysis of the mtDNA cytochrome b. Symposium on Conservation of the Atlantic Sturgeon Acipenser sturio. L., 1758 in Europe, Boletín. Instituto Español de Oceanografía, 16(1-4), 17-27.

16. Gasent-Ramírez, J.M. Godoy, J.A. and Jordano, P. (2001) Identificación de esturiones procedentes del Guadalquivir mediante análisis de ADN en especimenes de museo. Publicaciones de la Consejería de medio Ambiente de la Junta de Andalucía, 36, 44-49.

17. Ludwig, A., Arndt, U., Debus, L., Roselló, E. and Morales, A. (2009) Ancient mitochondrial DNA analyses of Iberian sturgeons. Journal Applied Ichthyology, 25(1), 5-9.
18. Aguayo, P., Carrilero, M., Martínez, G., Afonso, J.A., Garrido, O. and Radial, B. (1989) Excavaciones Arqueológicas en el yacimiento de Ronda la Vieja (Acinipo). Campaña de 1988. Anuario Arqueológico de Andalucía 88/II, Consejería de Cultura, Junta de Andalucía, Sevilla, 309-314.
19. Carrilero, M., Aguayo, P., Garrido, O. and Padial, B. (2002) Autóctonos y fenicios en la Andalucía Mediterránea. en XVI Jornadas de Arqueología fenicio-púnica (Eivissa, 2001), Treballs del Museu Arqueològic d´ Eivissa i Formentera, Ibiza, 50, 69-125.
20. Simón Vallejo, M.D. (2003) Una secuencia con mucha prehistoria: la Cueva de Nerja. Mainake, 25, 249-274.
21. Cortés, M. (2004) Del Magdaleniense al Neolítico en la costa de Malaga. Novedades y perspectivas. Actas Jornadas Temáticas Andaluzas de Arqueología, Sociedades recolectoras y primeros productores, Junta de Andalucía, 109-122.
22. Martínez-Espín, E., Martínez-González, L.J., Álvarez, J.C., Roby, R.K. and Lorente, J.A. (2009) Forensic strategies used for DNA extraction of ancient and degraded museum sturgeon specimens. In: Carmona, et al., Ed., Biology, Conservation and Sustainable Development of Sturgeons, Fish & Fisheries, Series 29, Springer, 85- 96.
23. Donoghue, H.D., Spigelman, M., Greenblatt, C.L., Lev-Maor, G., Bar-Gal, G.K., Matheson, C., Vernon, K., Nerlich, A.G. and Zink, A.R. (2004) Tuberculosis: from prehistory to Robert Koch, as revealed by ancient DNA. Lancet Infectious Diseases, 4(9), 584-592.
24. Hagelberg, E. and Clegg, J.B. (1991) Isolation and characterization of DNA from archaeological bone. Proceedings of the Royal Society B: Biological Sciences, 244 (1309), 45-50.
25. De la Herrán, R., Fontana, F., Lanfredi, M., Congiu, L., Leis, M., Rossi, R., Ruiz-Rejón, C., Ruiz-Rejón, M. and Garrido-Ramos, M.A. (2001) Slow rates of evolution and sequence homogenization in an ancient satellite DNA family of sturgeons. Molecular Biology and Evolution, 18(3), 432-436.
26. Robles, F., De la Herrán, R., Ludwig, A., Ruiz-Rejón, C., Ruiz-Rejón, M. and Garrido-Ramos, M.A. (2004) Evolution of ancient satellite DNAs in sturgeon genomes. Gene, 338(1), 133-142.

27. Tagliavini, J., Williot, P., Congiu, L., Chicca, M., Lanfredi, M., Rossi, R. and Fontana, F. (1999) Molecular cytogenetic analysis of the karyotipe of the European Atlantic sturgeon, Acipenser sturio. Heredity, 83(5), 520-525.
28. King, T.L., Lubinski, B.A. and Spidle, A.P. (2001) Microsatellite DNA variation in Atlantic sturgeon (Acipenser oxyrinchus oxyrinchus) and cross-species amplification in the Acipenseridae. Conservation Genetics, 2(2), 103-119.
29. Ludwig, A. and Kirschbaum, F. (1998) Comparison of mitochondrial DNA sequences between the European and the Adriatic sturgeon. Journal of Fish Biology, Vol. 52(6), 1289-1291.
30. Ludwig, A., May, B., Debus, L. and Jenneckens, I. (2000) Heteroplasmy in the mtDNA control region of sturgeon (Acipenser, Huso and Scaphirhynchus). Genetics, 156(4), 1933-1947.
31. Thompson, J.D., Gibson, T.J., Plewniak, F., Jeanmougin, F. and Higgins, D.G. (1997) The ClustalX windows interface: flexible strategies for multiple sequence alignment aided by quality analysis tools. Nucleic Acids Research, 25(24), 4876-4882.
32. Tamura, K., Dudley, J., Nei, M. and Kumar, S. (2007) MEGA4: Molecular Evolutionary Genetics Analysis (MEGA) software version 4.0. Molecular Biology and Evolution, 24(8), 1596-1599.
33. Michener, D. and Sokal, R.R. (1957) A quantitative approach to a problem of classification. Evolution, 11(2), 130-162.
34. Saitou, N. and Nei, M. (1987) The neighbor-joining method: a new method for reconstructing phylogenetic trees. Molecular Biology and Evolution, 4(4), 406-425.
35. Fontana, F., Tagliavini, J. and Congiu, L. (2001) Sturgeon genetics and cytogenetics: Recent advancements and perspectives. Genetica, 111(1-3), 359-373.
36. Arefjev, V.A. (1997) Cytogenetics of interploid hybridization of sturgeons. Proceedings of the 3rd International Symposium on Sturgeon, Piacenza, Italy, 277.
37. Ludwig, A., Lippold, S., Debus, L. and Reinartz, R. (2009) First evidence of hybridization between endangered starlets (Acipenser ruthenus) and exotic Siberian sturgeons (Acipenser baerii) in the Danube River. Biological Invasions, 11, 753-760.
38. Ludwig, A., Congiu, L., Pitra, C., Fickel, J., Gessner, J., Fontana, F., Patarnello, T. and Zane, L. (2003) Nonconcordant evolutionary history of

maternal and paternal lineages in Adriatic sturgeon. Molecular Ecology, 12(12), 3253-3264.

39. Tiedemann, R., Moll, K., Paulus, K.B., Scheer, M., Williot, P., Bartel, R., Gessner, J. and Kirschbaum, F. (2007) Atlantic sturgeons (Acipenser sturio, Acipenser oxyrinchus): American females successful in Europe. Naturwissenschaften, 94(3), 213-217.

40. Pagès, M. Desse-Berset, N. Tougard, C. Brosse, L. Hänni, C. and Berebi, P. (2008) Historical presence of the sturgeon Acipenser sturio in the Rhône basin determined by the analysis of ancient DNA cytochrome b sequences. Conservation Genetics, 10(1), 217-224.

41. Rochard, E., Castelnaud, G. and Lepage, M. (1990) Sturgeons (Pisces: Acipenseridae), threats and prospects. Journal of Fish Biology, 37(Suppl A), 123-132.

42. Lepage, M. and Rochard, E. (1995) Threatened fishes of the world: Acipenser sturio Linnaeus, 1758 (Acipenseridae). Environmental Biology of Fishes, 43(1), 28.

43. Williot, P., Rochard, E., Castelnaud, G., Rouault, T., Brun, R., Lepage, M. and Elie, P. (1997) Biological and ecological characteristics of European Atlantic sturgeon, Acipenser sturio, as foundations for a restoration programme in France. Environmental Biology of Fishes, 48 (1-4), 359-370.

44. Williot, P., Arlati, G., Chebanov, M., Gulyas, T., Kasimov, R., Kirschbaum, F., Patriche, N., Pavlovskaya, L., Poliakova, L., Pourkazemi, M., Yu, K., Zhuang, P. and Zholdasova, I.M. (2002) Status and management of Eurasian sturgeon: An overview. International Review of Hydrobiology, 87(5-6), 483-506.

45. Choudhury, A. and. Dick, T.A (1998) The historical biogeography of sturgeons (Osteichthyes: Acipenseridae): A synthesis of phylogenetics, palaeontology and palaeogeography. Journal Biogeography, 25(4), 623-640.

46. Ludwig, A., Debus, L., Lieckfeldt, D., Wirgin, I., Benecke, N., Jenneckens, I., Williot, P., Waldman, J.R. and Pitra, C. (2002) When the American sea sturgeon Swam east. Nature, 419(6906), 447-448.

47. Ludwig, A., Arndt, U., Lippold, S., Benecke, N., Debus, L., King T.L. and Matsumura, S. (2008) Tracing the first steps of American sturgeon pioneers in Europe. BMC Evolutionary Biology, 8(1), 221-252.

Chapter 4

THE EFFICACY OF MOLECULAR MARKERS ANALYSIS WITH INTEGRATION OF SENSORY METHODS IN DETECTION OF AROMA IN RICE

H. Y. Yeap[1], G. Faruq[1], H. P. Zakaria[1], and J. A. Harikrishna[1,2]

[1]Institute of Biological Sciences, Faculty of Science, University of Malaya, 50603 Kuala Lumpur, Malaysia

[2]Centre for Research in Biotechnology for Agriculture, University of Malaya, 50603 Kuala Lumpur, Malaysia

ABSTRACT

Allele Specific Amplification with four primers (External Antisense Primer, External Sense Primer, Internal Nonfragrant Sense Primer, and Internal Fragrant Antisense Primer) and sensory evaluation with leaves and grains were executed to identify aromatic rice genotypes and their F_1 individuals derived from different crosses of 2 Malaysian varieties with 4 popular land races and 3 advance lines. Homozygous aromatic (fgr/fgr) F_1 individuals demonstrated better aroma scores compared to both heterozygous nonaromatic (FGR/fgr) and homozygous nonaromatic (FGR/FGR) individuals, while, some F_1 individuals expressed aroma in both leaf and grain aromatic tests without possessing the fgr allele. Genotypic analysis of F_1 individuals for the fgr gene represented homozygous aromatic, heterozygous nonaromatic and homozygous nonaromatic genotypes in the ratio 20:19:3. Genotypic and phenotypic analysis revealed that aroma in F_1 individuals was successfully inherited from the parents, but either molecular analysis or sensory evaluation alone could not determine aromatic condition completely. The integration of molecular analysis with sensory methods was observed as rapid and reliable for the screening of aromatic genotypes because molecular analysis could distinguish aromatic homozygous, nonaromatic homozygous and nonaromatic heterozygous individuals, whilst the sensory method facilitated the evaluation of aroma emitted from leaf and grain during flowering to maturity stages.

INTRODUCTION

Aroma is the most important quality trait of aromatic rice which commands a higher price than nonaromatic rice. Thus, aromatic or scented rice plays a vital role in global rice trading [1–3]. Several chemical constituents including different volatile compounds are the major causes of aroma in cooked rice [4–6]. Moreover, Bradbury et al. [7] reported that a recessive gene (fgr) on chromosome 8 of rice which contains an 8 bp deletions and 3 Single Nucleotide Polymorphism (SNPs) produced a nonfunctional Betaine Aldehyde Dehydrogenase 2 (BADH2) enzyme resulting in aroma in rice. Many molecular markers such as RFLPs, RAPDs, STSs, and iso-enzymes have been developed for fragrant rice selection and identification [8]. Meanwhile, two types of molecular marker that is, Simple Sequence Repeat (SSR) and Single Nucleotide Polymorphism (SNP) were identified as promising marker, because they are genetically linked to aroma [6,9–11]. In Addition, a perfect marker technique named Allele Specific Amplification (ASA) was developed by Bradbury et al. [12] for aroma genotyping and discriminating aromatic and nonaromatic rice. This technique was considered useful for selection of aromatic and nonaromatic rice genotypes in rice breeding programs [1]. In Malaysia, some constraints including high temperature during grain filling and ripening stage are slowing down the effectiveness of Maker-assisted selection for the improvement of aromatic rice varieties. So, aroma analysis throughout the life cycle using a combined sensory and molecular marker approach, may overcome these constraints by facilitating selection of the most appropriate parental materials for breeding programs [13, 14]. In this study, we evaluated the efficacy of molecular markers and integration of sensory methods with these molecular markers for the detection of aroma in different rice genotypes.

MATERIALS AND METHODS

Plant Materials

Six globally popular land races and eleven advance lines from International Rice Research Institute (IRRI) and three Malaysian cultivars from the Malaysian Agricultural Research Development Institute (MARDI) were used in this investigation (Table 1).

Table 1: Description of plant materials.

Source	Name	Type
	Khau Dau Mali, Rato Basmati, Ranbir Basmati, Sadri, Gharib, Kasturi	Land races
International Rice Research Institute (IRRI)	Entry-7 (IR 77734-93-2-3-2), Entry-11 (IR 78554-145-1-3-2) Entry-13 (IR 77512-2-1-2-2) Entry-14 (IR 77629-72-2-1-3) Entry-15 (M1-10-29 UL) Entry-16 (TOX 3226-5-2-2-2-2) Entry-18 (WAB 272-B-B-5-H5) Entry-19 (WAB 99-84) Entry-20 (WAB 337-B-B-15-H1) Entry-37 (PSB RC2 = IR 32809-26-3-3) Entry-38 (PSB RC18 = IR51672-62-2-1-1-2-3)	Advance lines
Malaysian Agricultural Research Development Institute (MARDI)	MRQ 72, MRQ 50 and MR 219	Malaysian cultivars

Crossing and Development of F_1 Seeds

Crosses were made at the experimental field of the Genetic and Molecular Biology Division of the Institute of Biological Science, University of Malaya (71.43°E, 30.2°N & 122 meter above the sea level) from 15th June 2010 to 30th July 2010. Among twenty genotypes, two local genotypes MR 219 (Homozygous nonfragrant) and MRQ 50 (Homozygous fragrant) were used as the female and seven fragrant genotypes (Entry-7, Entry-11, Entry-13, Sadri, Gharib, Rato Basmati, and Ranbir Basmati) were used as male to produce F_1 genotypes as they clearly demonstrated homozygous conditions by both ASA and sensory test.

Aroma Evaluation

The assessment of aroma in leaf was done according to Yeap et al. [15] which is a modified method of Sood and Siddiq [16] by using 0.2 g of leaf samples and cut into tiny pieces (<2 mm), but for grain aromatic test five (5) grains of each genotype were used following the method of Faruq et al. [14].

Extraction of DNA, PCR, and Genotyping

Young leaves were used for extracting total genomic DNA using Quick Extract plant DNA extraction solution from Epicentre biotechnologies

(USA). Primers were designed as TTGTTTGGAGCTTGCTGATG (ESP), CATAGGAGCAGCTGAAATATATACC (IFAP), CTGGTAAAAAGATTATGGCTTCA (INSP), and AGTGCTTTACAAAGTCCCGC (EAP) based on Bradbury et al. [12] which were synthesised by Medigene (Malaysia). PCR was performed using 2.0 µL of 10X reaction buffer (with 20 mM Mg^+), 0.2 µL of 10 mM dNTPs mix, 0.25 µL of YEAtaq DNA Polymerase (Yeastem Biotech Co. Ltd., Taiwan), 5.0 µL of DNA template, 0.4 µL of each primer EAP, and ESP, 0.5 µL of primer INSP and 0.5 µL of IFAP, the total volume were 20 µL. Amplification was carried out using a thermal cycler (C1000, BioRad, USA). Cycling conditions were performed 5 min at 94°C, followed by 35 cycles of 30 second at 94°C, 30 second at 53°C and 1 minutes at 72°C concluding with the final extension of 7 min at 72°C and hold at 4°C until recovery. Electrophoresis in 1.0% agarose gel and staining in ethidium bromide was done to analyse PCR products. PCR fragment size was estimated through 100 bp ladder (Vivantis, USA). The bands representing homozygous aromatic, homozygous nonaromatic, and heterozygous for fgr gene were analyzed by Allele Specific Amplification technique.

RESULTS AND DISCUSSION

Allele Specific Amplification (ASA) of Parental Materials

A set of 20 genotypes including 3 local checks (MRQ 50, MRQ 72, and MR 219) was chosen for aroma analysis of parental genotypes by Allele Specific Amplification. Among them Entry-11 and Gharib which scored 4 (Leaf & Grain aromatic test) for aroma were identified as homozygous for the fragrance gene (fgr) and genotype Sadri which also scored 4 (Leaf & Grain aromatic test) was identified as homozygous for the fgr gene. Moderate aroma (Mean aroma score 3) was found in Entry-7, Entry-13, Rambir Basmati, Rato Basmati, MRQ 50, and MRQ 72 which were also identified as homozygous for the fgr gene. On the other hand, Entry-14, Entry-15, Entry-16, Entry-18, Entry-19, Entry-20, Entry-37, Entry-38, and MR 219 which scored 1 (mean aroma) were identified as homozygous nonfragrant through ASA analysis. Surprisingly, Kasturi, with an aroma score 3 in the sensory test was scored as homozygous nonfragrant by ASA (Figure 1). Bounphanousay et al. [17] observed the same incident in a popular aromatic rice variety named Kai Noi Leuang from Laos.

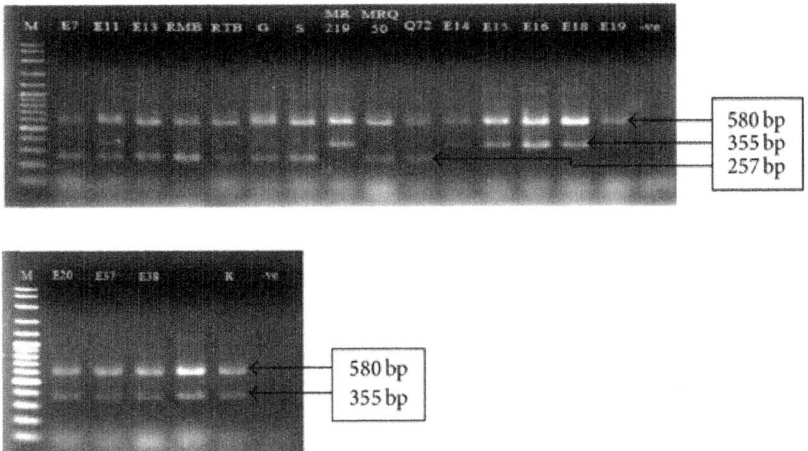

Figure 1: Aroma analysis for parental genotypes through single tube Allele Specific Amplification. E, Entry; RMB, Ranbir Basmati; RTB, Rato Basmati; G, Gharib; S, Sadri; K, Kasturi.

ASA analysis of parent materials resulted in 580 bp sized bands representing the positive control, amplified by both EAP and ESP external primers, while 355 bp bands indicated a PCR product amplified from the nonfragrant allele by the External Antisense Primer (EAP) and Internal Nonfragrant Sence Primer (INSP). The 257 bp bands indicated a PCR product amplified from the fragrant allele (fgr) by the External Sence Primer (ESP) and Internal Fragrant Antisence Primer (IFAP). All genotypes produced a 580 bp band but only 9 genotypes (Entry-7, Entry-11, Entry-13, Rambir Basmati, Rato Basmati, Garib, Sadris, MRQ 50, and MRQ 72) showed bands of 257 bp indicating fragrant genotypes. The remaining genotypes produced 355 bp bands indicating nonfragrant genotypes. Previously, Bradbury et al. [12] mentioned that it is possible to differentiate nonfragrant from fragrant rice varieties and to identify fragrant homozygous, nonfragrant homozygous and nonfragrant heterozygous genotypes by using this method.

Allele Specific Amplification (ASA) of F_1 Hybrids

Among the twenty parental genotypes, seven aromatic genotypes (Entry-7, Entry-11, Entry-13, Rambir Basmati, Rato Basmati, and Garib and Sadris) were crossed with nonaromatic (MR 219) and aromatic (MRQ 50) local cultivars. The F_1s derived from aromatic (7 genotypes) with nonaromatic (MR 219) crosses were slightly aromatic (Mean aroma score 2) and nonaromatic

(Mean aroma score 1) represented heterozygous nonaromatic and homozygous nonaromatic individuals respectively. On the other hand, aromatic (7 genotypes) with aromatic (MRQ 50) crosses were produced homozygous aromatic (Mean aroma score 3) and slightly aromatic (Mean aroma score 2) F_1 individuals.

In Figure 2, Lanes 1–14 represented the F_1 individuals derived from 14 different crosses with 3 replications, that is, 42 individuals. The 580 bp band was amplified from all individuals (positive control from ESP and EAP external primers). Bands of 355 bp (from amplification from the nonfragrant allele of fgr gene by primers EAP and INSP) and 257 bp (indicating presence of the fgr allele by the ESP and IFAP primers) were both amplified for 18 F_1 individuals. The presence of only the 580 bp with the 335 bp band was observed for 3 F_1 individuals whilst 21 individuals had only the 580 bp and 257 bp bands. The presence of 355 bp band indicated homozygous nonfragrant (without fgr allele) while 257 bp bands were represented the individuals as homozygous fragrant (homozygous for fgr gene). The individuals that represented both bands (355 bp and 257 bp) were identified as heterozygous nonfragrant individuals. Similar amplification pattern of fragrance (fgr) gene was observed by Bradbury et al. [12].

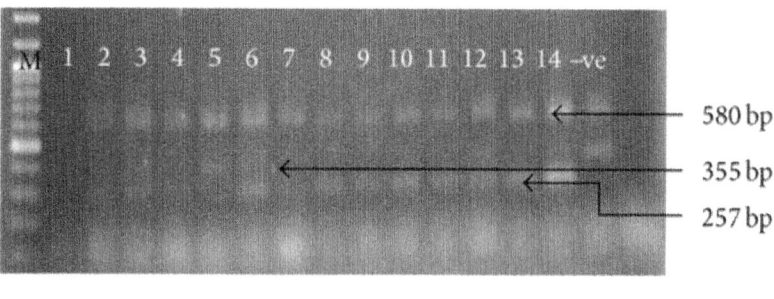

(a)

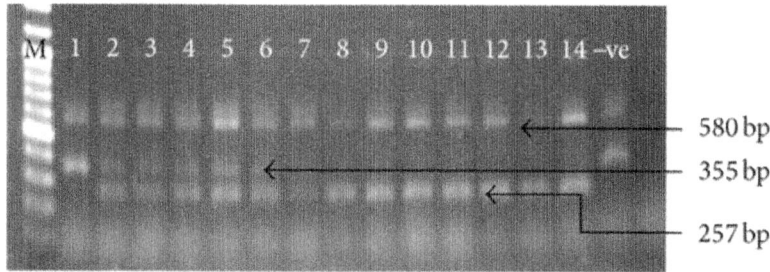

(b)

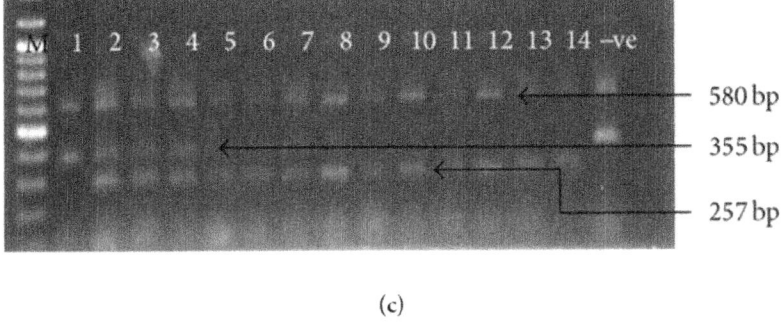

(c)

Figure 2: Fragrance analysis in F_1 individuals using single tube Allele Specific Amplification (ASA). Lane 1–7, Gharib, Entry-7, Rambir Basmati, Entry-13, Rato Basmati, Entry-11, and Sadri crossed with MR 219, respectively; Lane 8–14, Rambir Basmati, Rato Basmati, Sadri, Gharib, Entry-11, Entry-13, and Entry-7 crossed with MRQ 50, respectively; (a), (b), and (c) representing 3 biological replicates (total of 42 seeds).

During this screening process, homozygous aromatic, heterozygous nonaromatic, and homozygous nonaromatic genotypes appeared in the ratio 20:19:3, which suggests that there are 20 aromatic and 22 (19 + 3) nonaromatic F_1 individuals (Figure 2). Mohamad et al. [18] also observed a similar amplification pattern while used EAP, ESP, INSP, and IFAP primers (STS markers) in multiplex PCR condition and they identified 28 homozygous aromatic: 2 heterozygous nonaromatic: 45 homozygous nonaromatic, indicated 28 aromatic and 47 nonaromatic rice individuals. Meanwhile, another group of researchers, Bounphanousay et al. [17], detected 36 homozygous aromatic: 3 heterozygous nonaromatic: 17 homozygous nonaromatic whilst Sarhadi et al. [1] found 10 aromatic: 18 nonaromatic, also demonstrating the efficiency of these markers and 100% accuracy to detect this aroma allele. The results also confirmed the previous findings of Bradbury et al. [7] who demonstrated that the fragrance of basmati or jasmine rice were associated with the presence of a gene (fgr) on chromosome 8 of rice encoding nonfunctional BADH2.

Aroma Evaluation through Sensory Methods

In this investigation, leaf and grain of selected parents (9 genotypes) and their F_1 hybrid individuals were used for aroma evaluation (Table 2). From the leaf aromatic test, the highest mean aroma score was 3 while the lowest was 1. Individuals from five different crosses scored 3, from eight crosses scored 2 and from one cross scored 1. From grain aromatic test, the highest scoring was 2 and the lowest was 1. Individuals from 9 different crosses scored 2 and from 6 different crosses scored 1. Comparing the mean aroma scores from both methods, most produced a better leaf aroma score (Score 3) than the grain (score 2), except for the hybrids from MR 219/Rato basmati, where leaf aroma score was 1 and grain aroma score 2. F_1s from MRQ 50/Gharib, MR 219/Entry-7, MR 219/Rambir Basmati, MR 219/Sadri had the same aroma score (score 2) in both leaf and grain. Genotypic analysis within F_1 hybrids revealed that homozygous fragrant individuals produced better mean aroma score in leaf and grain than both heterozygous nonfragrant individuals and homozygous nonfragrant individuals, except for F_1 from MRQ 50/Entry-11, MRQ 50/Sadri. Through comparison of both genotypic and phenotypic characteristics of aroma in F_1 rice individuals, it was observed that aroma character from parents was successfully inherited to F_1 individuals.

Table 2: Aroma performance of parents and F_1 individuals in aromatic test (leaf and grain) and their genotypic expression.

Crosses	Leaf aroma			Grain aroma			Genotypic expression		
	♀ Parent	♂ Parent	F_1	♀ Parent	♂ Parent	F_1	♀ Parent	♂ Parent	F_1
MR 219/Gharib	1	3	2	1	4	1	N	H	N
MR 219/RMB	1	3	2	1	3	2	N	F	H
MR 219/RTB	1	3	1	1	2	2	N	F	H
MR 219/Sadri	1	4	2	1	4	2	N	F	H
MR 219/Entry-7	1	3	2	1	3	2	N	F	H
MR 219/Entry-11	1	4	2	1	4	1	N	H	H
MR 219/Entry-13	1	3	2	1	2	1	N	F	H
MRQ 50/Gharib	3	4	2	2	4	2	F	H	F
MRQ 50/RMB	3	3	3	2	3	2	F	F	F
MRQ 50/RTB	3	3	3	2	2	2	F	F	F
MRQ 50/Sadri	3	4	2	2	4	1	F	F	N
MRQ 50/Entry-7	3	3	3	2	3	2	F	F	F
MRQ 50/Entry-11	3	4	3	2	4	1	F	H	H
MRQ 50/Entry-13	3	3	3	2	2	2	F	F	F

Genotypic expression F, N, and H represented homozygous fragrance, homozygous nonfragrance and heterozygous nonfragrance, respectively. The number 1 to 4 represented aroma condition such as 1: absence of aroma, 2: slight aroma, 3: moderate aroma, and 4: strong aroma and RMB for Ranbir Basmati, RTB for Rato Basmati.

Integration of Sensory Methods and Molecular Markers for Detection of Aroma in Rice

Aromatic rice varieties emit aroma from their leaves, grains, and flowering organs at various stages of maturity [19]. There are many approaches used by researchers to determine the presence or absence of aroma in rice, such as evaluating aroma from leaves and grains with dilute KOH [16], analyzing the aroma using gas chromatography [20], and molecular markers related to rice aroma [12]. Sensory method facilitate the identification of aromatic and nonaromatic genotypes while molecular markers assist to identify specific allele but single tube allele specific amplification guides to identify zygosity (Homozygous or heterozygous for fgrgene) of individuals. In this study, we combined sensory tests and molecular marker methods for the detection of the presence or absence of aroma in parents and F_1 hybrids. The F_1 individuals, which were classified as having the aroma alleles (fgr gene) through molecular marker analysis, also showed presence of aroma in sensory tests. However, in less than 40% of the individuals which possessed aroma alleles and showed presence of aroma in sensory tests, the variation was from light aroma to strong aroma in leaf and grain aromatic tests. Less than 10% of the F_1 individuals did not exhibit aroma in grain aromatic test while carrying fgr gene and producing leaf aroma. In another cases, around 30–40% of F_1 individuals that were classified as homozygous nonfragrant (without fgr allele) but produced aroma in both the leaf and grain aromatic tests and <5% produced only grain aroma. Therefore, less than 50% of the F_1 individuals that were classified as aromatic or nonaromatic rice by ASA were scored the same way in both leaf and grain using sensory detection. While more than 50% of the individual's demonstrated aroma by leaf or grain or ASA or in combination of the tests. These results indicated that only molecular marker analysis or sensory methods could not represent the complete aromatic conditions. Bounphanousay et al. [17] reported that the molecular marker results agreed well with chemical analysis in most of the rice varieties, except some contrasting results such as in a local aromatic rice variety, Kai Noi Leuang, which produced aroma but was identified as homozygous nonaromatic by molecular marker analysis. They suggested that different gene location might be responsible for the observed aroma or the presence of another major aromatic compound. Sarhadi et al. [1] reported coincidence between results from 1.7% KOH sensory testing and molecular marker analysis for the classification of aromatic and nonaromatic rice, but occasionally molecular markers could not classify heterozygous and homozygous genotypes. Yi et al. [2] also reported that variation in the sensory score may arise from minor genes or environmental factors and that some rice varieties may carry minor QTLs which have an influence on rice aroma.

CONCLUSION

Aroma evaluation of rice genotypes is complicated in the tropical environment (countries like Malaysia) because of the large effects of environment and low sense of heritability. The integration of molecular markers and sensory tests can make the evaluation more effective. In allele specific amplification method, Entry-11, Gharib and Sadri was identified as homozygous for the fragrance allele (fgr gene), while the aroma scores were 4 in the sensory test. Genotypes Entry-13, Rato Basmati, Entry-7, Rambir Basmati, and two local checks MRQ 50 and MRQ 72 which scored 3, were also identified as homozygous for fragrance gene, but Kasturi with an aroma score of 3 in the sensory test was found as homozygous nonfragrant. However, homozygous aromatic F_1 individuals possessed higher mean aroma score in leaf and grain compared to heterozygous and homozygous nonaromatic individuals. High aroma score was observed in F_1s of MRQ 50/Entry-13, MRQ 50/Rambir Basmati and MRQ 50/Rato Basmati (leaf: 3; grain: 2). So, integration of sensory methods (Grain and Leaf aromatic test) along with allele specific amplification of 3 SNPs with 4 primers (ESP, EAP, INSP and IFAP) were observed as reliable, fast, and cost effective techniques in identifying parental materials and F_1 individuals to evaluate rice aroma in this investigation.

ACKNOWLEDGMENT

The authors wish to express their gratitude to the International Rice Research Institute (IRRI) and Malaysian Agricultural Research Development Institute (MARDI) for supplying rice genotypes and the Ministry of Higher Education Malaysia (MOHE) and the University of Malaya, Kuala Lumpur, Malaysia for IPPP (PV044-2011A) and UMRG (no. RG 033/10 BIO) Grants during this research.

REFERENCES

1. W. A. Sarhadi, N. L. Hien, M. Zanjani, W. Yosofzai, T. Yoshihashi, and Y. Hirata, "Comparative analyses for aroma and agronomic traits of native rice cultivars from Central Asia," Journal of Crop Science and Biotechnology, vol. 11, pp. 17–22, 2011.

2. M. Yi, K. T. Nwea, A. Vanavichit, W. Chai-arree, and T. Toojinda, "Marker assisted backcross breeding to improve cooking quality traits in Myanmar rice cultivar Manawthukha," Field Crops Research, vol. 113, no. 2, pp. 178–186, 2009.

3. K. Sakthivel, R. M. Sundaram, N. S. Rani, S. M. Balachandran, and C. N. Neeraja, "Genetic and molecular basis of fragrance in rice," Biotechnology Advances, vol. 27, no. 4, pp. 468–473, 2009.

4. I. Yajima, T. Yanai, M. Nakamura, H. Sakakibara, and K. Hayashi, "Volatile flavor components of cooked kaorimai (scented rice, O. sativa japonica)," Agricultural and Biological Chemistry, vol. 43, no. 12, pp. 2425–2429, 1979.

5. L. M. Nijssen, C. A. Visscher, H. Maarse, L. C. Willemsens, and M. H. Boelens, "Volatile compounds in foods," in Qualitative and Quantitative Data, L. M. Nijssen, C. A. Visscher, H. Maarse, L. C. Willemsens, and M. H. Boelens, Eds., pp. 1.1–1.18, 13.1–13.5, TNO Nutrition and Food Research Institute, Zeist, The Netherlands, 7th edition, 1996.

6. G. M. Cordeiro, M. J. Christopher, R. J. Henry, and R. F. Reinke, "Identification of microsatellite markers for fragrance in rice by analysis of the rice genome sequence," Molecular Breeding, vol. 9, no. 4, pp. 245–250, 2002.

7. L. M. Bradbury, T. L. Fitzgerald, R. J. Henry, Q. Jin, and D. L. Waters, "The gene for fragrance in rice,"Plant Biotechnology Journal, vol. 3, no. 3, pp. 363–370, 2005.

8. M. Lorieux, M. Petrov, N. Huang, E. Guiderdoni, and A. Ghesquière, "Aroma in rice: genetic analysis of a quantitative trait," Theoretical and Applied Genetics, vol. 93, no. 7, pp. 1145–1151, 1996.

9. Q. Jin, D. Waters, G. M. Cordeiro, R. J. Henry, and R. F. Reinke, "A single nucleotide polymorphism (SNP) marker linked to the fragrance gene in rice (Oryza sativa L.)," Plant Science, vol. 165, no. 2, pp. 359–364, 2003.

10. S. Chen, J. Wu, Y. Yang, W. Shi, and M. Xu, "The fgr gene responsible for rice fragrance was restricted within 69 kb," Plant Science, vol. 171, no. 4, pp. 505–514, 2006.

11. K. Kibria, M. M. Islam, and S. N. Begum, "Screening of aromatic rice lines by phenotypic and molecular markers," Bangladesh Journal of Botany, vol. 37, no. 2, pp. 141–147, 2008.

12. L. M. Bradbury, R. J. Henry, Q. Jin, R. F. Reinke, and D. L. Waters, "A perfect marker for fragrance genotyping in rice," Molecular Breeding, vol. 16, no. 4, pp. 279–283, 2005.

13. F. Golam, K. NorZulaani, A. H. Jennifer et al., "Evaluation of kernel elongation ratio and aroma association in global popular aromatic rice cultivars in tropical environment," African Journal of Agricultural Research, vol. 5, no. 12, pp. 1515–1522, 2010.

14. F. Golam, Y. H. Yin, A. Masitah et al., "Analysis of aroma and yield components of aromatic rice in Malaysian tropical environment," Australian Journal of Crop Science, vol. 5, no. 11, pp. 1318–1325, 2011.

15. H. Y. Yeap, G. Faruq, and J. A. Harikrisna, "Aroma analysis in few rice genotypes," in Proceedings of the 9th Genetics Congress, Kuching Sarawak, Malaysia, September 2011.

16. B. C. Sood and E. A. Siddiq, "A rapid technique for scent determination in rice," Indian Journal of Genetics and Plant Breeding, vol. 38, no. 2, pp. 268–275, 1978.

17. C. Bounphanousay, P. Jaisil, J. Sanitchon, M. Fitzgerald, and N. S. Hamilton, "Chemical and molecular characterization of fragrance in black glutinous rice from Lao PDR," Asian Journal of Plant Sciences, vol. 7, no. 1, pp. 1–7, 2008.

18. O. Mohamad, N. Amiran, K. Hadzim et al., "Molecular screening for aroma in rice," in Plant Breeding News, C. H. Hershey, Ed., 1.34, FAO, New York, NY, USA, 195th edition, 2008, http://www.fao.org/ag/agp/agpc/doc/services/pbn/pbn-195.htm#a134.

19. S. Wongpornchai, T. Sriseadka, and S. Choonvisase, "Identification and quantitation of the rice aroma compound, 2-acetyl-1-pyrroline, in bread flowers (Vallaris glabra Ktze)," Journal of Agricultural and Food Chemistry, vol. 51, no. 2, pp. 457–462, 2003.

20. V. Laksanalamai and S. Ilangantileke, "Comparison of aroma compound 2-acetyl-1-pyrroline in leaves from pandan," Cereal Chemistry, vol. 70, no. 4, pp. 381–384, 1993.

Chapter 5

HOMOGENEOUS NATURE OF MALAYSIAN MARINE FISH EPINEPHELUS FUSCOGUTTATUS(PERCIFORMES; SERRANIDAE): EVIDENCE BASED ON MOLECULAR MARKERS, MORPHOLOGY AND FOURIER TRANSFORM INFRARED ANALYSIS

A'wani Aziz Nurdalila[1], Hamidun Bunawan[1], Subbiah Vijay Kumar[2], Kenneth Francis Rodrigues[2], and Syarul Nataqain Baharum[1]

[1]Institute of Systems Biology, Universiti Kebangsaan Malaysia, UKM Bangi, 43600 Selangor, Malaysia

[2]Biotechnology Research Institute, Universiti Malaysia Sabah, Jalan UMS, 88400 Kota Kinabalu Sabah, Malaysia

ABSTRACT

Taxonomic confusion exists within the genus *Epinephelus* due to the lack of morphological specializations and the overwhelming number of species reported in several studies. The homogenous nature of the morphology has created confusion in the Malaysian Marine fish species *Epinephelus fuscoguttatus* and *Epinephelus hexagonatus*. In this study, the partial DNA sequence of the 16S gene and mitochondrial nucleotide sequences of two gene regions, Cytochrome Oxidase Subunit I and III were used to investigate the phylogenetic relationship between them. In the phylogenetic trees, *E. fuscoguttatus* was monophyletic with *E. hexagonatus* species and morphology examination shows that no significant differences were found in the morphometric features between these two taxa. This suggests that *E. fuscoguttatus* is not distinguishable from *E. hexagonatus* species, and that *E. fuscoguttatus* have been identified to be *E. hexagonatus* species is likely attributed to differences in environment and ability to camouflage themselves under certain conditions. Interestingly, this finding was also supported by Principal Component Analysis on Attenuated Total Reflectance–Fourier-transform Infrared (ATR-FTIR) data analysis. Molecular, morphological and meristic characteristics were combined with

ATR-FTIR analysis used in this study offer new perspectives in fish species identification. To our knowledge, this is the first report of an extensive genetic population study of *E. fuscoguttatus* in Malaysia and this understanding will play an important role in informing genetic stock-specific strategies for the management and conservation of this highly valued fish.

INTRODUCTION

Groupers (*Epinephelinae* spp., Serranidae) are an important marine fish in the marine culture industry since they are easily bred in captivity. Groupers are highly prized and sought after since they have a higher market value compared to other marine fishes [1]. The fish are wildly distributed and can be found in the in the Atlantic, Mediterranean and Indo-Pacific region, including the Red Sea [2].

Grouper farming appears to have great promise for commercialization, and coastal cage culture has the potential for continued development. However, the unreliable and limited supply of grouper fry has hindered the growth of the grouper industry. For sustainable growth, the production of the fish fry in a controlled hatchery setting is needed to lessen the demand for supply from the wild [3]. Furthermore, precise taxonomic and species identification is essential for the appropriate managing of fishery resources.

Incorrect identifications of waste resources have cost the agricultural sector many millions of pounds, and have invalidated the results of entire research programmes. Kuo *et al.*, [4] also reported that inbreeding was expected to cause defective recessive alleles that will reduce trait qualities and also survival rates. These can result in growth depression due to the sensitivity to environmental stress, which can lead to trait depression, compromising the animals' fitness and disease resistance. As a result, the farmers may observe poor survival and growth of the seed and wild brood stock since some fishes are a low value species that were formerly not considered as food fish [5].

Under field conditions and natural habitat, the identification and diagnostic methods to distinguish the grouper species are typically based on the colour pattern and morphological characters. In the genus *Epinephelus*, taxonomic confusion often occurred to identify individual species due to lack of morphological distinctive characters and specializations as well as the overwhelming number of the species [6]. The identification of the grouper species in this genus is normally based on colour configuration and geographic zone, even if it is not really effective to discriminate among grouper genera by geographic locality [6].

Genetic markers such as the Cytochrome b (Cyt b) sequence and 16s rRNA have proven to be informative to facilitate the investigation of deeper evolutionary relationships in the genetic structure and gene flow and can improve species identification within members of the genus *Epinephelus* [6,7,8]. The first comprehensive molecular and phylogenetic relationship study on the genus *Epinephelus* was reported by Craig *et al.* [6] using 16S rDNA. The study suggested that the genus *Epinephelus* is paraphyletic by forming two distant clades. Further study on molecular phylogeny of the subfamily Epinephelinae was reported by Craig and Hasting [9] using two nuclear and two mitochondrial genes and confirmed the monophyly of the genera *Cephalopholis*, *Epinephelus* and *Mycteroperca*.

Previously, we reported a study identifying *E. fuscoguttatus* and *E. hexagonatus* using the Cyt b gene to construct phylogenetic tree. Based on the previous results, the phylogenetic tree of Maximum Parsimony (MP), Molecular Evolution (ME) and Neighbor-Joining (NJ) showed that the group cluster of the trees have a mix of individuals of those two different species [10]. In this study, we used the 16s rRNA, Cytochrome Oxidase Subunit I (COI) and III (COIII) genes as molecular markers to study the Malaysian marine fishes *Epinephelus fuscoguttatus* and *Epinephelus hexagonatus*. *Epinephelus fuscoguttatus*(Forsskal, 1775), also known as Brown-marbled grouper, locally known as the Tiger grouper and *Epinephelus hexagonatus*(Forster, 1801), known as Starspotted grouper. *Epinephelus* hexagonatus have been observed in several locations in Peninsular Malaysia and Borneo. Both of them have similar characteristics such as a black spot on the upper caudal peduncle, dark brown circular spots, and head, jaws, and gills cover with spots, however we believe that this is probably due to environment or the homogeneous nature of the morphology. In order to identify these two taxa, we performed molecular identification, morphology characterization and Attenuated Total Reflectance-Fourier-transform Infrared (ATR-FTIR) data analysis to solve this mystery.

RESULTS

The 16S and Cytochrome Oxidase Subunit I and III genes were partially sequenced from 125 samples. The number of base substitutions per site is based on the pairwise analysis of 16S, COI and COIII haplotypes. The evolutionary history was determined using the ME (data not shown), and MP (data not shown) also NJ methods (data are shown). The ME, MP, and NJ trees showed similar pattern for the 16S, COI and COIII genes.

Phylogenetic Trees

For the 16S gene, the bootstrap consensus NJ tree evolutionary relationships (Figure 1) determined from 1000 replicates to represent the evolutionary history of the taxa analysed. From the total of 125 samples that were successfully sequenced, six unique haplotypes were found in the analysis. The consistency index is 0.782609; the retention index is 0.871795; and the composite indices are 0.815911 and 0.782274 for all sites and parsimony-informative sites, respectively. Branches corresponding to partitions reproduced in less than 70% of the bootstrap replicates are collapsed [11]. The tree is drawn to scale. Branch lengths were calculated using the average pathway method and are in units of the number of changes over the entire sequence. The statistical values in genetic distances among the grouper samples ranged from 0.009 to 0.015. Out of 125 samples, 54 were identified as *E. hexagonatus* and 71 were identified as *E. fuscoguttatus*.

The sequences of COI were deposited in GenBank with accession numbers JN674159 to JN674166. The number of base substitutions per site is based on the pairwise analysis of 76 sequences. The evolutionary history was inferred using the MP, ME, and NJ methods. The ME optimal tree branch length is 0.94837547. The percentage of replicate trees in which the associated taxa clustered together in the bootstrap test (1000 replicates) is shown next to the branches. The phylogenetic tree was linearised assuming equal evolutionary rates in all lineages. We did not observe major changes in the genealogy of the *E. fuscoguttatus* and *E. hexagonatus* sequences.

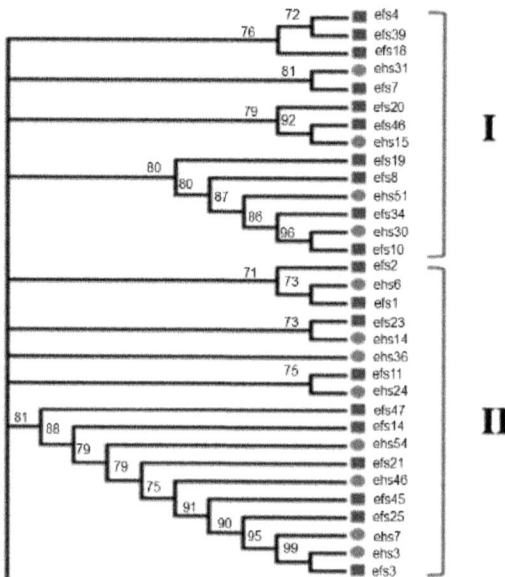

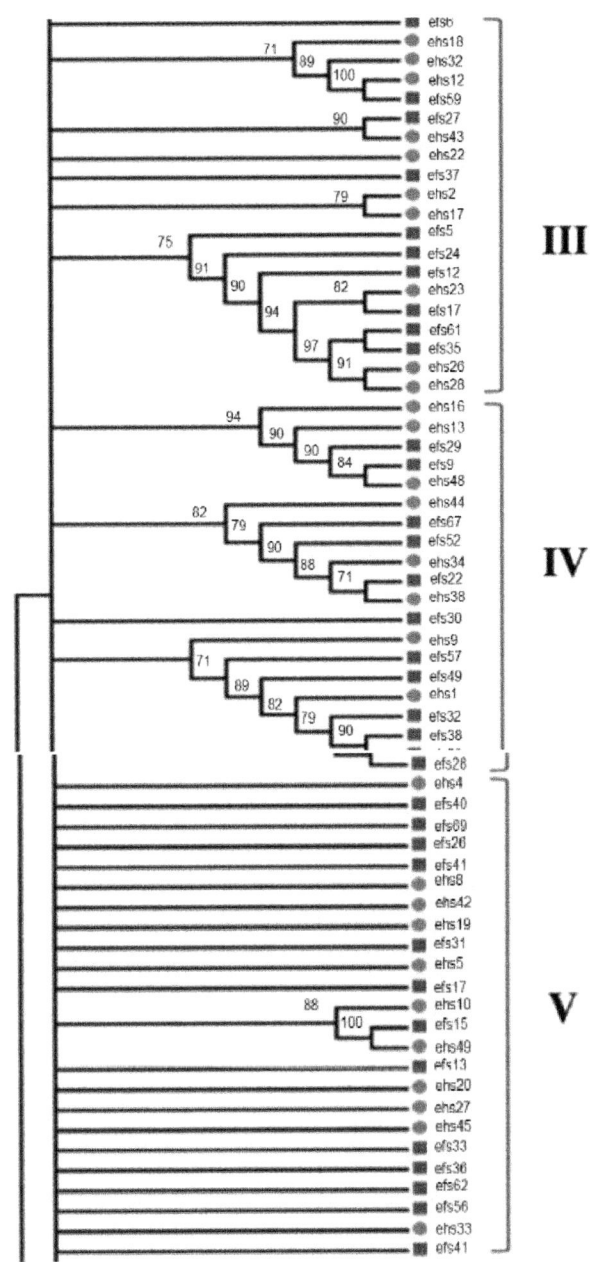

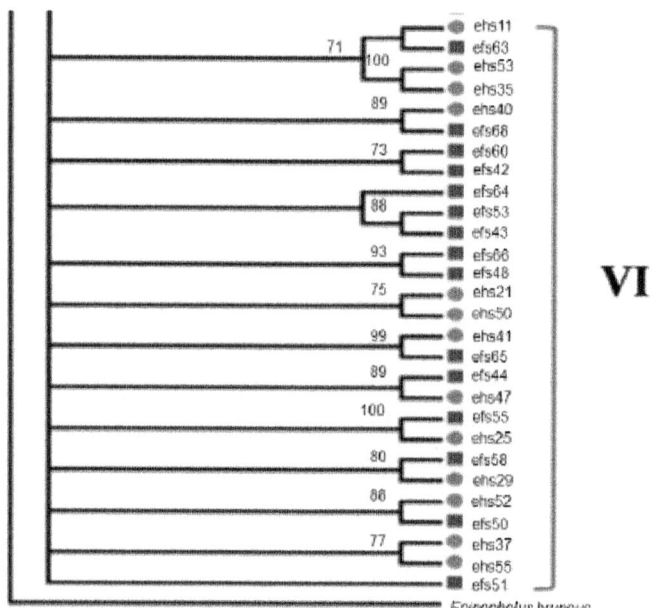

Figure 1. NJ evolutionary of 16S gene. The square shapes represent *E. fuscoguttatus*, and the round shapes represent *E. hexagonatus*. Numbers represent grouper cluster. Branches corresponding to partitions reproduced in less than 70% of the bootstrap replicates are collapsed. I–VI represent six unique of haplotypes.

For the MP method (data not shown), the consensus tree inferred from the 65 most parsimonious trees is shown. Branches corresponding to partitions reproduced in less than 70% of trees are collapsed. For parsimony-informative sites, the consistency index is 0.917730, the retention index is 0.803582, and the composite index is 0.716039. For all sites, the composite index is 0.748753. The percentage of parsimonious trees in which the associated taxa clustered together is shown next to the branches. The evolutionary history was inferred using the NJ method. The NJ optimal tree branch length is 0.91644275 (data not shown). The percentage of replicate trees in which the associated taxa clustered together in the bootstrap test (1000 replicates) is shown next to the branches. The phylogenetic tree was linearised assuming equal evolutionary rates in all lineages.

An average of 1500 nucleotide base pairs per taxon was amplified from DNA of *E. fuscoguttatus* and *E. hexagonatus* by employing the COIII primer pairs. No insertions/deletions or stop codons were observed with high sequence similarity (98%–100%) to their respective species data in GenBank. Thus, the correct identity of these species was confirmed. The sequences of COIII were deposited in GenBank with accession numbers JN859014 to JN859041. The evolutionary history of the COIII gene was also inferred using the MP, ME, and NJ methods. The ME optimal tree branch length is 0.75335376. The percentage of replicate trees in which the associated taxa clustered together in the bootstrap test (1000 replicates) is shown next to the branches. The tree is drawn to scale, with branch lengths in the same units as those of the evolutionary distances used to infer the phylogenetic tree. The transition/transversion ratio, base frequencies, and gamma distribution shape parameter as estimated were Ti/Tv = 8.7879; A = 0.2885, C = 0.3589, G = 0.1481, and T = 0.2044; and α = 2.5876, respectively. For the MP tree, the consistency index is 0.725490, the retention index is 0.810902, and the composite index is 0.835294 for the parsimony-informative sites. The composite index is 0.764407 for all sites (data not shown).

The optimal neighbour-joining tree a branch length is 0.75584609 (data not shown). The percentage of replicate trees in which the associated taxa clustered together in the bootstrap test (1000 replicates) is shown next to the branches. These results indicate that the topological structures of the two trees were nearly identical. They showed relatively close relationships among the species groups and were strongly supported by bootstrap values.

Morphological Analysis

Ten specimens of *E. fuscoguttatus* and *E. hexagonatus* (Figure 2) from Terengganu and Borneo, ranging from 58.0 to 61.0 cm total length (TL) and 6.0 to 8.5 kg body weight (BW), were used for morphometric and meristic characteristic analyses (Figure 2). The main morphometric and meristic data are reported in Table 1 and Table 2, respectively. Fork length (FL), standard length (SL), head length (HL), caudal fin length (CFL), dorsal fin length (DFL), pectoral fin length (PFL), anal fin length (AFL), mouth length (ML), snout length (SnL), body width length (BwL), pelvic fin ray, first dorsal fin ray, second dorsal fin ray, anal fin ray, caudal fin ray and pectoral fin ray were examined. Body depth was 2.6 to 2.9 times in standard length and the lateral-body scales of fish is approximately more than 10 cm standard length smooth, with auxiliary scales of shape.

Figure 2. Morphology, morphometric and meristic characteristics (**A**) *Epinephelus hexagonatus*; (**B**) *Epinephelus fuscoguttatus*; (**C**) Bones of *Epinephelus hexagonatus* and (**D**) Bones of *Epinephelus fuscoguttatus*.

Table 1. Definitions of morphometric measurements and meristic counts used in this study.

Characters	Description
11 Morphometrics	
Total length (TL)	Tip of the lower jaw to the end of caudal fin
Fork length (FL)	Tip of the upper jaw to the tail base

Standard length (SL)	Tip of the upper jaw to the end of caudal fin
Head length (HL)	From the front of the upper lip to the posterior end of the opercular
Caudal fin length (CFL)	From tail base to tip of the caudal fin
Dorsal fin length (DFL)	Front of the upper lip to the origin of the dorsal fin
Pectoral fin length (PFL)	From base to tip of the pectoral fin
Anal fin length (AFL)	Front of the upper lip to the origin of the anal fin
Mouth length (ML)	Straight line measurement between the snout tip and posterior
Snout length (SnL)	The front of the upper lip to the flesh anterior edge of the orbit
Body width length (BwL)	The greatest width just posterior to the gill opening
6 Meristic	
Pelvic fin ray	Number of soft fin rays in the pelvic fin
1st dorsal fin ray	Number of soft fin rays in 1st dorsal fin
2nd dorsal fin ray	Number of soft fin rays in 2nd dorsal fin
Anal fin ray	Number of soft fin rays in anal fin Caudal fin ray
Anal fin ray	Number of soft fin rays in anal fin Caudal fin ray
Pectoral fin ray	Number of soft fin rays in pectoral fin

Table 2. Morphometric and meristic measurements.

Morphometrics								Meristics					
Measurements (cm)	Min		Max/Mean		Mean ± SD		TL (%)		Measurement	Range		Mean ± SD	
	EF	EH	EF	EH	EF	EH	EF	EH		EF	EH	EF	EH
Total length (TL)	58	58	61	59	59.5 ± 1.13	58.5 ± 1.21							
Fork length (FL)	50.5	49.4	53.7	53.3	52.3 ± 1.31	51.4 ± 0.89	87.8	87.8	Pelvic fin ray	6 to 7	6 to 7	6.5 ± 0.53	6.5 ± 0.5
Standard length (SL)	48.3	45.2	51.7	52	50.1 ± 1.18	48.6 ± 1.03	84	83.1	1st dorsal fin ray	11 to 12	11 to 12	11.6 ± 0.52	11.5 ± 0.5

Head length (HL)	14	14	15.5	14.8	14.8 ± 0.47	14.4 ± 0.58	24.9	24.6	2nd dorsal fin ray	9 to 13	9 to 14	11.1 ± 1.73	11.5 ± 1.78
Caudal fin length (CFL)	5.7	5.5	6.4	7.1	6.07 ± 0.25	6.3 ± 0.31	10.2	10.8	Anal fin ray	7 to 9	7 to 8	8 ± 0.82	7.5 ± 0.73
Dorsal fin length (DFL)	13.2	14.2	16.4	15.8	15.0 ± 1.13	15 ± 1.03	25.3	25.6	Caudal fin ray	14 to 15	14 to 15	14.5 ± 0.53	14.5 ± 0.42
Pectoral fin length (PFL)	8.2	8.2	9.5	9.5	8.85 ± 0.43	8.85 ± 0.4	14.9	15.1	Pectoral fin ray	18 to 20	17 to 20	18.8 ± 0.92	18.5 ± 1.13
Anal fin length (AFL)	7.9	7.5	11.3	10.8	9.62 ± 1.21	9.15 ± 1.33	16.1	15.6					

EF: *Epinephelus fuscoguttatus*; EH: *Epinephelus hexagonatus*.

Attenuated Total Reflectance-Fourier Transform Infrared Analysis (ATR-FTIR)

Quantitative FT-IR data for each sample are given in Figure 3. The PCA of these FT-IR data is displayed in a two-dimensional plot using the first two principal components (Figure 3). Three replicate samples of each individual were grouped in discrete clusters, indicating that PCA is able to discriminate individuals by sample and location. The FTIR spectrum (Figure 3) resulted in six functional groups consisting of amines, carboxylic acids, alkenes, and alcohols (Table 3). The patterns of absorptions are similar for both samples; however, the FTIR spectrum shows different intensities at the 3500–3300 cm^{-1} (Amines), 1700–1500 cm^{-1} (Carboxylic acids), and 1430–1290 cm^{-1} (Alkenes) regions.

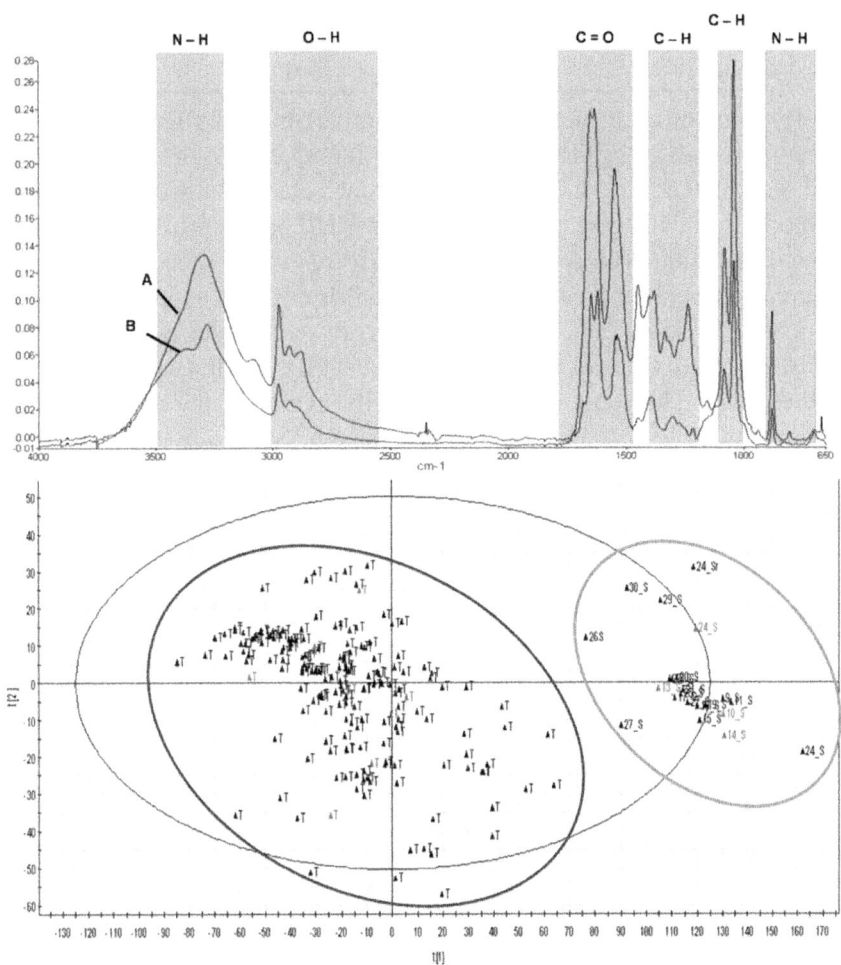

Figure 3. Two-dimensional plot using the first two principal components. Green cluster represents location from Sabah (A) and blue cluster represents location from Terengganu (B). Red plot represents *E. hexagonatus* and black plot represents *E. fuscoguttatus*.

Table 3. Functional Group of FTIR Spectrum.

Functional Group		Spectrum Region (cm^{-1})
Amines	N–H Stretch	3500–3300
Carboxylic acids	O–H Stretch	3000–2500
Carboxylic acids	C=O Stretch	1700–1500
Alkenes	C–H Bend	1430–1290

| Alcohols | C–O | 1260–1000 |
| Amines | N–H Bend | ~800 |

The score scatter plot divided the samples into two groups by location along PC1 and PC2. The samples from Terengganu were clustered on the right side of the plot, and the samples from Sabah (Borneo) were clustered on the left, which indicates that the ATR-FTIR is able to separate samples based on location. The separation of individuals in the score scatter plot was in agreement with known locations of the fish and also supports the findings of a previous study [10]. However, the score scatter plot could not separate *E. fuscoguttatus* from *E. hexagonatus*, indicating that the samples should be in one group. The score scatter plot results supported the phylogenetic tree results from molecular analysis. The non-separation of individuals in the score scatter plot was in agreement with the known molecular analysis of the fish. These results confirm the same putative introgressed sequences indicated by the phylogenetic reconstructions.

DISCUSSION

Molecular Analysis

The congruent results by the MP and NJ analyses further reinforce our preliminary investigation that *E. fuscoguttatus* and *E. hexagonatus* form a monophyletic clade and that the population is phylogenetically closer to each other. We found no significant differences among these types of calibration/estimated nodes. The molecular data demonstrate that there is little genetic distance between the *E. fuscoguttatus* and *E. hexagonatus* (e.g., Kimura 2-parameter distance is 0.008 and 0.029) indicating a very close relationship on the molecular level. Phylogenetic inference by neighbor-joining and maximum-parsimony analysis revealed that *E. fuscoguttatus* and *E. hexagonatus* representing a similar taxon analysed either in COI, COIII or in 16S. This study also supported our previous phylogenetic relationships of *E. fuscoguttatus* and *E. hexagonatus* using cytochrome b mtDNA markers [10].

This study has revealed a remarkable level of similarity between the *E. fuscoguttatus* and *E. hexagonatus*, using several molecular markers. Our molecular analysis, supported by high bootstrap values for the *E. hexagonatus* node, clearly demonstrated that the species *E. fuscoguttatus* is misidentified as *E. hexagonatus* for which sequence divergence is very low. It has been demonstrated in other groups of fishes that some morphological characters are clearly adaptative to nature and that this may be affected by recurrent evolution in different lineages and caused confusion in identifying species that resulting in misidentification [12,13,14].

Moreover, several studies based on molecular data have indicated that phylogenetic relationships in fish are not always congruent with the phylogeny based on morphological data [15]. Molecular data examined suggest that the previous practice of identifying grouper species by morphological data may not be effective in species recognition and in discerning the true relationships within Epinephelinae spp. These 125 samples with six unique haplotypes of 16S, COI and COIII are identifiable using PCR techniques on a segment of mitochondrial 16S. The ability to distinguish different species based on muscle tissues when the external characteristics have been removed is of great commercial importance. The mitochondrial 16S, COI and COIII are commonly targeted for phylogenetic analysis, and several unique nucleotide positions that are useful to identify the respective species have been described.

Morphological Analysis

The biometric analysis, including the meristic and morphometric characteristics, has been adopted by many studies to identify and relate different fish species and populations [16]. This trend in biometric analysis reflects its validity in stock identification in different fisheries throughout the world. The present findings confirmed the validity of this biometric approach. Moreover, the type of allometry was used to study intra- and inter-specific variations in some fish species [17]. The present work confirms this statement and emphasises the taxonomic significance of allometric criterion in revealing intra- and inter-specific variations of morphometric characters of *E. fuscoguttatus* and *E. hexagonatus*. In the present study, the total ray counts and length measurements were reliable for the differentiation between *E. fuscoguttatus* and *E. hexagonatus*. All patterns of morphometric variations referred to *E. fuscoguttatus* as not clearly separated from *E. hexagonatus* because they shared the same number of rays and length. The dorsal, pectoral, pelvic and anal fin rays and spines were found to be constant on the generic level.

Attenuated Total Reflectance-Fourier Transform Infrared Analysis (ATR-FTIR)

The IR spectrum is measured by calculating the intensity of the IR radiation before and after it passes through a sample and the spectrum is traditionally plotted with the Y-axis units as absorbance or transmittance and the X-axis units as wave number units. For quantitative purposes it is necessary to plot the spectrum in absorbance units [18]. The intensity and, more accurately, the areas of the absorption peaks in the FTIR spectrum, were directly related to the concentration of the molecules. The spectrum is quite complex and contains several peaks arising from the contribution of different functional groups belonging to proteins, lipids and carbohydrates. Analysis was performed using a scanning process conducted on wavenumbers 4000–400 cm^{-1} with a resolution of 4 cm^{-1}. The scanning results were percentage absorbance on the specific wavenumber for every functional group in each sample [19].

Each functional group has a specific marker group. Carbonyl groups as fatty acid markers were detected on wavenumber 1746 cm^{-1}. The fatty acid absorbencies in every sample were used to compare relative protein levels in samples. The functional group was determined by comparing the wave numbers of functional groups. Various types of functional groups exist in each sample. All of the functional groups were identified by FTIR because of the specific mechanism that FTIR offers. It has been previously demonstrated that tissue proteins, carbohydrates and lipids play a role in the sensitivity of environmental effects. The infrared of the protein is characterised by a set of absorption regions: the amide region and the C–H region. The peaks arise principally from the C=O stretching vibration of the peptide group which is primarily N–H bending with a contribution from C–N stretching vibrations. This C=O peak is sensitive to the environment of the peptide linkage and also depends on the proteins overall secondary structure. The ratio of the peak intensities observed (1541 and 3297 cm^{-1}) due to N–H bending and O–H stretching, respectively, could be used as indicators of the relative protein concentration and elucidate environmental effects on biological tissues [20].

The overall spectral profile is similar except for the variation in the intensities of the peak. The lower peak intensities represent samples from Terengganu and the higher peak intensities represent samples from Sabah. The peak at 3297 cm^{-1} is assigned to O–H stretching. The peaks observed at 2923 and 2853 cm^{-1} are due to the asymmetric and symmetric stretching of the membrane lipids, respectively. The other peak observed at 1653 cm^{-1} is assigned to the peak of C=O respectively. FT-IR analyses are often chosen by various studies because they are relatively fast, simple and require little or no sample preparation for spectral acquisition. Furthermore, the machine is non-destructive towards the sample cells/tissues because the samples remain intact during analysis. According to Chen *et al.* [21], FT-IR analysis is a universal method where the instruments and software are readily available and can be used for routine analysis as well as quantitative and qualitative analyses. The spectra provide information on the cell/tissue composition and the number of functional groups present.

Moreover, the FT-IR allows a multiple sample environment where samples can be in the form of liquid, gas, powder, solid, or film. Additionally, it is relatively less expensive for animals, plants and bacteria identification compared to several commonly used methods [22]. Multivariate statistical analysis of FT-IR spectra (chemometrics) can be divided into two types: supervised methods and unsupervised methods. The objective of unsupervised methods is to extrapolate the spectral data without prior knowledge of *E. fuscoguttatus* and *E. hexagonatus*. Principal component analysis (PCA) is an example of an unsupervised method.

PCA is used to reduce the multidimensionality of the data set into its most dominant components or scores while maintaining the relevant variation between the data points. PCA identifies the natural clusters in the data set with the first principal component (PC) expressing the largest amount of variation, followed by the second PC that conveys the second most important factor of the remaining analysis [23] and so forth. Score plots can be used to interpret the similarities and differences between *E. fuscoguttatus* and *E. hexagonatus*. The closer the samples are within a score plot, the more similar they are with respect to the principal component score evaluated. Studies using FT-IR have been utilised in plants [24,25,26], but few studies have been conducted with animals [27,28]. To date, no studies have been conducted on marine fish. This is the first study on marine fish using FT-IR to differentiate the origin and location of marine fish populations.

Misidentification due to homogenous nature leading to unrelated individuals being assigned to a full-sib family would not increase the inbreeding in the progeny. *E. hexagonatus*, *E. spilotoceps*, *E. macrospilos*, *E. howlandi*, *E. faveatus*, and *E. melanostigma* have always been frequently confused in the literature. They were often misidentified as other species since we know that grouper are highly stressed fish and when they become stressed, they tend to change their characteristics and camouflage to the surrounding environment [29].

If breeding with different species, especially with one that was not used as a food source, the outcome will be unfortunate for the farmers. Cross-breeding with wrong species will produce all sorts of wrong traits including growth retardation, slow feeding rate, low disease defenses and higher cannibalism amongst each other. This will result in death and high loss to the farmers. Criteria for choosing individuals that will be founders should be essentially the same as those used when the selection response is optimized under restricted co-ancestry when pedigree information is available. It is necessary to avoid mating between close relatives for managing existing quantitative genetic variation at the start of the programme [30]. Once it is clear that two species are conspecific, the cost for farm management and the rate of breeding between different species that will produce low value fish, low eggs production and low survival rate of the fry, is reduced [29]. Here, *E. fuscoguttatus* was misidentified as *E. hexagonatus* due to homogenous nature.

EXPERIMENTAL SECTION

Collection and Preservation

A piece of caudal fin area (fin clips at the edge of the fin) (20–50 mg) was collected from 125 individuals of grouper representing two taxa *i.e.*, *E. fuscoguttatus* and *E. hexagonatus*. The specimens were collected from five different locations of the Fisheries Research Institute, in Kampung Raja: 5.812213° N 102.589055° E, Kampung Seberang Timur: 5.991414° N 104.561976° E, Kampong Sentosa of Terengganu and 6.293416° N 105.61385° E Tanjung Badak: 6.201995° N 116.195526° E and Tuaran: 6.950465° N 121.144854° E of Sabah (Borneo) Malaysia were used for the molecular analysis. The samples were obtained from hatcheries and open sea cages. The specimens were then preserved in 95% ethanol at ambient temperature while in the field and at −20 °C under laboratory conditions. The specimens were homogenised in chilled nuclei lysis solution Wizard DNA isolation Kit (Promega, Madison, WI, USA) and kept at −20 °C. Samples for morphology and meristics were also collected at the same five locations of the same species.

DNA Extraction

DNA extraction was conducted using the Wizard DNA isolation Kit (Promega, USA) adapted for fin clip genomic DNA extraction.

Amplifying and Sequencing

Polymerase Chain Reaction (PCR) was used to amplify an approximately 450-bp fragment of the cyt b gene and an approximately 650-bp fragment of the 16S gene. Amplification reactions of 100 μL were prepared with 10–100 ng of DNA, 1.5 mM $MgCl_2$, 2.5 U of *Taq* DNA polymerase, 200 mM dNTPs, and 0.1 mM of each primer.

The following step procedure was used to amplify the 16S gene following a 5 min denaturation at 94 °C: 94 °C for 1 min 30 s, 54.8 °C for 2 min and 72 °C for 1 min 30 s. A total of 30 cycles were performed using an Eppendorf MasterCycler (Hamburg, Germany). The following primer set was used to amplify the 16S gene: S16M: 5'-CGCCTGTTTATCAAAAACAT-3' and S16KM: 5'-CCGGTCTGAACTCAGATCACGT-3'. Two portions of the mitochondrial genome were also amplified by polymerase chain reaction (PCR): an approximately 800 base pair (bp) region within the COI gene and an approximately 1500 bp region within the COIII gene. Primer sequences for COI were Co1FR 5'-CGCCTGTTTATCAAAAACAT-3' and Co1RV 5'-GATATAAGAAGTCTAGCCTG-3'. Primer sequences for COIII were Co3FR 5'-GGAGGATTTGGAAATTGATTAGTTC-3' and Co3RV 5'-GGGATAGCAATATTATGT-3'. These primers were designed based on sequenced samples using Primer-BLAST, NCBI (Bethesda, MD, USA) Primer3 [31]. PCR amplification was performed in 50 µL reactions containing 5 µL of 10× PCR buffer (Promega), 2.5 mM MgCl2, 0.2 µM dNTPs, 2–6 µL of template DNA, 0.5 µM of each primer, and 1.25 units of Taq polymerase. Reactions were performed with an initial denaturation step at 94 °C for 3 min, then 30–40 cycles of 94 °C for 30 s, 53.2 °C for 30 s, 50 °C for 90 s, 72 °C for 120 s, and a final extension step at 72 °C for 10 min. Negative controls were performed for all amplification reactions. In addition, PCR and sequencing were repeated to confirm putative introgressed sequences and to exclude the possibility that they were the result of PCR contamination. No identification problems were found.

PCR amplification products were separated on a 1% agarose gel stained with ethidium bromide (EtBr). Products of interest were identified using a *Hin*dIII DNA ladder as a reference marker (Promega, USA). PCR fragments were purified from the gel using a Wizard-Prep PCR Purification Kit (Promega). Automated fluorescent dideoxy sequencing of both strands was carried out using an ABI Prism 3100 DNA sequencer (Applied Biosystems, Foster City, CA, USA) at the available sequencing centre and we provide our own 16S, COXI and COXIII primers for the sequencing services. Sequences from both strands of all the individuals from each species were assembled into a single consensus sequence using the assembly editor option in the Biology Workbench (San Diego, CA, USA) (ver. 3.2) [32]. The consensus sequences were aligned using the alignment program Clustal W with the default settings [33].

Data Analysis

The DNA sequence data for all samples were analysed for ambiguities and the nucleotide sequences obtained were aligned by ClustalX2 [34] with default settings. Only unique haplotype was chosen for the phylogenetic analysis. Because variation within the nuclear loci was low, relationships between sequences for each locus were determined, and the data set was analysed with maximum parsimony (MP), molecular evolution (ME), and neighbour-joining (NJ) methods using the program MEGA 4.0 [35]. Sites with missing data were removed, and the mitochondrial region sequences were used to test models of evolution. The Kimura 2-parameter model of evolution was employed to test the MP, ME and NJ trees. Kimura 2-parameter model corrects for multiple hits, taking into account transitional and transversional substitution rates, the four nucleotide frequencies are the same and that rates of substitution do not vary among sites [36].

The haplotype were analysed with MP, ME and NJ as implemented DNAsp Version 5 [37]. Support for nodes was estimated using the bootstrap technique with 1000 replicates using AY950700 and DQ067314 from *Epinephelus bruneus* as an outgroup. All new sequences were deposited in GenBank, as detailed in Supplementary Table S1. The haplotype and the nucleotide diversity were calculated in order to examine the levels of gene variability within samples and gene genealogy when dealing with large data sets of closely related species; for this purpose Arlequin Ver 3.5 [38] and TCS v1.21 [39] were used. The neutrality value in the samples was assessed by calculating Tajima's D value using 1000 permutations in the Arlequin Ver 3.5.

Morphological Analysis

Wild species of *E. fuscoguttatus* and *E. hexagonatus* specimens were caught by fishing nets in ponds in the coastal and open South China Sea. Ten specimens (four males and six females) were collected at the five locations, two specimens from each location. The total length of the analysed specimens ranged from 58.0 to 61.0 cm and the weights ranged from 6.0 to 8.5 kg. All morphometric and meristic characteristics were examined according to Heemstra and Randall [40] by colour patterns and the scale of rays, fins, and spines as diagnostic characters for identification. The specimens were measured with a digital slide caliper (24 inches measuring range) and weighed with a high capacity electric balance (12,000 g × 0.1 g).

Attenuated Total Reflectance-Fourier Transform Infrared Analysis (ATR-FTIR)

Fin tissues of samples were subjected to FTIR analysis. The freeze-dried fin samples were placed directly on the diamond window. All samples were oriented in the north-south configuration and aligned with the probing beam to minimise unwanted spectral differences due to sample placement. The micrometer accessory had a straight-edged metal tip attachment and each sample was collected while applying approximately 800 psi (55 bar) of pressure (PerkinElmer, Boston, MA, USA). The analysis was conducted using an FT-IR spectrometer (PerkinElmer, USA) equipped with a DTGS detector. MIR spectra were recorded from an accumulation of 4 scans in the 4000–400 cm^{-1} range.

CONCLUSIONS

This study has provided important molecular, morphological and spectroscopy information that can be used to identify the *E. fuscoguttatus* species more precisely. Additionally, the study has successfully proven the utility of molecular, morphological and spectroscopy techniques in identifying *E. fuscoguttatus*. This study found that *E. fuscoguttatus* has been misidentified to *E. hexagonatus*. The variation in color pattern and morphology of *E. hexagonatus* that have been observed are likely attributed to differences in environment or nature of the fish's habitat such as the coral reef; it is well known that the grouper family can adapt and camouflage in order to protect themselves [9]. It could be that these two taxa are in fact undergoing speciation, however the process is not complete and, therefore, they are currently occupying separate environmental niches, leading to difference in morphology, but are capable of interbreeding, and that *E. fuscoguttatus* was misidentified as *E. hexagonatus*. The former name takes precedence over *E. hexagonatus* as it was described first. They may form two separate species in the future if the populations become reproductively isolated. The potential of molecular genetics as one of the techniques in identification of grouper species should be considered for establishing proper management and breeding strategies for the valuable *Epinephelus fuscoguttatus* species in Malaysia.

SUPPLEMENTARY MATERIALS

Table S1. List of taxa sequenced of Cytochrome b and 16S and their GenBank accession numbers.

	Species	Label	GenBank ID
Cytochrome b	*Epinephelus hexagonatus*	3C11	GU591708
	Epinephelus hexagonatus	3C27	GU591711
	Epinephelus hexagonatus	4C7	GU591702
	Epinephelus hexagonatus	4C10	GU591703
	Epinephelus hexagonatus	4C13	GU591704
	Epinephelus hexagonatus	4C17	GU591705
	Epinephelus hexagonatus	5C18	GU591718
	Epinephelus hexagonatus	6C6	GU591720
16S	*Epinephelus hexagonatus*	3s11	HQ840441
	Epinephelus hexagonatus	3s20	HQ840443
	Epinephelus hexagonatus	4s1	HQ840444
	Epinephelus hexagonatus	4s7	HQ840445
	Epinephelus hexagonatus	4s17	HQ840446
	Epinephelus hexagonatus	4s18	HQ840447
	Epinephelus hexagonatus	5s18	HQ840449
	Epinephelus hexagonatus	6s9	HQ840450

ACKNOWLEDGMENTS

This work was supported by research grant number 07-05-ABI-AB005 from the Ministry of Science, Technology and Innovation of Malaysia (MOSTI). We thank Micheal Meyrick Burrell from University of Sheffield, and Thelma Patricia Burrell as well as Kathryn Ford from University of Bristol for providing feedback and critical reading on the manuscript.

AUTHOR CONTRIBUTIONS

Syarul Nataqain Baharum conceived and designed the experiments; A'wani Aziz Nurdalila performed the experiments; Syarul Nataqain Baharum, Subbiah Vijay Kumar, Kenneth Francis Rodrigues, A'wani Aziz Nurdalila and Hamidun Bunawan performed the data analysis; A'wani Aziz Nurdalila, Hamidun Bunawan and Syarul Nataqain Baharum wrote the paper.

REFERENCES

1. James, C.M.; Al-Thobaiti, S.A.; Rasem, B.M.; Carlos, M.H. Potential of Grouper Hybrid *Epinephelus fuscoguttatus* X *E. polyphekadion* for Aquaculture Naga. *ICLARM Q.* 1999, *22*, 19–23.
2. Hseu, J.R.; Huang, W.I.; Yeong, T.C. What causes cannibalization-associated suffocation in cultured brown-marbled grouper, *Epinephelus fuscoguttatus* (Forsskal, 1775)? *Aquac. Res.* 2007, *38*, 1056–1060.
3. Yeh, S.L.; Kuo, C.M.; Ting, Y.Y.; Chang, C.F. Androgens stimulate sex change in protogynous grouper, *Epinephelus coioides*: spawning performance in sex-changed males. *Comp. Biochem. Physiol. C* 2003, *135*, 375–382.
4. Kuo, H.-C.; Hsu, H.-H.; Chua, C.S.; Wang, T.-Y.; Chen, Y.-M.; Chen, T.-Y. Development of pedigree classification using microsatellite and mitochondrial markers for giant grouper broodstock (*Epinephelus lanceolatus*) management in Taiwan. *Mar. Drugs* 2014, *12*, 2397–2407.
5. Nedwman, S.J.; Cappo, D.; Williams, M.B. Age, growth, mortality rates and corresponding yield estimates using otoliths of the tropical red snappers, *Lutjanus erythropterus*, *L. malabaricus* and *L. sebae*, from the central Great Barrier Reef. *Fish. Res.* 2000, *48*, 1–14.
6. Craig, M.T.; Pondella, D.J.; Franck, J.P.C.; Hafner, J.C. On the status of the serranid fish genus *Epinephelus*: Evidence for paraphyly based upon 16S rDNA sequence. *Mol. Phylogenet. Evol.* 2001, *19*, 121–130.
7. Maggio, T.; Andaloro, F.; Hemida, F.; Arculeo, M. A molecular analysis of some Eastern Atlantic grouper from the Epinephelus and Mycteroperca

genus. *J. Exp. Mar. Biol. Ecol.* 2005, *321*, 83–92.

8. Zhu, Z.Y.; Yue, G.H. The complete mitochondrial genome of red grouper *Plectropomus leopardus* and its applications in identification of grouper species. *Aquaculture* 2008, *276*, 44–49.

9. Craig, M.T.; Hastings, P.A. A molecular phylogeny of the groupers of the subfamily Epinephelinae (Serranidae) with a revised classification of the Epinephelini. *Ichthyol. Res.* 2007, *54*, 1–17.

10. Baharum, S.N.; Nurdalila, A.A. Phylogenetic Relationships of *Epinephelus fuscoguttatus* and *Epinephelus hexagonatus*Inferred from Mitochondrial Cytochrome b Gene Sequences using Bioinformatic Tools. *Int. J. Biosci. Biochem. Bioinform.* 2011, *1*, 47–52.

11. Felsenstein, J. Confidence limits on phylogenies: An approach using the bootstrap. *Evolution* 1985, *39*, 783–791.

12. Chisholm, L.A.; Morgan, J.A.T.; Adlard, R.D.; Whittington, I.D. Phylogenetic analysis of the Monocotylidae (Monogenea) inferred from 28S rDNA sequences. *Int. J. Parasitol.* 2001, *31*, 1537–1547.

13. Meyer, A. Phylogenetic relationship and evolutionary processes in east African cichlid fishes. *Trends Ecol. Evol.* 1993, *8*, 279–284.

14. De-La, H.R.; Ruiz, R.C.; Ruiz, R.M.; Garrido-Ramos, M.A. The molecular phylogeny of the Sparidae species (Pieces, Perciformes) based on two satellite DNA families. *Heredity* 2001, *87*, 691–697.

15. Birstein, V.J.; Desalle, R. Molecular phylogeny of Acipenserinae. *Mol. Phylogenet. Evol.* 1998, *9*, 141–155.

16. Turan, C. Stock identification of Mediterranean horse mackerel (Trachurus mediterraneus) using morphometric and meristic characters. *ICES J. Mar. Sci.* 2004, *61*, 774–781.

17. Meyer, X. Morphometrics and allometry in the trophically polymorphic cichlid fish, Cichlasoma citrinellum: Alternative adaptations and ontogenic changes in shape. *J. Zool.* 1990, *221*, 237–260.

18. Kuhm, A.E.; Suter, D.; Felleisen, R.; Rau, J. Application of Fourier transform infrared spectroscopy (FT-IR) for the identification of Yersinia enterocolitica on species and subspecies level. *Appl. Environ. Microbiol.* 2009, *75*, 5809–5813.

19. Khairudin, K.; Sukiran, N.A.; Goh, H.-H.; Baharum, S.N.; Noor, M.N. Direct discrimination of different plant populations and study on temperature effects by Fourier transform infrared spectroscopy. *Metabolomics* 2014, *10*, 203–211.

20. Jagadeesan, G.; Kavitha, A.V.; Subasini, J. FTIR study of the influence of Tribulus terrestris on mercury intoxicated mice, Mus musculus, liver. *Trop. Biomed.* 2005, *22*, 15–22.
21. Chen, L.; Carpita, N.C.; Reiter, W.D.; Wilson, R.H.; Jeffries, C.; McCann, M.C. A rapid method to screen for cell-wall mutants using discriminant analysis of Fourier transformation infrared spectra. *Plant J.* 1998, *16*, 385–392.
22. Wilson, R.H.; Slack, P.T.; Appleton, G.P.; Sun, L.; Belton, P.S. Determination of the fruit content of jam using Fourier transform infrared spectroscopy. *Food Chem.* 1993, *47*, 303–308.
23. Rodriguez-Saona, L.E.; Khambaty, F.; Fry, F.; Dubois, J.; Calvey, E.M. Detection and identification of bacteria in a juice matrix with Fourier transform-near infrared spectroscopy and multivariate analysis. *J. Food Prot.* 2004, *67*, 2555–2559.
24. Evans, P.A. Differentiating "hard" from "soft" woods using Fourier transform infrared and Fourier transform spectroscopy. *Spectrochim. Acta* 1991, *47*, 1441–1447.
25. Müller, G.; Schöpper, C.; Vos, H.; Kharazipour, A.; Polle, A. FTIR-ATR spectroscopic analysis of chages in wood properties during particle- and fibreboard production of hard- and softwood trees. *Bioresources* 2009, *4*, 49–71.
26. Smith, A.R.; Johnson, H.E.; Hall, M. Metabolic Fingerprinting of Salt-Stressed Tomatoes. *Bulg. J. Plant Physiol.* 2003, 153–163. Available online: http://www.bio21.bas.bg/ippg/bg/wp-content/uploads/2011/06/03_essa_153-163.pdf(accessed on 25 June 2015).
27. Espinoza, E.O.; Baker, B.W.; Moores, T.D.; Voin, D. Forensic identification of elephant and giraffe hair artifacts using HATR FTIR spectroscopy and discriminant analysis. *Endang. Spec. Res.* 2008, *9*, 239–246.
28. Boskey, A.; Camacho, N.P. FT-IR imaging of native and tissue-engineered bone and cartilage. *Biomaterials* 2007, *28*, 2465–2478.
29. Heemstra, P.C.; Randall, J.E. Grouper of the world. *FAO Fish. Synop.* 1993, *125*, 104–276.
30. Young, N.D. A cautiously optimistic vision for marker-assisted breeding. *Mol. Breed.* 1999, *5*, 505–510.
31. Rozen, S.; Skaletsky, H.J. Primer3 on the WWW for General Users and for Biologist Programmers. In *Bioinformatics Methods and Protocols: Methods in Molecular Biology*; Misener, S., Krawetz, S.A., Eds.; Humana Press Inc.: Totowa, NJ, USA, 2000; pp. 365–386.

32. Subramaniam, S. The Biology Workbench-a seamless database and analysis environment for the biologist. *Proteins* 1998, *32*, 1–2.
33. Thompson, J.D.; Higgins, D.G.; Gibson, T.J. CLUSTAL W: Improving the sensitivity of progressive multiple sequence alignment through sequence weighting, position-specific gap penalties and weight matrix choice. *Nucleic Acids Res.* 1994, *22*, 4673–4680.
34. Larkin, M.A.; Blackshields, G.; Brown, N.P.; Chenna, R.; McGettigan, P.A.; McWilliam, H.; Valentin, F.; Wallace, I.M.; Wilm, A.; Lopez, R.; *et al*. Clustal W and Clustal X version 2.0. *Bioinformatics* 2007, *23*, 2947–2948.
35. Tamura, K.; Dudley, J.; Nei, M.; Kumar, S. MEGA4: Molecular Evolutionary Genetics Analysis (MEGA) software version 4.0. *Mol. Biol. Evol.* 2007, *24*, 1596–1599.
36. Nei, M.; Kumar, S. *Molecular Evolution and Phylogenetics*; Oxford University Press: New York, NY, USA, 2000.
37. Librado, P.; Rozas, J. DnaSP v5: A software for comprehensive analysis of DNA polymorphism data. *Bioinformatics* 2009, *25*, 1451–1452.
38. Excoffier, L. Computational and Molecular Population Genetics Lab CMPG. 2010. Available online: http://cmpg.unibe.ch/software/arlequin3 (accessed on 1 July 2012).
39. Clement, M.; Posada, D.; Crandall, K.A. TCS: A computer program to estimate gene genealogies. *Mol. Ecol.* 2000, *9*, 1657–1660.
40. Heemstra, P.C.; Randall, J.E. Groupers of the world (Family Serranidae, Subfamily Epinephelinae). In *An Annotated and Illustrated Catalogue of the Grouper, Rockcod, Hind, Coral Grouper and Lyretail Species Known to Date*; Food and Agriculture Organization of the United Nations: Rome, Italy, 1993; p. 382.

Chapter 6

SYNTHESIS, MOLECULAR DOCKING AND BIOLOGICAL EVALUATION OF GLYCYRRHIZIN ANALOGS AS ANTICANCER AGENTS TARGETING EGFR

Yong-An Yang[1], Wen-Jian Tang[2], Xin Zhang[1], Ji-Wen Yuan[1], Xin-Hua Liu[1,2], and Hai-Liang Zhu[1]

[1]State Key Laboratory of Pharmaceutical Biotechnology, Nanjing University, Nanjing 210093, China

[2]School of Pharmacy, Anhui Medical University, Hefei 230032, China

ABSTRACT

Glycyrrhizin (GA) analogs in the form of 3-glucuronides and 18-epimers were synthesized and their anticancer activities were evaluated. Alkaline isomerization of monoglucuronides is reported. In vitro and in vivo studies showed that glycyrrhetinic acid monoglucuronides (GAMGs) displayed higher anticancer activities than those of bisglucuronide GA analogs, while anticancer activity of the 18α-epimer was superior to that of the 18β-epimer. 18α-GAMG was firstly nicely bound to epidermal growth factor receptor (EGFR) via six hydrogen bonds and one charge interaction, and the docking calculation proved the correlation between anticancer activities and EGFR inhibitory activities. Highly active 18α-GAMG is thus of interest for the further studies as a potential anticancer agent.

INTRODUCTION

The epidermal growth factor receptor (EGFR) is a transmembrane glycoprotein that defines a family of tyrosine kinase receptors (TKRs) including ErbB2/HER2, ErbB3/HER3 and ErbB4/HER4 [1,2]. As a cell surface protein that binds to epidermal growth factor, its binding to a ligand induces receptor dimerization and tyrosine autophosphorylation and leads to cell proliferation, of which altered activity has been implicated in the development and growth of many tumors [3]. EGFR is highly expressed in adult hepatocytes and the EGFR family plays a central hepatoprotective and pro-regenerative role in the liver [4,5]. Mice lacking EGFR or heparin-binding EGF show delayed regeneration after partial hepatectomy (PH), which demonstrates that EGFR is a critical regulator of hepatocyte proliferation during liver regeneration [6,7]. The treatment of hepatocellular carcinoma (HCC) cells with EGFR-specific tyrosine kinase inhibitors or neutralizing antibodies induces cell cycle arrest and apoptosis and increases chemosensitivity [8,9]. Hence, EGFR has long been an attractive candidate as anticancer drug target. Over the past 30 years, much effort has been directed at developing anticancer agents that can interfere with EGFR activity, such as, monoclonal antibodies and small-molecule inhibitors.

Natural products play a major role in drug discovery, and nearly half of the new drugs introduced into the market over the past two decades are natural products or their derivatives [10]. The roots and rhizomes of licorice (Glycyrrhiza) species have long been used worldwide as a herbal medicine and natural sweetener. Glycyrrhizin (Glycyrrhizic acid, GA, 18β-GA), the major bioactive compound in licorice, is developed as a drug with multi-pharmacological effects such as anti-inflammation, antivirus, anti-tumor, and immuno-modulating properties, among others [11,12,13]. GA has been used in Japan for more than 60 years as a treatment for chronic hepatitis C, thus long-term administration was effective in preventing hepatic cirrhosis and HCC [14,15,16]. GA exhibits hepatoprotective activity by decreasing serum liver enzyme levels and improving tissue pathology in hepatitis patients, while in vitro studies showed that its anticancer activity is achieved by inhibiting abnormal cell proliferation, tumor formation and growth [17,18,19,20].

GA is a conjugate of an 18β-H-oleanane-type aglycone and two glucuronic acids at the C-3 position, and it could be transformed into 18β-glycyrrhetinic acid monoglucuronide (18β-GAMG) by removing one terminal glucuronic acid [21,22]. Compared to GA, 18β-GAMG showed similar (or stronger) pharmacological activities, such as antitumor, antivirus, anti-allergic, and anti-inflammatory activities [23,24]. 18α-Glycyrrhizin (18α-GA), a D/E-trans-epimer, was prepared by alkaline isomerization of 18β-GA [25]. 18α-GA also showed similar anti-inflammatory and anticancer activity [26,27]. Researches showed that GA analogs are primary hepatocyte mitogens that bind to EGFRs and subsequently stimulate the receptor tyrosine kinase mitogenactivated protein kinase pathway to induce hepatocyte DNA synthesis and proliferation [28]. Herein, we prepared glycyrrhizin analogs (Figure 1) and further evaluated their anticancer activities in vitro and in vivo.

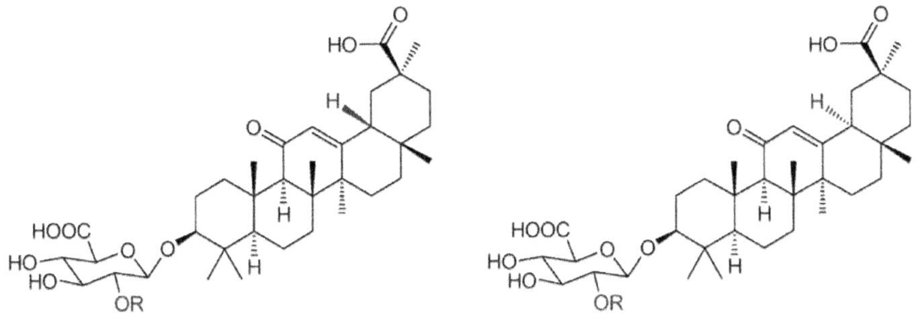

18β-GAMG: R = H
18β-GA: R = glucuronide

18α-GAMG: R = H
18α-GA: R = glucuronide

Figure 1. Glycyrrhizin analogs with 3-glucuronides and 18-epimers.

Based on the EGFR complex structure (PDB Code 1M17) [29], computer-generated docking molecular models of GA analogs were analyzed using the Discovery Studio 3.5, and we also initiated an effort to leverage molecular modeling in combination with available data to study the effect of structure of glycyrrhizin on anticancer activity.

RESULTS AND DISCUSSION

GA (18β-GA) was provided by Jiangsu Tian Sheng Pharmaceutical Co. Ltd., Jiangsu, China. Aspergillus sp Ts-1, a kind of β-glucuronidase, selectively hydrolyzed the terminal–glucuronyl linkage of 18β-GA to produce 18β-GAMG in 54% yield. 18α-GA and 18α-GAMG were respectively synthesized in yields of 63% and 71% after recrystallization from 18β-GA and 18β-GAMG by alkaline isomerization. The isomerization reaction was monitored by ^{13}C-NMR spectroscopy, and the structure of 18α-GAMG was elucidated by comparison with ^{1}H-NMR and ^{13}C-NMR data of 18β-GAMG [30] (see Supplementary Information).

Anti-proliferative activities of glycyrrhizin analogs and erlotinib against the HepG2 (hepatocellular carcinoma), HeLa (cervix of uterus adenocarcinoma) and A549 (lung carcinoma), were evaluated by CCK8 dye assays. The results, summarized in Table 1, revealed that four glycyrrhizin analogs exhibited significant antitumor activities. Two monoglucuronide compounds, 18α-GAMG and 18β-GAMG, exhibited more significant antitumor activities than those with bisglucuronide GAs, while the antitumor activity of the 18α-epimer was superior to that of 18β-epimer for identical cell lines. Among the four analogs, 18α-GAMG displayed the most potent activity, with IC_{50} values of 6.67, 7.43 and 15.76 μM against HepG2, Hela and A549, respectively.

Table 1. In vitro anticancer activities (IC_{50}, μM) of title compounds against human tumor cell lines.

Compd.	IC_{50}, (μM) [a]		
	HepG2 [b]	HeLa [b]	A549 [b]
18α-GAMG	6.67	7.43	15.76
18β-GAMG	33.60	8.39	21.55
18α-GA	54.24	15.13	41.57
18β-GA	63.59	18.93	51.92
Erlotinib	0.12	0.20	0.13

[a] Antiproliferation activity was measured using the CCK-8 assay. Values are the average of three independent experiments run in triplicate. Variation was generally 5%–10%; [b] Cancer cells kindly supplied by State Key Laboratory of Pharmaceutical Biotechnology, Nanjing University.

To generate data concerning the broad spectrum potential of these compounds in Table 2, the IC_{50} values of synthesized compounds against EGFR enzymes are summarized in Table 2. Reference data for erlotinib had also been included for comparison with the compounds reported in this study. For the majority of the compounds, we found that compound 18α-GAMG, with an IC_{50} of 0.028 μM, was a better inhibitor than the positive control erlotinib with an IC_{50} of 0.030 μM, suggesting that, at least in part, inhibition of proliferation of the these lines may be the result of EGFR inhibition.

Table 2. Data of the in vitro EGFR (IC_{50}, μM) enzyme inhibition assay of the synthesized compounds.

Compd.	EGFR (IC_{50}, μM) [a]	Compd.	EGFR (IC_{50}, μM) [a]
18α-GAMG	0.028	18β-GA	0.092
18β-GAMG	0.069	Erlotinib	0.030
18α-GA	0.081		

[a] Minimum cytotoxic concentration required to cause a microscopically detectable alteration of normal cell morphology.

In order to gain more understanding of the structure–activity relationships observed at the EGFR, molecular docking of the most potent inhibitor 18α-GAMG and EGFR was performed on the binding model based on the EGFR complex structure (PDB Code 1M17) using the Discovery Studio 3.5 software [29]. The docking calculation of the analogs was depicted in Table 3, and as shown in Table 3, all analogs had nice binding affinity to EGFR and four analogs› -EDOCKER_ INTERACTION_ ENERGY had the same trend as the anti-proliferative activities, which further proved the correlation between the anti-proliferative activities and EGFR inhibitory activities of the analogs.

Table 3. -EDOCKER_INTERACTION_ENERGY of title compounds and 1M17.

Compd.	-EDOCKER_INTERACTION_ENERG ΔG (kcal/mol)
18α-GAMG	72.0274
18β-GAMG	66.9106
18α-GA	58.7009
18β-GA	58.6731
Erlotinib	44.3732

In the result of molecular docking, 18α-GAMG showed maximum -EDOCKER_ INTERACTION_ENERGY, which suggested it was mostly easy to bind to EGFR. The 2D and 3D binding models of 18α-GAMG with EGFR are depicted inFigure 2. The amino acid residues which had interactions with EGFR as well as bond lengths were labeled. In the binding models, 18α-GAMG was nicely bound to EGFR via six hydrogen bonds with ASP831 (angle = 120.49°, distance = 2.14 Å), GLU738 (angle = 142.28°, distance = 2.2 Å), THR766 (three bonds: angle = 140.28°, distance = 2.1 Å; angle = 117.6°, distance = 2.3 Å; angle = 152.94°, distance = 2.0 Å) and LYS721 (angle = 179.02°, distance = 1.7 Å). In addition, compound 18α-GAMG was also nicely bound to EGFR via one charge interaction. The end group of LYS692 formed one charge interaction with a carboxyl which strengthened the binding affinity, leading to the increased anticancer activities of 18α-GAMG. Besides, the hydrogens of LYS692, LYS692 and PRO770 formed three hydrogen bonds interaction with the amino group nitrogen atom of 18β-GAMG (angle H-N_{LYS692} O35 = 151.7°, distance = 1.98 Å, angle H_{LYS692} O36 = 123.6°, distance = 2.47 Å, angle H96 O_{PRO770} = 113.4°, distance = 2.22 Å). Furthermore, compound 18β-GAMG was also nicely bound to EGFR via three charge interactions. The end group of LYS692, LYS704 and LYS721 respectively formed three charge interactions with two carboxyls. These molecular docking results, along with the biological assay data, suggest that compound 18α-GAMGpossesses higher anticancer activity than 18β-GAMG, which will help us carry out structure optimization based on computer-aided design.

Recently, 18β-GA has been recognized as a hepatoprotective high-mobility group protein 1 (HMGB1) inhibitor, which binds directly to both HMG boxes in HMGB1 and attenuates HMGB1-induced hepatocyte apoptosis, thus leading to induce hepatocyte DNA synthesis and proliferation [31,32]. As it is, 18β-GA induced hepatocyte proliferation, while we got the opposite results in cancer cells. The binding moiety of 18α-GAMG with EGFR was mainly the glucuronide unit, but GA could inhibit HMGB1 by binding of its triterpene ring directly to the two HMG boxes. These results showed that the protein targets and molecular pathways affected by GA may be complicated and heterogeneous.

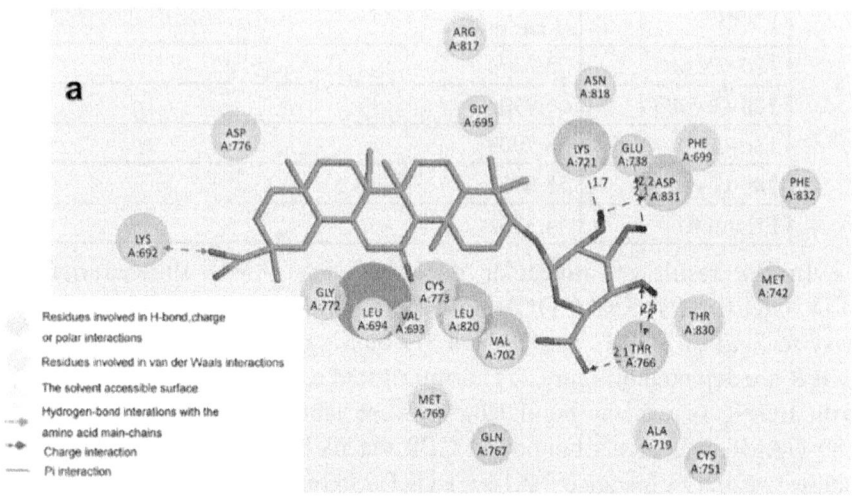

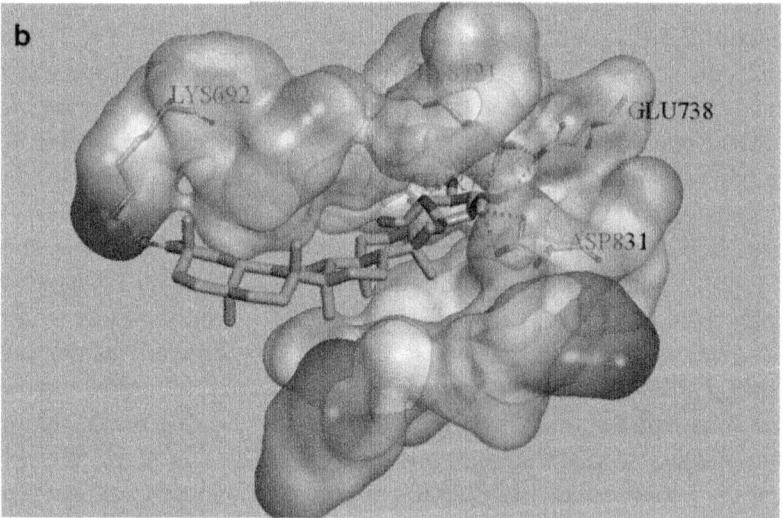

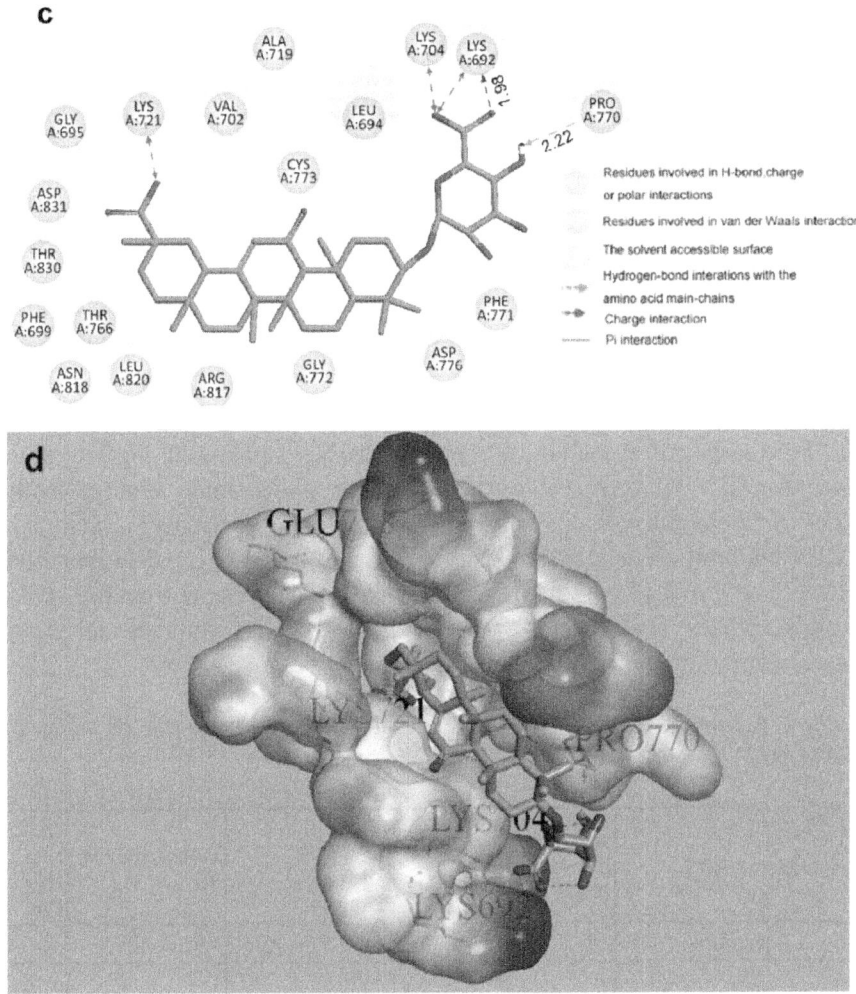

Figure 2. (**a**) 2D molecular docking modeling of 18α-GAMG with 1M17. (**b**) 3D model of the interaction between 18α-GAMG and 1M17 site. (**c**) 2D molecular docking modeling of 18β-GAMG with 1M17. (**d**) 3D model of the interaction between 18β-GAMG and 1M17 site.

To further verify the inhibitory effect of glycyrrhizin analogs on the growth of tumor cells in vivo, sarcoma cells S180, hepatoma cells HepG2 and Ehrlich ascites cells EAC were selected to evaluate in vivo antitumor effects. The inhibitory effects of glycyrrhizin analogs on the growth of the transplanted S180 or HepG2 carcinoma are presented in Table 4 andFigure 3. The results revealed that glycyrrhizin analogs significantly decreased the tumor weights of S180 and HepG2 tumor-bearing mice. The inhibitory rates of GAMGs were higher than those of GAs, while the inhibitory rate of the 18α-epimer was higher than that of corresponding 18β-epimer. The most potent activity was showed by 18α-GAMG with inhibitory rates 39.8% and 49.7% for S180 and HepG2 tumor-bearing mice at the dosage of 60 mg/kg/day, respectively.

EAC tumor-bearing mice were observed for mean survival time. The effect of glycyrrhizin analogs on percentage increases in life span was calculated on the basis of mortality of the experimental mice. Survival response of untreated EAC-bearing mice died within 16.4 days (Table 5). A similar phenomenon was observed: mice administered monoglucuronide and 18α-epimer displayed longer survival times. The 18α-GAMG group was observed to enhance the survival rate to 45.4%.

Based on in vitro and in vivo experiments, followed by molecular docking, we here demonstrated that the protein target Epidermal Growth Factor Receptor (EGFR) was also sensitive to four glycyrrhizin analogs in three types of carcinoma cells, indicative of their potential anticancer activity as the EGFR inhibitors. The result was significant and intriguing, but further studies needs to be provided to systematically elucidate the direct correlation between the glycyrrhizin analogs and the EGFR target, which would reveal the new mechanism of glycyrrhizin action.

Table 4. Antitumor effects of glycyrrhizin analogs against tumor growth on the S180 and HepG2 xenograft mice. [a]

Models	Groups	Animal number (End, n)	Body weight (g)		Tumor weight (g)	Inhibition rate (%)
			Beginning	End		
S180	Control	10	19.80 ± 1.32	23.97 ± 2.23	1.91 ± 0.29	
	18α-GAMG	9	19.40 ± 1.07	25.25 ± 1.80	1.15 ± 0.50 **	39.8
	18β-GAMG	10	19.80 ± 1.39	25.02 ± 2.58	1.25 ± 0.19 **	34.6
	18α-GA	10	20.00 ± 1.49	25.20 ± 1.11	1.29 ± 0.47 **	32.5
	18β-GA	9	19.70 ± 1.25	24.48 ± 2.37	1.33 ± 0.67 **	30.4
HepG2	Control	10	19.4 ± 1.35	25.05 ± 1.89	1.95 ± 0.22	
	18α-GAMG	10	19.6 ± 1.51	27.02 ± 2.10	0.98 ± 0.43 **	49.7
	18β-GAMG	10	19.00 ± 0.94	27.8 ± 1.57	1.20 ± 0.35 **	38.4
	18α-GA	10	20.1 ± 1.45	26.08 ± 1.26	1.22 ± 0.46 **	37.4
	18β-GA	10	19.4 ± 1.08	26.94 ± 2.05	1.26 ± 0.65 **	35.4

[a] Mice were inoculated with S180 or HepG2 subcutaneously into the right front armpit and randomly divided into five test groups. The mice were daily treated by 18α-GAMG, 18β-GAMG, 18α-GA, 18β-GA (60 mg/kg/day), or

normal saline (NS, 10 mL/kg) by oral gavage for ten consecutive days. Data were analyzed using SPSS11.0. Significant difference between each treatment and the control are shown as P < 0.05 (*) and P < 0.01 (**).

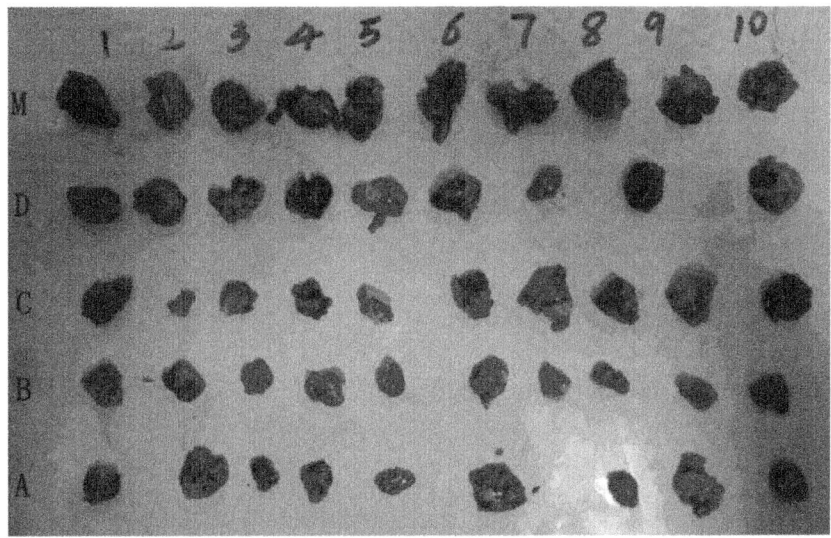

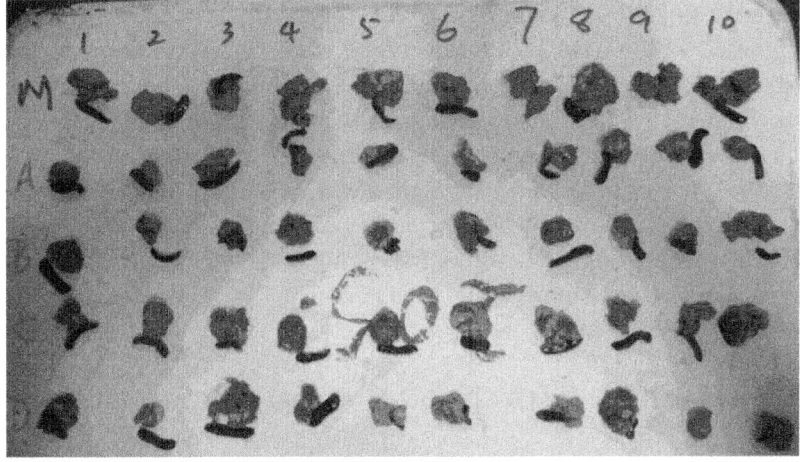

Figure 3. Solid tumors from S180 (above) and HepG2 (below) tumor-bearing mice. A: 18α-GAMG; B: 18β-GAMG; C: 18α-GA; D: 18β-GA; M: Normal saline.

Table 5. Effects of glycyrrhizin analogs against the survival of EAC-bearing mice.

Groups	Animal number (n)	Body weight (g)	Survival time [a] (d)	Survival rate (%)
Control	10	19.0 ± 1.25	16.40 ± 2.07	
18α-GAMG	10	19.5 ± 1.27	23.85 ± 5.41*	45.4
18β-GAMG	10	19.0 ± 0.82	21.05 ± 4.65*	28.4
18α-GA	10	19.4 ± 0.97	19.75 ± 3.08*	20.4
18β-GA	10	19.4 ± 1.27	19.40 ± 3.77*	18.3

[a] Time denoted by number of days. $P < 0.05$ (*)

EXPERIMENTAL SECTION

Synthesis of Glycyrrhizin Analogs

General Methods

Aspergillus sp. Ts-1 was isolated from soil collected in Kashi of the Xinjiang Uygur Autonomous Region (China) and selectively hydrolyzed the terminal–glucuronyl linkage of 18β-GA to yield glycyrrhetic acid 3-O-mono-β-D-glucuronide (18β-GAMG). Its subculture and 18β-GA were provided by Jiangsu Tian Sheng Pharmaceutical Co. Ltd. (Nanjing, China). All materials were obtained from commercial suppliers, were of analytical reagent grade and used without further purification. Melting points were uncorrected. Silica gel (200–300 mesh, Huanghai, Qingdao, China). TLC: pre-coated silica gel F_{254} plates. Optical rotations: polar 3002 polarimeter. NMR spectra: Bruker AV NMR spectrometer (^{1}H: 500 or 300 and ^{13}C: 125 or 75 MHz), the residual solvent peaks used as an internal standard, J in Hz. TOF- HR MS: Agilent 1260-6221 TOF LC/MS.

Preparation of 18β-GAMG from 18β-GA via Biotransformation

18β-GA: white powder. Mp 234−236°C; $[\alpha]_{20}^{D}$ = +52 (c = 1.0, MeOH); ^{13}C-NMR (75 MHz, DMSO-d_6): Table S1. TOF-HRMS:m/z [M + Na]$^+$ calcd for $C_{42}H_{62}NaO_{16}$: 845.3930; found: 845.3935.

Aspergillus sp. Ts-1 on glucose yeast agar slant was inoculated into a 250 mL Erlenmeyer flask containing 100 mL of seed medium consisting of 1.0 g glucose, 0.2 g yeast, 1.0 g agar, 0.1 g KH_2PO_4 and 0.025g $MgSO_4$ in distilled water (pH 7.0). The culture media were sterilized at 121°C for 20 min and the fermentation was carried out at 30°C on a rotary shaker at 200 rpm. After 24 h of inoculation, 30 mL sterilized medium was inoculated into a 1,000 mL Erlenmeyer flask containing 300 mL pre-culture sample consisting of 15 g GA, 0.30 g KH_2PO_4, 3.0 g urea and 0.24 g $MgSO_4$ in distilled water and the pH value was adjusted to 6.0. The culture media were sterilized at 121°C for 20 min and the fermentation was carried out at 30°C on a rotary shaker at 250 rpm.

After 72 h of inoculation, the culture solution was filtered and the filtrate was extracted with ethyl acetate. The extract was concentrated under the reduced pressure. The residue (14.5 g) was applied to a silica gel column (800 g, 5.0 × 100 cm) and eluted with $CHCl_3$–MeOH in a gradient manner from 100:1 to 1:1. By TLC analysis, fractions I–IX was obtained. Fractions VI–VIII was concentrated in vacuo and recrystallization from aqueous MeOH to give 18β-GAMG (6.35 g, 54% yield) as a white crystalline powder. Mp 237−239°C; $[α]_{20}^{D}$ = +91 (c = 1.0, MeOH); ^{1}H-NMR (500 MHz, DMSO-d_6) δ (ppm): 0.76 (s, 3H, 24-CH_3), 0.77 (s, 3H, 28-CH_3), 0.99 (s, 3H, 23-CH_3), 1.06 (s, 2 × 3H, 25-CH_3, 26-CH_3), 1.10 (s, 3H, 29-CH_3), 1.34 (s, 3H, 27-CH_3), 2.34 (s, 1H, 9-H), 3.01 (m, 1H, 4'-H), 3.08 (dd, 1H, J_1= 4.8 Hz, J_2= 11.2 Hz, 3-H), 3.15 (t, 1H, J= 9.0 Hz, 3'-H), 3.30 (m, 1H, overlapped, 2'-H), 3.58 (d, 1H, J= 9.7 Hz, 5'-H), 4.25 (d, 1H, J= 7.8 Hz, 1'-H), 5.40 (s, 1H, 12-H); ^{13}C-NMR (125 MHz, DMSO-d_6): Table S1. TOF-HRMS: m/z $[M + Na]^+$ calcd for $C_{36}H_{54}NaO_{10}$: 669.3609; found: 669.3608.

General Procedure of Alkaline Isomerization of the 18β-isomer to the 18α-isomer

A solution of 18β-isomer (6.0 mmol) in 5.0 M NaOH solution (100 mL) was heated and stirred for 12 h at 90°C. After the reaction mixture was cooled to <5°C, the pH was adjusted to 2.5 with concentrated HCl. After 12 h, the mixture was filtrated, washed with water, dried. The product (18α-isomer) was obtained by crystallization from ethanol/EtOAc. (Scheme S1 and Figure S1). 18α-GA: According to the above procedure, diammonium 18α-GA (3.17 g, 63% yield) was obtained from 18β-GA (5.00 g) as a white crystalline powder. Mp 211−216°C; $[α]_{20}^{D}$ = +20 (c = 1.0, MeOH); ^{1}H-NMR (300 MHz, DMSO-d_6) δ(ppm): 0.65 (s, 3H, 28-CH_3), 0.73 (s, 3H, 24-CH_3), 0.95 (s, 3H, 23-CH_3), 1.04 (s, 3H, 26-CH_3), 1.10 (s, 3H, 25-CH_3), 1.16 (s, 3H, 29-CH_3), 1.33 (s, 3H, 27-CH_3), 4.31 (d, 1H, J= 7.3 Hz, 1'-H), 4.49 (d, 1H, J= 7.6 Hz, 1''-H), 5.33 (s, 1H, 12-H); ^{13}C-NMR (75 MHz, DMSO-d_6): Table S1. TOF-HRMS: m/z $[M + Na]^+$ calcd for $C_{42}H_{62}NaO_{16}$: 845.3930; found: 845.3938.

18α-GAMG: 18α-GAMG (2.83 g, 71% yield) was obtained from 18β-GAMG (4.00 g) as a white crystalline powder. Mp 229−231°C; $[α]_{20}^{D}$ =

+24 (c = 1.0, MeOH); ^{1}H-NMR (300 MHz, DMSO-d_6) δ (ppm): 0.65 (s, 3H, 28-CH_3), 0.77 (s, 3H, 24-CH_3), 0.92 (s, 3H, 23-CH_3), 0.98 (s, 3H, 25-CH_3), 1.04 (s, 3H, 26-CH_3), 1.16 (s, 3H, 29-CH_3), 1.33 (s, 3H, 27-CH_3), 2.27 (overlapped, 9-H), 3.01 (t, 1H, J= 8.4 Hz, 4'-H), 3.07 (dd, 1H, J_1= 6.5 Hz, J_2= 9.7 Hz, 3-H), 3.15 (t, 1H, J= 9.0 Hz, 3'-H), 3.30 (t, 1H, J= 9.8 Hz, 2'-H), 3.58 (d, 1H, J= 9.7 Hz, 5'-H), 4.24 (d, 1H, J= 7.8 Hz, 1'-H), 5.33 (s, 1H, 12-H); ^{13}C-NMR (75 MHz, DMSO-d_6): Table S1. TOF-HRMS: m/z [M + Na]$^+$ calcd for $C_{36}H_{54}NaO_{10}$: 669.3609; found: 669.3600.

Biological Assay of in Vitro Anticancer Activities

CCK8 is much more convenient and helpful than MTT for analyzing cell proliferation, because it can be reduced to soluble formazan by dehydrogenase in mitochondria and has little toxicity to cells. Cell proliferation was determined using CCK8 dye (BeyotimeInst Biotech, Shanghai, China) according to manufacturer's instructions. Briefly, 1–5 × 10^3 cells per well were seeded in a 96-well plate, grown at 37°C for 12 h, Subsequently, cells were treated with compounds at increasing concentrations in the presence of 10% FBS for 24 or 48 h. After 10 μL CCK8 dye was added to each well, cells were incubated at 37°C for 1–2 h and Plates were read in a Victor-V multilabel counter (Perkin-Elmer, Waltham, MA, USA) using the default europium detection protocol. Percent inhibition or IC_{50} values of compounds were calculated by comparison with DMSO-treated control wells.

General Procedure for Preparation, Purification of EGFR, and Inhibitory Assay

A 1.6 kb cDNA encoded for the EGFR cytoplasmic domain (EGFR-CD, amino acids 645–1186) were cloned into baculoviral expression vectors pBlueBacHis2B and pFASTBacHTc (Huakang Company, Changsha, China), separately. A sequence that encodes (His)$_6$ was located at the 5' upstream to the EGFR sequences. Sf-9 cells were infected for 3 days for protein expression. Sf-9 cell pellets were solubilized at 0°C in a buffer at pH 7.4 containing 50 mM HEPES, 10 mM NaCl, 1% Triton, 10 μM ammonium molybdate, 100 μM sodium vanadate, 10 μg/mL aprotinin, 10 μg/mL leupeptin, 10 μg/mL pepstatin, and 16 μg/mL benzamidine HCl for 20 min followed by 20 min centrifugation. Crude extract supernatant was passed through an equilibrated Ni-NTA superflow packed column and washed with 10 mM and then 100 mM imidazole to remove nonspecifically bound material. Histidine tagged proteins were eluted with 250 and 500 mM imidazole and dialyzed against 50 mM NaCl, 20 mM HEPES, 10% glycerol, and 1 μg/mL each of aprotinin, leupeptin and pepstatin for 2 h. The entire purification procedure was performed at 4°C or on ice [29].

EGFR kinase assays were set up to assess the level of autophosphorylation based on DELFIA/Time-Resolved Fluorometry. All compounds were dissolved in 100% DMSO and diluted to the appropriate concentrations with 25 mM HEPES at pH 7.4. In each well, 10 µL compound was incubated with 10 µL (5 ng for EGFR) recombinant enzyme (1:80 dilution in 100 mM HEPES) for 10 min at room temperature. Then, 10 µL of 5 mM buffer (containing 20 mM HEPES, 2 mM $MnCl_2$, 100 µM Na_3VO_4 and 1 mM DTT) and 20 µL of 0.1 mM ATP–50 mM $MgCl_2$ were added for 1 h. Positive and negative controls were included in each plate by incubation of enzyme with or without ATP–$MgCl_2$. At the end of incubation, liquid was aspirated, and plates were washed three times with wash buffer. A 75 µL (400 ng) sample of europium labeled anti-phosphotyrosine antibody was added to each well for another 1 h of incubation. After washing, enhancement solution was added and the signal was detected by Victor (Wallac Inc., Gaithersburg, MD, USA) with excitation at 340 nm and emission at 615 nm. The percentage of auto-phosphorylation inhibition by the compounds was calculated using the following formula: 100% − [(negative control)/(positive control − negative control)]. The IC_{50} was obtained from curves of percentage inhibition with eight concentrations of compound. As the contaminants in the enzyme preparation are fairly low, the majority of the signal detected by the anti-phosphotyrosine antibody is from EGFR.

Evaluation of the in Vivo Antitumor Activities

Animals and Cell Lines

Kunming mice (SPF, male or female, 20 ± 2 g) were purchased from the experimental animal center of China Pharmaceutical University. Animals were housed in a temperature (22 ± 2°C) and relatively humidity (50%)-controlled room on a 12 h light/dark cycle, given free access to food and water, and acclimatized for at least one week prior to use. All the animal experiments were performed in accordance with the Regulations of the Experimental Animal Administration issued by the State Committee of Science and Technology of China.

Cell lines used for evaluation of the in vivo antitumor activity in this study included three tumor cell lines, namely S180 (sarcoma tumer cell line), HepG2 (liver carcinoma cell line), EAC (Ehrlich ascites carcinoma cell line). All of cell lines were purchased by the Shanghai Institutes for Biological Sciences, Chinese Academy of Sciences, and the cells were cultured in RPMI-1640 medium, which was supplemented with 10% heat-inactivated fetal bovine serum, 100 U/mL penicillin and 100 U/mL streptomycin and cultured in an atmosphere of 5% CO_2 at 37°C. Cells were collected for the experiments in the logarithmic growth phase.

To establish the tumor-bearing mouse model, the cell lines were harvested and inoculated subcutaneously into the right armpit region of the mice. On the 7th day, the tumor ascrites were obtained and washed with sterile PBS. Under sterile condition, the tumor ascrites were diluted with sterile nomal saline to 1×10^{10} /L cell suspension. Tumor ascites were maintained in vivo in mice by transplantation of 0.2 mL of ascites (2×10^6 cells) from the infected mice to the non-infected mice.

In Vivo Tumor Xenograft Model

Each Kunming mouse (male or female, weight 20 ± 2 g) were inoculated with seven-day-old ascrite (0.2 mL, 2×10^6 cells) subcutaneously into the right front armpit. 24 h after implantation of tumor cells, the mice were randomly divided into five test groups with 10 mice per group. Each mouse was weighed immediately after inoculation. The mice were treated by oral gavage with test samples (60 mg/kg/day) or normal saline (NS, 10 mL/kg) for ten days once daily. On day 11, the mice were sacrificed via cervical dislocation, and the mouse and tumor were excised and weighed for evaluating the tumor growth inhibition. The tumor inhibitory rate was calculated by the following formula:

$$Tumor\ inhibitory\ rate\ rate\ (\%) = \left(\frac{W_{control} - W_{treated}}{W_{control}} \right) \times 100\% \qquad (1)$$

where $W_{control}$ and $W_{treated}$ were the average tumor weights of the control and treated mice, respectively. EAC tumor-bearing mice were observed for mean survival time. The effect of glycyrrhizin analogs on percentage increases in life span was calculated on the basis of mortality of the experimental mice:

$$Mean\ survival\ time = \frac{\Sigma Survival\ time\ (days)\ of\ each\ mouse\ in\ a\ group}{Total\ number\ of\ mice} \qquad (2)$$

$$\%ILS = \frac{MST\ of\ treated\ group}{MST\ of\ control\ group} \times 100 \qquad (3)$$

CONCLUSIONS

In summary, we synthesized glycyrrhizin analogs by glucuronidase biotransformation and alkaline isomerization, and evaluated their biological activities in vitro and in vivo. Anticancer activities of monoglucuronide GAMGs were higher than those of bisglucuronide GAs, while 1the 8α-epimer showed better activity than the 18β-epimer. Among them, 18α-GAMG displayed the most potent in vitro activity, with IC_{50} values of 6.67, 7.43 and 15.76 μM against HepG2, Hela and A549, and in vivo activity with inhibitory rates 39.8% and 49.7% for S180 and HepG2 tumor-bearing mice, respectively, and it significantly enhanced the survival rate of EAC tumor-bearing mice to 45.4%. The docking calculations showed that four analogs had better binding affinity to EGFR than the reference compound erlotinib and their binding energy had the same trend as anticancer activities, which further proved the correlation between anticancer activities and EGFR inhibitory activities of these compounds. In the binding model, high active compound 18α-GAMG was nicely bound to EGFR via six hydrogen bonds with ASP831, GLU738, THR766 (three bonds), LYS721 and one charge interaction, leading to the increased anticancer activities of 18α-GAMG. Therefore, 18α-GAMG is of interest for further studies as a potential anticancer agent. Further structural optimization of 18α-GAMG is ongoing using a variety of rational design strategies [33].

SUPPLEMENTARY MATERIALS

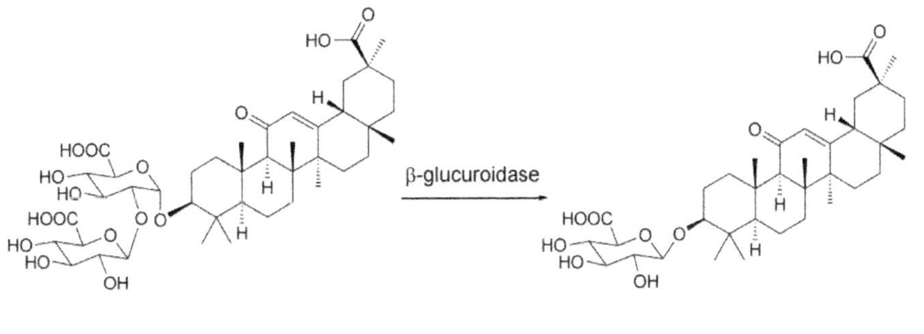

Scheme S1. Synthesis of 18β-GAMG, 18α-GAMG and 18α-GAMG.

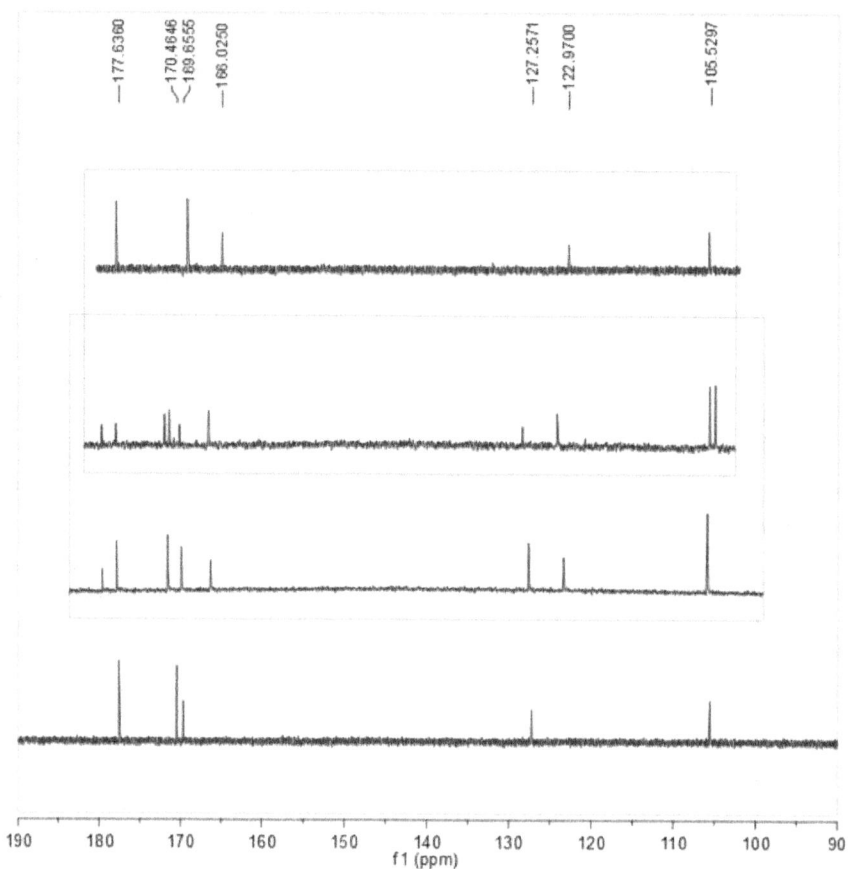

Figure S1. The isomerization reaction was monitored by ^{13}C-NMR spectroscopy.

Table S1. ^{13}C-NMR data for 18α-GA, 18α-GAMA, 18β-GA and 18β-GAMG in DMSO-d$_6$ (δ, ppm, J in Hz).

Position	18α-GA	18α-GAMG	18β-GA	18β-GAMG
1	38.6	38.3	38.7	38.4
2	25.5	25.7	25.8	25.6
3	88.3	87.9	88.3	87.7
4	39.0	39.1	39.1 ^{a}	39.1 ^{a}
5	54.4	54.3	54.4	54.1
6	17.1	17.1	17.0	16.9
7	33.2	33.2	32.2	32.1
8	43.4	43.4	44.9	44.9
9	59.9	59.9	61.1	61.0
10	36.2	36.2	36.4	36.3
11	198.8	198.7	199.1	198.9
12 *	123.1	123.1	127.4	127.3
13 *	166.2	166.1	169.8	169.7
14	44.7	44.7	43.0	42.9
15	26.3	26.3	26.1	26.1
16 *	36.8	36.7	25.9	25.8
17 *	35.4	35.3	31.6	31.5
18 *	39.6 ^{a}	39.8 ^{a}	48.1	48.0
19 *	31.5	31.4	40.7	40.6
20 *	41.7	41.6	43.2	43.1
21 *	28.5	28.4	30.4	30.4
22	35.2	35.1	37.6	37.5
23	27.3	27.5	27.2	27.4
24	16.2	16.45	16.0	16.2
25	16.4	16.49	16.3	16.4
26	18.2	18.2	18.4	18.3
27 *	20.4	20.4	23.0	23.0
28 *	15.7	15.7	28.5	28.4
29 *	20.7	20.6	27.9	27.8
30	179.7	179.5	177.8	177.6
1'	103.5	105.6	104.8	105.5
2'	81.0	75.6	82.7	76.1
3'	74.2	73.7	76.3	75.6
4'	72.0 or 71.9	71.6	71.6 or 71.3	71.5
5'	76.6	76.1	74.9	73.7
6'	172.4	170.6	172.4	170.5
1"	103.7		103.5	
2"	75.0		75.7	
3"	75.2		75.9	
4"	72.0 or 71.9		71.6 or 71.3	
5"	76.0		75.2	
6"	172.0		172.0	

* Carbon atoms of chemical shifts with significant difference; ^{a} Overlapped with solvent.

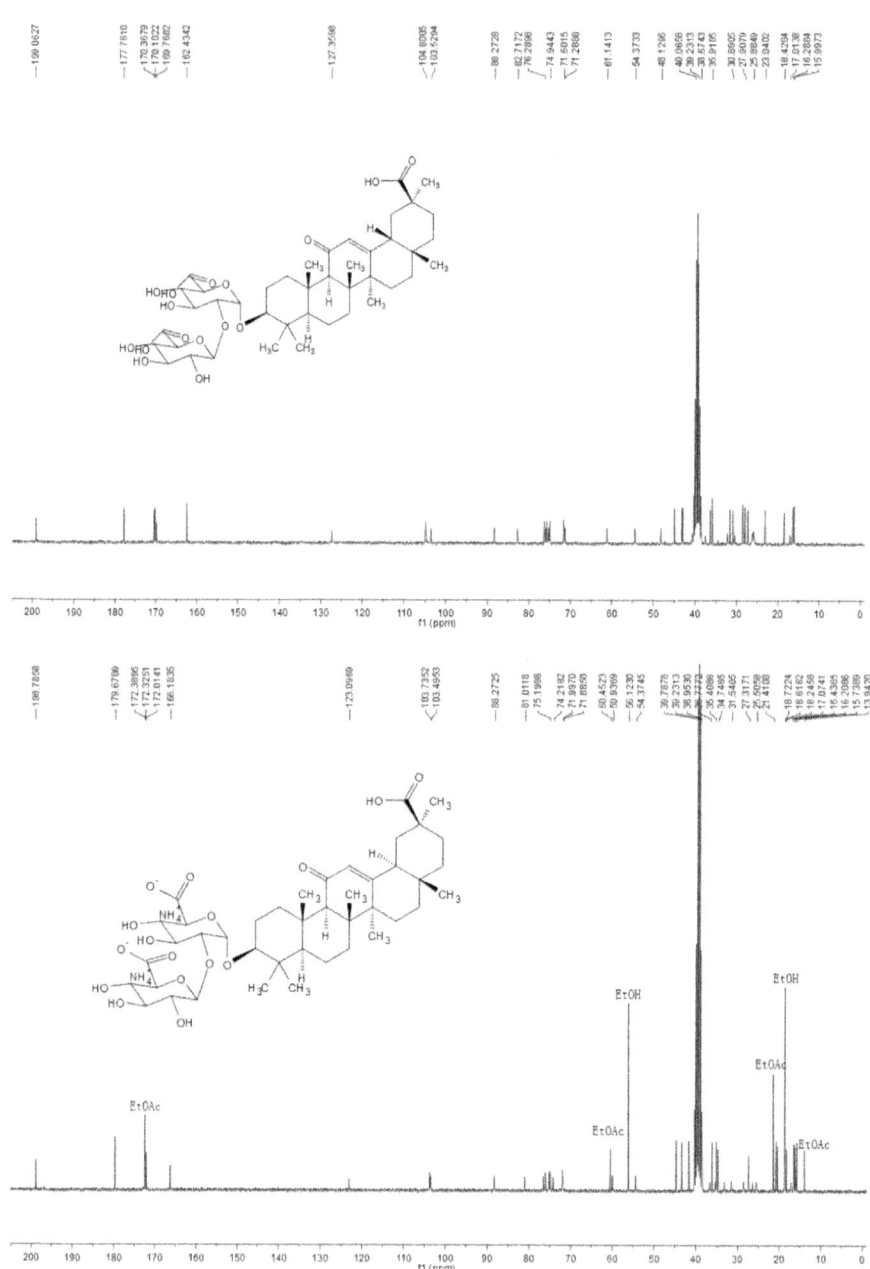

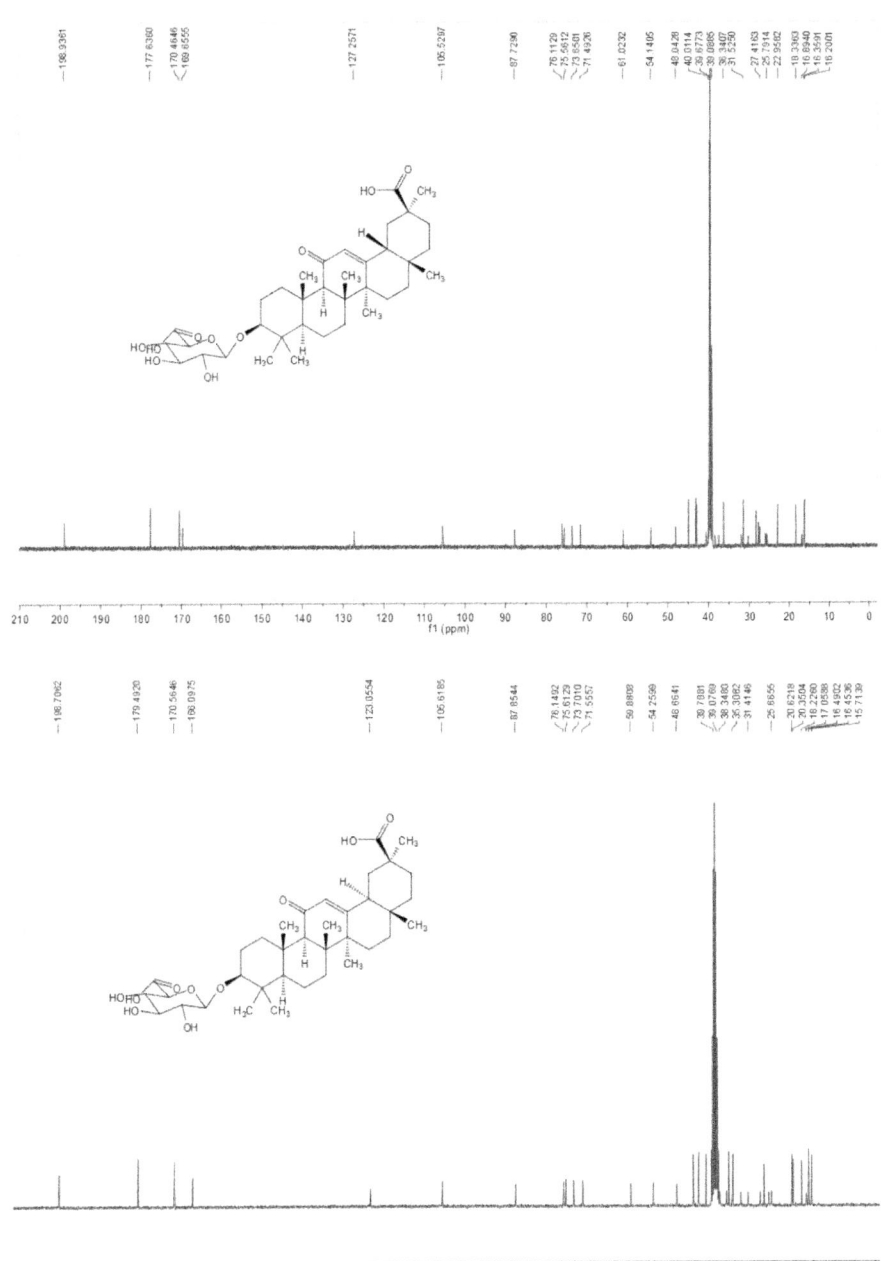

Copies of ^{13}C-NMR spectra of all compounds

ACKNOWLEDGMENTS

The authors wish to thank the National Natural Science Foundation of China (Nos. 21272008, 20802003), Science and Technological Fund of Anhui Province for Outstanding Youth (1408085J04).

AUTHOR CONTRIBUTIONS

YongAn Yang and HaiLiang Zhu designed research; YongAn Yang, WenJian Tang, Xin Zhang, JiWen Yuan and XinHua Liu performed research and analyzed the data; YongAn Yang and WenJian Tang wrote the paper. All authors read and approved the final manuscript.

REFERENCES

1. Citri, A.; Yarden, Y. EGF-ERBB signaling: Towards the systems level. Nat. Rev. Mol. Cell Biol. 2006, 7, 505–516.
2. Mendelsohn, J. Targeting the epidermal growth factor receptor for cancer therapy. J. Clin. Oncol. 2002, 20, 1S–13S.
3. Zandi, R.; Larsen, A.B.; Andersen, P.; Stockhausen, M.T.; Poulsen, H.S. Mechanisms for oncogenic activation of the epidermal growth factor receptor. Cell Signal. 2007, 19, 2013–2023.
4. Berasain, C.; Castillo, J.; Prieto, J.; Avila, M.A. New molecular targets for hepatocellular carcinoma: The ErbB1 signaling system. Liver Int. 2007, 27, 174–185.
5. Berasain, C.; Ujue, L.M.; Urtasun, R.; Goñi, S.; Elizalde, M.; Garcia-Irigoyen, O.; Azcona, M.; Prieto, J.; Avila, M.A. Epidermal growth factor receptor (EGFR) crosstalks in liver cancer. Cancers 2011, 3, 2444–2461.
6. Natarajan, A.; Wagner, B.; Sibilia, M. The EGF receptor is required for efficient liver regeneration. Proc. Natl. Acad. Sci. USA 2007, 104, 17081–17086.
7. Mitchell, C.; Nivison, M.; Jackson, L.F.; Fox, R.; Lee, D.C.; Campbell, J.S.; Fausto, N. Heparin-binding epidermal growth factor-like growth factor links hepatocyte priming with cell cycle progression during liver regeneration. J. Biol. Chem. 2005, 280, 2562–2568.
8. Hopfner, M.; Sutter, A.P.; Huether, A.; Schuppan, D.; Zeitz, M.; Scherubl, H. Targeting the epidermal growth factor receptor by gefitinib for treatment of hepatocellular carcinoma. J. Hepatol. 2004, 41, 1008–1016.

9. Huether, A.; Hopfner, M.; Sutter, A.P.; Schuppan, D.; Scherubl, H. Erlotinib induces cell cycle arrest and apoptosis in hepatocellular cancer cells and enhances chemosensitivity towards cytostatics. J. Hepatol. 2005, 43, 661–669.

10. Newman, D.J.; Cragg, G.M. Natural products as sources of new drugs over the 30 years from 1981−2010. J. Nat. Prod.2012, 75, 311–335.

11. Baltina, L.A. Chemical modification of glycyrrhizic acid as a route to new bioactive compounds for medicine. Curr. Med. Chem. 2003, 10, 155–171.

12. Morgan, A.G.; McAdam, W.A. Glycyrrhiza glabra. (Monograph). Altern. Med. Rev. 2005, 10, 230–237.

13. Asl, M.N.; Hosseinzadeh, H. Review of pharmacological effects of Glycyrrhiza sp. and its bioactive compounds.Phytother. Res. 2008, 22, 709–724.

14. Van Rossum, T.G.; Vulto, A.G.; de Man, R.A.; Brouwer, J.T.; Schalm, S.W. Glycyrrhizin as a potential treatment for chronic hepatitis C. Aliment Pharmacol. Ther. 1998, 12, 199–205.

15. Arase, Y.; Ikeda, K.; Murashima, N.; Chayama, K.; Tsubota, A.; Koida, I.; Suzuki, Y.; Saitoh, S.; Kobayashi, M.; Kumada, H. The long-term efficacy of glycyrrhizin in chronic hepatitis C patients. Cancer 1997, 79, 1494–1500.

16. Chayama, K. Management of chronic hepatitis C and prevention of hepatocellular carcinoma. J. Gastroenterol. 2002, 37, 69–73.

17. Van Rossum, T.G.; Vulto, A.G.; Hop, W.C.; Schalm, S.W. Glycyrrhizin-induced reduction of ALT in European patients with chronic hepatitis C. Am. J. Gastroenterol. 2001, 96, 2432–2437.

18. Thirugnanam, S.; Xu, L.; Ramaswamy, K.; Gnanasekar, M. Glycyrrhizin induces apoptosis in prostate cancer cell lines DU-145 and LNCaP. Oncol. Rep. 2008, 20, 1387–1392.

19. Tripathi, M.; Singh, B.K.; Kakkar, P. Glycyrrhizic acid modulates t-BHP induced apoptosis in primary rat hepatocytes.Food Chem. Toxicol. 2009, 47, 339–347.

20. Kim, K.J.; Choi, J.S.; Kim, K.W.; Jeong, J.W. The anti-angiogenic activities of glycyrrhizic acid in tumor progression.Phytother. Res. 2013, 27, 841–846.

21. Feng, S.; Li, C.; Xu, X.; Wang, X. Screening strains for directed biosynthesis of β-d-mono-glucuronide-glycyrrhizin and kinetics of

enzyme production. J. Mol. Catal. B: Enzym. 2006, 43, 63–67.
22. Lu, L.; Zhao, Y.; Yu, H.S.; Huang, H.Z.; Kang, L.P.; Cao, M.; Cui, J.M.; Yu, L.Y.; Song, X.B.; Ma, B.P. Preparation of glycyrrhetinic acid monoglucuronide by selective hydrolysis of glycyrrhizic acid via biotransformation. Chin. Herb. Med. 2012, 4, 324–328.
23. Mizutani, K.; Kambara, T.; Masuda, H.; Tamura, Y.; Ikeda, T.; Tanak, O.; Tokuda, H.; Nishino, H.; Kozuka, M.; Konoshima, T.; et al. Glycyrrhetic acid monoglucuronide (MGGR): Biological activities. Int. Congr. Ser. 1998, 1157, 225–235.
24. Park, H.Y.; Park, S.H.; Yoon, H.K.; Han, M.J.; Kim, D.H. Anti-allergic activity of 18α-glycyrrhetinic acid-3-O-α-d-glucuronide. Arch. Pharm. Res. 2004, 27, 57–60.
25. Baltina, L.A.; Stolyarova, O.V.; Baltina, L.A.; Kondratenko, R.M.; Plyasunova, O.A.; Pokrovskii, A.G. Synthesis and antiviral activity of 18α-glycyrrhizic acid and its esters. Pharm. Chem. J. 2010, 44, 299–302.
26. Shetty, A.V.; Thirugnanam, S.; Dakshinamoorthy, G.; Samykutty, A.; Zheng, G.; Chen, A.; Bosland, M.C.; Kajdacsy-Balla, A.; Gnanasekar, M. 18α-glycyrrhetinic acid targets prostate cancer cells by down-regulating inflammation-related genes. Int. J. Oncol. 2011, 39, 635–640.
27. Qu, Y.; Chen, W.H.; Zong, L.; Xu, M.Y.; Lu, L.G. 18α-Glycyrrhizin induces apoptosis and suppresses activation of rat hepatic stellate cells. Med. Sci. Monit. 2012, 18, BR24–32.
28. Kimura, M.; Inoue, H.; Hirabayashi, K.; Natsume, H.; Ogihara, M. Glycyrrhizin and some analogues induce growth of primary cultured adult rat hepatocytes via epidermal growth factor receptors. Eur. J. Pharmacol. 2001, 431, 151–161.
29. Tsou, H.R.; Mamuya, N.; Johnson, B.D.; Reich, M.F.; Gruber, B.C.; Ye, F.; Nilakantan, R.; Shen, R.; Discafani, C.; DeBlanc, R.; et al. 6-Substituted-4-(3-bromophenylamino)quinazolines as putative irreversible inhibitors of the epidermal growth factor receptor (EGFR) and human epidermal growth factor receptor (HER-2) tyrosine kinases with enhanced antitumor activity. J. Med. Chem. 2001, 44, 2719–2734.
30. Li, W.; Sha, Y.; Chen, L.X.; Qiu, F.; Wu, L.J. NMR data assignments of two 18-epimers of diammonium glycyrrhizinate. J. Shenyang Pharm. Univ. 2005, 7, 273–278.
31. Mollica, L.; Marchis, F.D.; Spitaleri, A.; Dallacosta, C.; Pennacchini, D.; Zamai, M.; Agresti, A.; Trisciuoglio, L.; Musco, G.; Bianchi, M.E. Glycyrrhizin binds to high-mobility group box 1 protein and inhibits its cytokine activities. Chem. Biol. 2007, 14, 431–441.

32. Du, D.; Yan, J.; Ren, J.H.; Lv, H.; Li, Y.; Xu, S.; Wang, Y.D.; Ma, S.G.; Qu, J.; Tang, W.B.; et al. Synthesis, biological evaluation, and molecular modeling of glycyrrhizin derivatives as potent high-mobility group box-1 inhibitors with anti-heart-failure activity in vivo. J. Med. Chem. 2013, 56, 97–108.

33. Huggins, J.D.; Sherman, W.; Tidor, B. Rational approaches to improving selectivity in drug design. J. Med. Chem. 2012, 55, 1424–1444.

Chapter 7

MOLECULAR CLONING OF A CHITINASE GENE FROM THE OVOTESTIS OF KURODA'S SEA HARE APLYSIA KURODAI

Gaku Matsunaga, Syuuji Karasuda, Ryo Nishino, Hideto Fukushima, and Masahiro Matsumiya

Department of Marine Science and Resources, College of Bioresource Sciences, Nihon University, Kanagawa, Japan

ABSTRACT

In this study, we report that we successfully cloned and sequenced a chitinase gene from the ovotestis of Kuroda's sea hare Aplysia kurodai. By using reverse transcription-polymerase chain reaction (RT-PCR) and a system for the 5' and 3' rapid amplification of cDNA ends, we obtained a 1352 bp chitinase gene (AkChi) from the ovotestis of A. kurodai. AkChi contains a 1263 bp open reading frame that encodes 421 amino acids. The domain structure predicted from the deduced amino acid sequence was an N-terminal signal peptide and a catalytic domain of glycoside hydrolase (GH) family 18 chitinase. A comparative analysis of the deduced amino acid sequences of AkChi with those of the acidic mammalian chitinase of the California sea hare Aplysia californica revealed the highest homology at 83%. The purified chitinase from the ovotestis was digested by trypsin, and 119 residues of digested peptides were consistent with the deduced amino acid sequence of AkChi. We used RT-PCR to evaluate the expression of AkChi in various tissues of A. kurodai, and we observed that AkChi was expressed only in the ovotestis. A phylogenetic tree analysis, performed using the amino acid sequences of AkChi and known GH family 18 chitinases, showed that AkChi was separated from the molluscan chitinases with a chitin binding domain. To our knowledge, this is the first study demonstrating the cDNA cloning of an ovotestis chitinase from a sea hare.

INTRODUCTION

Chitin, a major molecular constituent of the exoskeleton of insects and crustaceans, is a straight-chain homopolymer of β-1,4-linked N-acetyl-D-glucosamine units [1]-[3]. Chitinases (EC 3.2.1.14) are enzymes that randomly hydrolyze the β-1,4 glycosidic bonds of chitin [4]. They have been found in various organisms, and they play important physiological roles in functions such as attack, defense, morphological changes, and digestion [5] [6].

The characterization and cDNA cloning of chitinases from several fishes have been reported [7]-[9]. The stomach chitinases of fish have been identified and are classified into two groups, acidic fish chitinase-1 (AFCase-1) and acidic fish chitinase-2 (AFCase-2) based on the differences in their primary structure and the activity toward short substrates [8]. Chitinases from molluscs play important physiological roles in the digestion of food [10] [11], attacking crustaceans [12], and shell formation [13] [14]. However, reports on the distribution, characterization, and cDNA cloning of molluscan chitinases are limited [10]-[16]. In this study, we were using the Kuroda's sea hare, Aplysia kurodai. A. kurodai is a kind of herbivorous gastropoda seen in the vicinity of the coast from April to June. In addition, this creature was allowed to degenerate shells despite the shellfish. In a previous study, we detected chitinase activity in the ovotestis and egg of A. kurodai [16], whereas lysozyme activity (antibacterial enzyme activity) was not detected in all of the organs [16]. We also reported the purification and properties of a chitinase from the ovotestis of A. kurodai [16]. Together the results indicated that the physiological role of this chitinase was as a defense against nematodes and fungus which had chitin in the body wall as a structural component [16].

In the present study, we cloned the cDNA encoding chitinase from the ovotestis of A. kurodai and determined the primary structure of the chitinase.

MATERIALS AND METHODS

Materials

Kuroda's sea hare Aplysia kurodai and laid egg were captured from the tide pools of Shimoda Bay (Shizuoka, Japan) in June.

Cloning of the Chitinase cDNA from A. Kurodai

The sequences of all primers are presented in Table 1. Total RNA was extracted from the ovotestis of A. kurodai using ISOGEN II reagent (Nippon Gene, Tokyo) according to the manufacturer's instructions. First-strand cDNA was synthesized using 500 ng of total RNA and oligo dT primers with Prime

Script Reverse Transcriptase (Takara Bio, Shiga, Japan) according to the manufacturer's instructions. Six degenerate primers were designed for the reverse transcriptase-polymerase chain reaction (RT-PCR) from conserved sequences of molluscan chitinase, including those from California sea hare (Aplysia californica; GenBank: XM_005112601), triangle sail mussel (Hyriopsis cumingii; GenBank: JN582038), Pacific oyster (Crassostrea gigas; GenBank: AJ971239), Hawaiian bobtail squid (Euprymna scolopes; GenBank: KF015222), and golden cuttlefish (Sepia esculenta; GenBank: AB986212).

The first PCR was performed using A. kurodai cDNA as a template and P1 and P2 as primers (Figure 1). The PCR parameters were as follows: 94°C for 2 min, followed by 30 cycles of 94°C for 30 s, 55°C for 30 s, and 72°C for 30 s. Nested PCR was performed using the products of the first PCR as templates and P3, P4, P5, and P6 as primers, with the same PCR parameters as described above. The nucleotide sequence analysis of the RT-PCR amplified chitinase cDNA fragments from the ovotestis of A. kurodai detected one nucleotide sequence (AkChi).

For the 3' rapid amplification of cDNA ends (RACE), we designed primers specific to AkChi (i.e., P7, P8, and P9, respectively; Table 1) based on the detected sequences. We amplified cDNA fragments encoding the 3' region of AkChi using A. kurodai cDNA as the template and the primer pairs P7 and 3R, P8 and 3R, and P9 and 3R (Figure 1). The PCR parameters were as follows: 94°C for 2 min, followed by 30 cycles of 94°C for 30 s, 56°C for 30 s, and 72°C for 30 s. For 5' RACE, specific primers (P10, P11, and P12 for AkChi; Table 1) were designed based on the nucleotide sequences obtained from RT-PCR. cDNA fragments encoding the 5' regions of AkChi were amplified using PCR. The first PCR was performed using the newly synthesized first-strand cDNA as a template and the primer pairs P10 and P11 for AkChi. Nested PCR was performed using the first PCR products as templates and the primer pairs P10 and P12 for AkChi.

Table 1. Primers used for PCR, RACE, and tissue-specific expression.

Primer	Sequence (5' → 3')	Purpose
P1*	TNGCNGCNTTYGARTGGAAYGA	Primary PCR
P2*	CATNCCNSWRAARTCRTCRTTRTC	Primary PCR
P3*	GGNGGNTGGAAYATGGG	Primary PCR
P4*	ACCCAYTGRTTNCCNARNACNA	Primary PCR
P5*	GNAAYTTYGAYGGNYTNGA	Primary PCR
P6*	TTDATCATYTCRCANACYTCRTARTA	Primary PCR

P7	GCCGGATACGAAGTGGAC	3' RACE
P8	GGAACTTAACGAGTACTT	3' RACE
P9	GACAGACGAGAGCGACTCTGGTCG	3' RACE
3R	CTGTGAATGCTGCGACTACGAT	3' RACE
P10	CACAATGACGTTGCAAG	5' RACE, Full-length PCR
P11	ATGGCCTGGGCTCATTTT	5' RACE
P12	TTATCCTCTGGAGGGCT	5' RACE
P13	CACGTTATGATTGCGAC	Full-length PCR
P14	TCTGCTGCTGTGAGTGCTGGCAAGG	tissue-specific expression
P15	GCATTTCGCACACCTCGTAGTAAGA	tissue-specific expression
β-actin-a*	GAYAAYGGNWSNGGNATGTG	tissue-specific expression
β-actin-b*	TCRAACATDATYTGNGTCAT	tissue-specific expression

Note: *Degenerate primers.

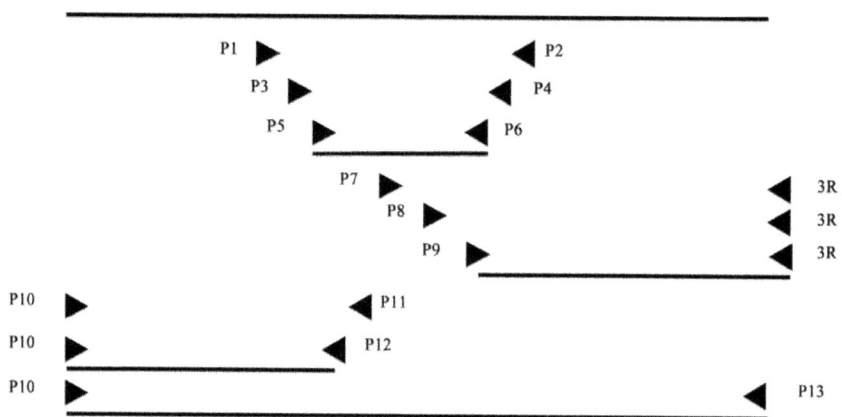

Figure 1. Schematic representation of the cDNA structure of AkChi and location of the primers. Arrowheads indicate the primers, and lines between the arrowheads indicate the amplified cDNA fragments.

The PCR parameters were as follows:

94°C for 1 min, followed by 30 cycles of 94°C for 30 s, 49°C for 30 s, and 72°C for 30 s.

The nucleotide sequences of cDNA fragments containing a full-length open reading frame (ORF) were confirmed by PCR using specific primers (P10 and P13 for AkChi; Table 1) and Platinum Pfx DNA Polymerase (Invitrogen, Carlsbad, CA).

Nucleotide Sequence Analysis

The RT-PCR, 3'RACE, and 5'RACE amplification products, and the full-length amplification products were subcloned into pGEM-T Easy Vector (Promega, Madison, WI), according to the manufacturer's instructions. Sequences were determined on an ABI PRISM 3130 genetic analyzer (Applied Biosystems, Foster City, CA) using a Big Dye Terminator v3.1 cycle sequencing kit (Applied Biosystems).

Amino Acid Sequence of the Peptide of the Purified Chitinase from the Ovotestis of A. kurodai

A chitinase from the ovotestis of A. kurodai was purified as described [16]. The purified chitinase was subjected to sodium dodecyl sulfate-polyacrylamide gel electrophoresis (SDS-PAGE) and stained with AE-1360 EzStain Silver (ATTO, Tokyo). A gel slice was cut into small pieces and destained by destaining solution (15 mM $K_3[Fe(CN)_6]$, 50 mM $Na_2S_2O_3$). Destained gel pieces were trypsinized as described in the manual of In-Gel Tryptic Digestion Kit manual (Thermo Scientific, Waltham, MA). The peptide mixtures thus obtained were subjected to a nano-scale liquid chromatography-electrospray ionization-tandem mass spectrometry (nanoLC- ESI-MS/MS) analysis using a Q Exactive mass spectrometer (Thermo Scientific) equipped with a captive spray ionization source (Michrom Bioresources, Auburn, CA) and an Advance UHPLC System (Michrom Bioresources).

Tissue-Specific Expression of AkChi

Total RNA was prepared from the ovotestis, egg, skin, gill, crop, anterior gizzard, and posterior gizzard as described in the cloning methods section (2.2) above. First-strand cDNA was pre-cloned from the RNA isolated from each tissue and egg as described in the RT-PCR section (2.2) above. For tissue-specific expression, we designed primers specific to AkChi (P14 and P15, respectively;Table 1) based on the detected sequences. AkChi was amplified using the first-strand cDNA as template and the primer pairs P14 and P15 (Table 1). The PCR parameters were as follows: 94°C for 1 min, followed by 35 cycles of 94°C for 30 s, 62°C for 30 s, and 72°C for 30 s. To determine the amount of total RNA in each tissue, we amplified β-actin mRNA fragments using specific primer pairs (Table 1).

Phylogenetic Tree Analysis of AkChi

In order to classify the chitinase from the ovotestis of A. kurodai among the GH family 18 chitinases, we constructed a phylogenetic tree based on the enzyme precursor sequences by the neighbor-joining method, using the ClustalW program (http://www.genome.jp/tools/clustalw/). A bacterial chitinase (GenBank: X03657) was used as the out group.

RESULTS AND DISCUSSION

Cloning of A. kurodai Chitinase cDNA

The structure of AkChi and the location of primer sequences are schematically represented inFigure 1. The internal sequence of the cDNA of A. kurodai ovotestis chitinase was amplified by RT-PCR using degenerate primers (from P1 to P6, respectively; Table 1); an amplified product of approx. 400 bp was obtained. The product was sequenced, and 86% homology with the acidic mammalian chitinase of A. californica was confirmed (accession no. XM_005112601). Because the sequence was part of ovotestis chitinase cDNA from A. kurodai, we used it to design gene-specific primers for 3' and 5' RACE (from P7 to P12; Table 1). An amplified product of approx. 430 bp was obtained by 3' RACE, and its sequence contained a stop codon. An amplified product of approx. 520 bp was also obtained by 5' RACE; its sequence contained a start codon. Based on these results, we designed full-length primers (P10 and P13; Table 1) to incorporate these start and stop codons. cDNA was amplified using the primers and the amplified product was sequenced.

The full-length cDNA of A. kurodai ovotestis chitinase (AkChi) was 1352 bp in length and contained an ORF of 1263 bp encoding 421 amino acids (Figure 2). The size of ORF of AkChi was smaller than it from H. cumingii [14], 1962 bp encoding 653 amino acids. A poly-A sequence in eukaryotes was detected at the 3' end of AkChi. AkChi, which encodes A. kurodai ovotestis chitinase, has been registered in the database of the DNA Data Bank of Japan (DDBJ) (accession no. LC085435). We compared the deduced amino acid sequence of AkChi with that of other organisms using BLAST, and the highest homology, 83%, was confirmed with the acidic mammalian chitinase of A. californica (accession no. XM_005112601).

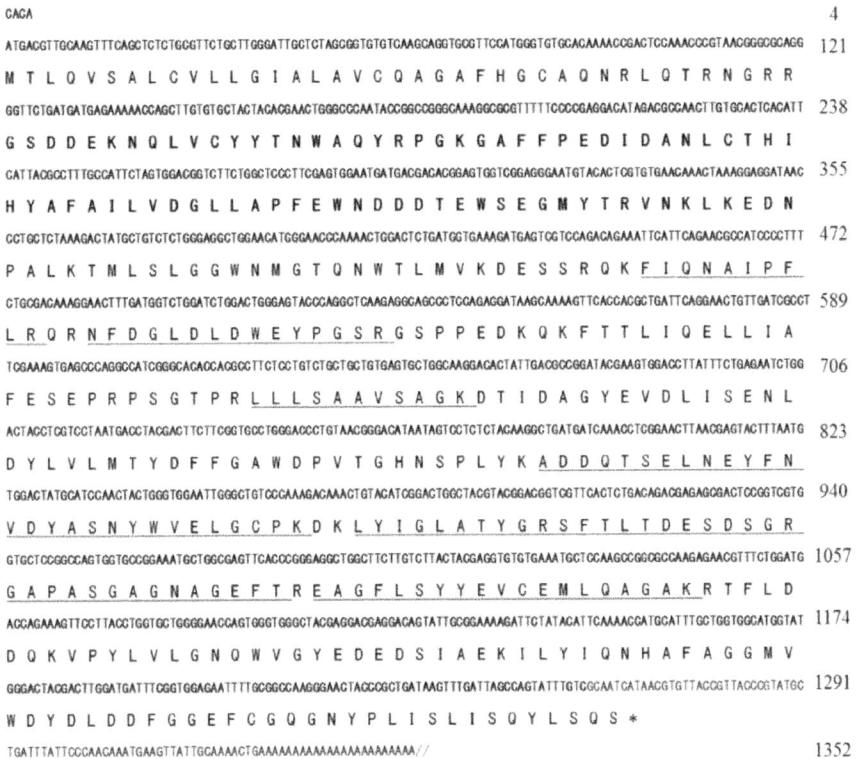

Figure 2. cDNA and deduced amino acid sequences of AkChi. Underlined sequences show matching with the peptide fragments of the purified and tripsinized enzyme (coverage: 35.39%, 119 residues).

Figure 3 compares amino acid sequences from AkChi and some other known molluscan chitinases (A. californica, H. cumingii, C. gigas, E. scolopes, and S. esculenta). The deduced amino acid sequence of AkChi was shown to have a structure of the GH family 18 chitinase, with an N-terminal signal peptide and a GH 18 catalytic domain. The catalytic domain also contained an active site that is a conserved sequence of GH family 18 chitinases (Figure 3). Though the chitinase of H. cumingii [14] and E. scolopes [15] had two chitin binding domains (CBDs) and the chitinase of S. esculenta had one CBD, AkChi lacked a CBD. It was reported that fish chitinases have one CBD [8]. This result suggests that the structure of molluscan chitinase is diverse compared to the fish chitinases.

Amino Acid Sequence of the Chitinase

We analyzed the sequences of the peptide fragments obtained by the tryptic treatment of the purified chitinase from the ovotestis of A. kurodai [16] were analyzed and compared them to the deduced amino acid sequence of AkChi. The obtained sequences from peptide fragments were consistent with the deduced amino acid sequence of AkChi (coverage: 35.39%, 119 residues) (Figure 2). This result suggests that AkChi is a gene coding the purified enzyme. In addition, trypsin is cut the C-terminal side of lysine and arginine. In this result, it was confirmed that the trypsin is working properly in the all of cleavage site.

Tissue-Specific Expression of AkChi

We investigated the tissue-specific expression of AkChi in A. kurodai by RT-PCR using the housekeeping β-ac- tin gene as a control (Figure 4). It is reported that fish express chitinase to the digestive organs for digestion of chitin from food [17]. The expression profile results indicated that AkChi was present only in the ovotestis. We previously detected chitinase activity in the ovotestis and egg from A. kurodai [16], whereas lysozyme activity (antibacterial enzyme activity) was not detected in any of the organs [16]. A. kurodai has to prey on seaweed.

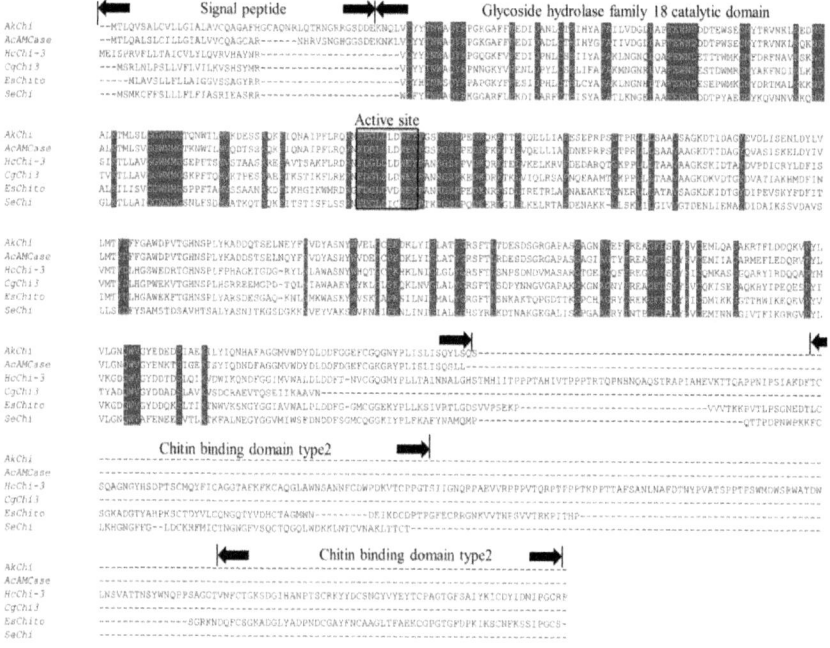

Figure 3. Multiple alignment of duduced amino acid sequences of A. kurodai chitinase (AkChi) with Aplysia californica acidic mammalian chitinase (AcAMCase), Hy-

riopsis cumingii chitinase-3 (HcChi-3), Crassostrea gigas Chit3 protein A (CgChi3), Euprymna scolopes chitotriosidase (EsChito), and Sepia esculenta chitinase (SeChi). GenBank accession nos.: AcAMCase, XM_005112601; HcChi-3, JN582038; CgChi3, AJ971239; EsChito, KF015222; SeChi, AB986212. Matched sequences are shown in black.

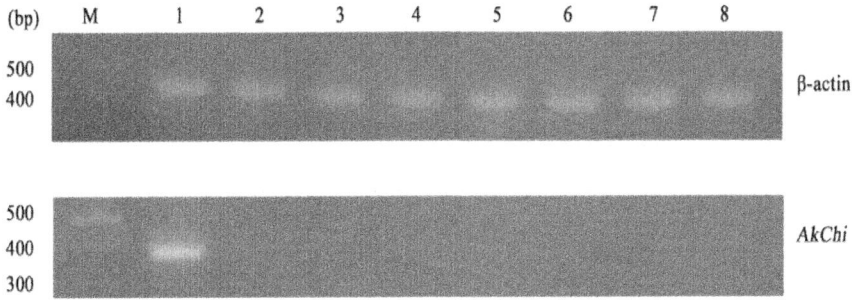

Figure 4. Expression profiles of AkChi and β-actin mRNA in tissue using RT-PCR. M, markers; 1, ovotestis; 2, egg; 3, skin; 4, gill; 5, buccal mass; 6, crop; 7, anterior gizzard; 8, posterior gizzard.

Thus, A. kurodai is not necessary chitinase in digestion and attack of food as squid [10] [11] and octopus [12], respectively. In addition, there is not necessary to shell formation because it does not even have shells. These results suggest that the role of this chitinase is as a defense against nematodes and fungus which have chitin in the body wall as a structural component.

Phylogenetic Tree Analysis of AkChi

We performed a phylogenetic tree analysis of GH family 18 chitinases and AkChi (Figure 5). Acidic mammalian chitinases (AMCases) have been found in the stomach of mammals.

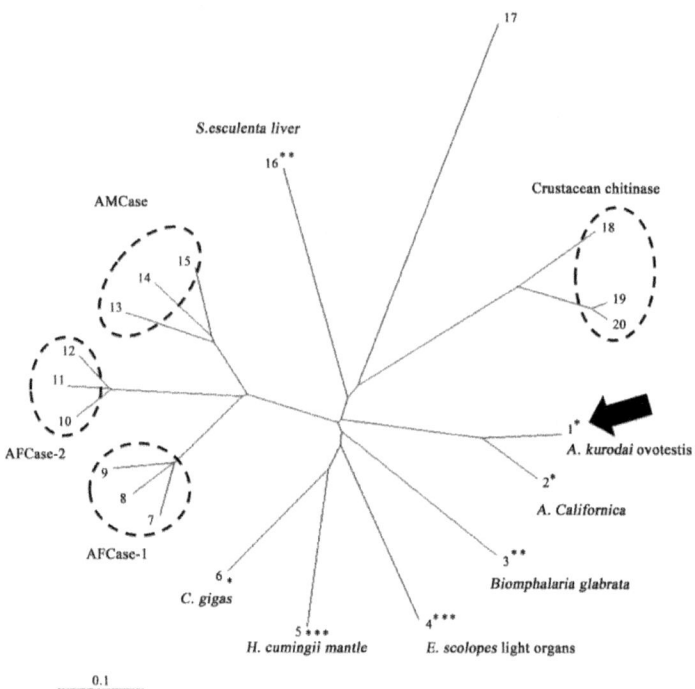

No.	Species	Genbank accession number
1	*Aplysia kurodai* (chitinase)	LC085435
2	*Aplysia californica* (acidic mammalian chitinase)	XM_005112601
3	*Biomphalaria glabrata* (chitinase-3-like protein 1)	XP_013090777
4	*Euprymna scolopes* (chitotriosidase)	KF015222
5	*Hyriopsis cumingii* (chitinase-3)	JN582038
6	*Crassostrea gigas* (Chit3 protein)	AJ971239
7	*Epinephelus coioides* (chitinase1)	AB686658
8	*Sebastiscus marmoratus* (chitinase1)	FJ169895
9	*Parapristipoma trilineatum* (chitinase1)	AB642677
10	*Epinephelus coioides* (chitinase2)	FJ169894
11	*Parapristipoma trilineatum* (chitinase2)	AB642678
12	*Sebastiscus marmoratus* (chitinase2)	AB686659
13	*Bos Taurus* (chitin binding protein b04)	AB051629
14	*Mus musculus* (acidic chitinase)	EF094027
15	*Homo sapiens* (acidic mammalian chitinase)	AF290004
16	*Sepia esculenta* (chitinase)	AB986212
17	*Serratia marcescens* (chiA protein precursor)	X03657
18	*Portunus trituberculatus* (chitinase1)	AB874469
19	*Portunus trituberculatus* (chitinase2)	AB890123
20	*Scylla serrata* (chitinase)	EU402970

Figure 5. Phylogenetic tree analysis of chitinase amino acid sequence by the neighbor-joining method of the program Clustal W. A bacterial chitinase, Serratia marcescens

chitinase, was used as the out group. The scale bar indicates the substitution rate per residue. The arrow shows AkChi obtained in the present study. * Molluscan chitinase without a CBD; ** Molluscan chitinase with one CBD; *** Molluscan chitinase with two CBDs.

Two chitinase groups with different structures and activity toward short substrates, AFCase-1 and AFCase-2, have been found in the stomach of fish [8]. Crustacean showed a chitinase group [18]. In contrast, molluscan chitinases did not show clear chitinase groups. The reason for this might be the differences in the chitinase domain structure that are due to the presence or absence of a CBD and the number of CBDs. We previously detected chitinase activity in the ovotestis and oviduct from the Walking sea hare Aplysia juliana [16]. If the success in cloning the chitinase from A. juliana, it will be conceivable to form a group of sea hare chitinase.

CONCLUSION

The cDNA of the ovotestis chitinase obtained from A. kurodai contained a 1263 bp open reading frame with a coding potential for 421 amino acid peptides. AkChi had the structural motifs of GH family 18 chitinase, but it did not have chitin binding domain. This study is the first report of the cloning of chitinase from the ovotestis of a sea hare.

ACKNOWLEDGEMENTS

This work was supported in part by a Grant-in-Aid for Scientific Research (C) (no. 25450309) and a College of Bioresource Science, Nihon-University Grant (2015).

REFERENCES

1. Khandeparker, L., Gaonkar, C.C. and Desai, D.V. (2013) Degradation of Barnacle Nauplii: Implications to Chitin Regulation in the Marine Environment. Biologia, 68, 696-706. http://dx.doi.org/10.2478/s11756-013-0202-6
2. Arbia, W., Arbia, L., Adour, L. and Amrane, A. (2013) Chitin Extraction from Crustacean Shells Using Biological Methods—A Review. Food Technology and Biotechnology, 51, 12-25.
3. Kramer, K.J. and Koga, D. (1986) Insect Chitin: Physical State, Synthesis, Degradation and Metabolic Regulation. Insect Biochemistry, 16, 851-877. http://dx.doi.org/10.1016/0020-1790(86)90059-4
4. Umemoto, N., Ohnuma, T., Mizuhara, M., Sato, H., Skriver, K.

and Fukamizo, T. (2013) Introduction of a Tryptophan Side Chain into Subsite +1 Enhances Transglycosylation Activity of a GH-18 Chitinase from Arabidopsis thaliana, AtChiC. Glycobiology, 23, 81-90. http://dx.doi.org/10.1093/glycob/cws125

5. Gooday, G.W. (1999) Aggressive and Defensive Roles for Chitinases, Chitin and Chitinases. Cellular and Molecular Life Sciences, 87, 157-169.

6. Henrissat, B. (1991) A Classification of Glycosyl Hydrolases Based on Amino Acid Sequence Similarities. Biochemical Journal, 280, 309-316. http://dx.doi.org/10.1042/bj2800309

7. Ikeda, M., Miyauchi, K. and Matsumiya M. (2012) Purification and Characterization of a 56 kDa Chitinase Isozyme (PaChiB) from the Stomach of the Silver Croaker, Pennahia argentatus. Bioscience, Biotechnology, and Biochemistry, 76, 971-979. http://dx.doi.org/10.1271/bbb.110989

8. Ikeda, M., Kondo, Y. and Matsumiya, M. (2013) Purification, Characterization, and Molecular Cloning of Chitinases from the Stomach of the Threeline Grunt Parapristipoma trilineatum. Process Biochemistry, 48, 1324-1334. http://dx.doi.org/10.1016/j.procbio.2013.06.016

9. Laribi-Habchi, H., Dziril, M., Badis, A., Mouhoub, S. and Mameri, N. (2012) Purification and Characterization of a Highly Thermostable Chitinase from the Stomach of the Red Scorpionfish Scorpaena scrofa with Bioinsecticidal Activity toward Cowpea Weevil Callosobruchus maculates (Coleoptera: bruchidae). Bioscience, Biotechnology, and Biochemistry, 76, 1733-1740. http://dx.doi.org/10.1271/bbb.120344

10. Nishino, R., Suyama, A., Ikeda, M., Kakizaki, H. and Matsumiya, M. (2014) Purification and Characterization of a Liver Chitinase from Golden Cuttlefish, Sepia esculenta. Journal of Chitin and Chitosan Science, 2, 238-243. http://dx.doi.org/10.1166/jcc.2014.1065

11. Matsumiya, M., Miyauchi, K. and Mochizuki, A. (2002) Characterization of 38 kDa and 42 kDa Chitinase Isozymes from the Liver of Japanese Common Squid Todarodes pacificus. Fisheries Science, 68, 603-609. http://dx.doi.org/10.1046/j.1444-2906.2002.00467.x

12. Ogino, T., Tabata, T., Ikeda, M., Kakizaki, H. and Matsumiya, M. (2014) Purification of a Chitinase from the Posterior Salivary Gland of Common Octopus Octopus vulgaris and Its Properties. Journal of Chitin and Chitosan Science, 2, 135-142. http://dx.doi.org/10.1166/jcc.2014.1049

13. Zhang, G., Fang, X., Guo, X., Li, L., Luo, R., Xu, F., Yang, P., Zhang, L., Wang, X., Qi, H., Xiong, Z., Que, H., Xie, Y., Holland, P.W.H., Paps, J., Zhu, Y., Wu, F., Chen, Y., Wang, J., Peng, C., Meng, J., Yang, L., Liu, J., Wen, B., Zhang, N., Huang, Z., Zhu, Q., Feng, Y., Mount, A., Hedgecock, D., Xu, Z., Liu, Y., Domazet-Loso, T., Du, Y., Sun, X., Zhang, S., Liu, B., Cheng, P., Jiang, X., Li, J., Fan, D., Wang, W., Fu, W., Wang, T., Wang, B., Zhang, J., Peng, Z., Li, Y., Li, N., Wang, J., Chen, M., He, Y., Tan, F., Song, X., Zheng, Q., Huang, R., Yang, H., Du, X., Chen, L., Yang, M., Gaffney, P.M., Wang, S., Luo, L., She, Z., Ming, Y., Huang, W., Zhang, S., Huang, B., Zhang, Y., Qu, T., Ni, P., Miao, G., Wang, J., Wang, Q., Steinberg, C.E.W., Wang, H., Li, N., Qian, L., Zhang, G., Li, Y., Yang, H., Liu, X., Wang, J., Yin, Y. and Wang, J. (2012) The Oyster Genome Reveals Stress Adaptation and Complexity of Shell Formation. Nature, 490, 49-54. http://dx.doi.org/10.1038/nature11413

14. Wang, G.-L., Xu, B., Bai, Z.-Y. and Li, J.-L. (2012) Two Chitin Metabolic Enzyme Genes from Hyriopsis cumingii: Cloning, Characterization, and Potential Functions. Genetics and Molecular Research, 11, 4539-4551. http://dx.doi.org/10.4238/2012.October.15.4

15. Kremer, N., Philipp, E.E.R., Carpentier, MC., Brennan, C.A., Kraemer, L., Altura, M.A., Augustin, R., H&aumlsler, R., Heath-Heckman, E.A.C., Peyer, S.M., Schwartzman, J., Rader, B., Ruby, E.G., Rosenstiel, P. and McFall-Ngai, M.J. (2013) Initial Symbiont Contact Orchestrates Host-Organ-Wide Transcriptional Changes that Prime Tissue Colonization. Cell Host & Microbe, 14, 183-194. http://dx.doi.org/10.1016/j.chom.2013.07.006

16. Karasuda, S., Ikeda, M., Miyauchi, K. and Matsumiya, M. (2011) Existence and Physiological Role of Chitinase in the Gonad of Two Species of Sea Hare, Kuroda's Sea Hare Aplysia kurodai and Walking Sea Hare Aplysia Juliana. Proceedings of the 9th Asia-Pacific Chitin and Chitosan Symposium, Vietnam, 3-6 August 2011, 169-172.

17. Kakizaki, H., Ikeda, M., Fukushima, H. and Masahiro, M. (2015) Distribution of Chitinolytic Enzymes in the Organs and cDNA Cloning of Chitinase Isozymes from the Stomach of Two Species of Fish, Chub Mackerel (Scomber japonicus) and Silver Croaker (Pennahia argentata). Open Journal of Marine Sciences, 5, 398-411. http://dx.doi.org/10.4236/ojms.2015.54032

18. Fujitani, N., Hasegawa, H., Kakizaki, H., Ikeda, M. and Masahiro, M. (2014) Molecular Cloning of Multiple Chitinase Genes in Swimming Crab Portunus trituberculatus. Journal of Chitin and Chitosan Science, 2, 149-156. http://dx.doi.org/10.1166/jcc.2014.1046

Chapter 8

MOLECULAR BIOLOGY AND GENETIC ENGINEERING

INTRODUCTION

Molecular genetics, or molecular biology, is the study of the biochemical mechanisms of inheritance. It is the study of the biochemical nature of the genetic material and its control of phenotype. It is the study of the connection between genotype and phenotype. The connection is a chemical one.

Control of phenotype is one of the two roles of DNA (transcription). You have already been exposed to the concept of the Central Dogma of Molecular Biology, i.e. that the connection between genotype and phenotype is DNA (genotype) to RNA to enzyme to cell chemistry to phenotype.

BIOTECHNOLOGY: BASIC CONCEPTS AND DEFINITIONS

Definition of Biotechnology

The term biotechnology was coined in 1919 by Karl Ereky, a Hungarian engineer. At that time, the term included all the processes by which products are obtained from raw materials with the aid of living organisms. Ereky envisioned a biochemical age similar to the stone and iron ages.

Nowadays, according to the Convention on Biological Diversity (CBD), Biotechnology is defined as "any technological application that uses biological systems, living organisms, or derivatives thereof, to make or modify products or processes for specific use" (CBD, 1992). The living organisms or derivatives thereof most frequently used include micro-organisms, animals and plants (or their isolated cells) as well as enzymes. They can be utilized to process substances, usually other natural, renewable materials, or serve themselves as sources for valuable substances or goods. Several branches of industry rely on biotechnological tools for the production of food, beverages, pharmaceuticals and biomedicals. The CBD definition is applicable to both "traditional" or "old" and "new" or "modern" biotechnology.

Long before the term biotechnology was coined for the process of using living organisms to produce improved commodities, people were utilizing living micro-organisms to obtain valuable products, for example through the process of fermentation.

A list of early biotechnological applications is given below in Table 1.

Table 1. Traditional applications of biotechnology

Providing bread with leaven	Prehistoric period
Fermentation of juices to alcoholic beverages	Prehistoric period
Knowledge of vinegar formation from fermented juices	Prehistoric period
Manufacture of beer in Babylonia and Egypt	3rd century BC
Wine manufacturing in the Roman Empire	3rd century AD
Production of spirits of wine (ethanol)	1150
Vinegar manufacturing industry	14th century AD
Discovery of the fermentation properties of yeast	1818
Description of the lactic acid fermentation by Pasteur	1857
Detection of fermentation enzymes in yeast by Buchner	1897
Discovery of penicillin by Fleming	1928
Discovery of many other antibiotics	≈1945

Since the middle of the twentieth century biotechnology has rapidly progressed and expanded. In the mid-1940s, scale-up and commercial production of antibiotics such as penicillin occurred.

The techniques used for this development were:

- Isolation of an organism producing the chemical of interest using screening/ selection procedures, and
- Improvement of production yields via mutagenesis of the organism or optimization of media and fermentation conditions. This type of biotechnology is limited to chemicals occurring in nature. It is also limited by its trial-and-error approach, and requires a lengthy procedure over years or even decades to improve yields (Rolinson, 1998).

About three decades ago, with the advance of molecular biology, biotechnology became more of a science than an art. Regions of deoxyribonucleic acid (DNA) (called genes) were found to contain information that directs the synthesis of specific proteins. Proteins can therefore be considered as the final product of a gene; they are the molecules that carry out almost all essential processes within a cell. Each protein has its own identity and function: many are so-called enzymes that catalyse (facilitate) chemical reactions, others are structural components of cells and organs (Morange and Cobb, 2000). Today it is possible to express a gene, regardless of its origin, in a simple bacterium such as Escherichia coli (E. coli), so that the bacterium produces large quantities of the protein coded for by the gene. The same principle can be applied to many other micro-organisms, as well as to higher organisms such as plants and animals.

The techniques used for this purpose include:
- Isolation of the gene coding for a protein of interest;
- Cloning (i.e. transfer) of this gene into an appropriate production host;
- Improving gene and protein expression by using stronger promoters, improving fermentation conditions etc. (Gellisen, 2005). Together, these techniques are known as recombinant DNA technology and will be discussed at some length throughout this resource book.

About two decades ago, protein engineering became possible as an offshoot of the recombinant DNA technology. Protein engineering differs from "classical" biotechnology in that it is concerned with producing new (engineered) proteins which have been modified or improved in some of their characteristics (Park and Cochran, 2009).

The techniques involved in protein engineering are essentially based on recombinant DNA technology and involve:
- Various types of mutagenesis (to cause changes in specific locations or regions of a gene to produce a new gene product);
- Expression of the altered gene to form a stable protein;
- Characterization of the structure and function of the protein produced;
- Selection of new gene locations or regions to modify for further improvement as a result of this characterization.

The commercial implications of the technical developments listed above are that a large number of proteins, existing only in tiny quantities in nature, can now be produced on an industrial scale. Furthermore, the yields of biochemical production can be increased much faster than what was originally possible with classical fermentation.

Importantly, the production of transgenic animals and plants that contain genetic elements from foreign sources and possess novel traits and characteristics is also based on the techniques outlined above. As all these approaches result in the creation of genetically modified organisms (GMOs) that can be potentially harmful to the environment and human health, the part of biotechnology that deals with GMOs is strictly regulated by biosafety laws and guidelines. The main thrust of this resource book is on the development and enforcement of such regulatory frameworks at domestic and international levels.

Biotechnology applications are developed by a collection of multidisciplinary research activities, commonly referred to as enabling technologies. Apart from fermentation and genetic engineering/recombinant DNA technology, other important enabling technologies are plant and animal cell culture technology and enzyme technology.

The basis of these enabling technologies are the scientific disciplines of molecular biology, genetics, microbiology, biochemistry, protein chemistry, chemical and process engineering and computer science. An overview of important events in the development of modern molecular biology and recombinant DNA technology is provided in Table 2.

Table 2. An overview of recombinant DNA-based biotechnology

Double helix structure of DNA is first described by Watson and Crick	1953
Cohen and Boyer, amongst others, develop genetic engineering	1973
The first human protein (somatostatin) is produced in a bacterium *(E. coli)*	1977
The first recombinant protein (human insulin) approved for the market	1982
Polymerase chain reaction (PCR) technique developed	1983
Launch of the Human Genome Project	1990
The first genome sequence of an organism *(Haemophilus influenzae)* is determined	1995
A first draft of the human genome sequence is completed	2000
Over 40 million gene sequences are deposited in GenBank, and genome sequences of hundreds of prokaryotes and dozens of eukaryotes are finished or in draft stage	2005

Overview of Applications of Biotechnology

Since the advance of recombinant DNA technology, several techniques and applications have been developed that are benefiting humankind in the areas of agriculture, medicine, environment, industry and forensics. The following sections briefly describe some of these applications and their potential benefits to society

Industry

Biotechnology can be used to develop alternative fuels; an example is the conversion of maize starch into ethanol by yeast, which is subsequently used to produce gasohol (a gasoline-ethanol mix). Bacteria are used to decompose sludge and landfill wastes (Soccol et al., 2003). Through biotechnology, micro-organisms or their enzymes can be adapted to convert biomass into feed stocks, or they can be used for manufacturing biodegradable plastics (bioplastics). Other organisms (micro-organisms, plants and mammals) are used as bioreactors for producing chemical compounds that are extracted from them and processed as drugs and other products. Plant and animal fibres are used for producing a variety of fabrics, threads and cordage. Biotechnology is applied to improve the quality and quantity of these products. Biopulping is a technique whereby a fungus is used to convert wood chips into pulp for papermaking (Gavrilescu and Chisti, 2005).

Health and Medicine

In the area of health and medicine, biotechnology has numerous and important functions. Biotechnologies are used to develop diagnostic tools for identifying diseases.

Biotechnology is also used to produce more effective and efficient vaccines, therapeutic antibodies, antibiotics, and other pharmaceuticals. Biotechnology is a USD 70 billion a year industry that has produced several blockbuster drugs and vaccines, i.e. drugs with sales volumes exceeding USD 1 billion per year (Lawrence, 2007). Furthermore, there are more than 370 drug products and vaccines obtained through biotechnology currently in clinical trials, targeting more than 200 diseases including various cancers, Alzheimer's disease, heart disease, diabetes, multiple sclerosis, AIDS and arthritis (Sullivan et al., 2008).

Through the biotechnology of gene therapy, scientists are making efforts at curing genetic diseases by attempting to replace defective genes with the correct version. A revolutionary strategy is being developed whereby staple foods such as potatoes, bananas, and others are used as delivery vehicles to facilitate the immunization of people in economically depressed regions of the world (Tacket, 2009).

Development and usage of alternative fuels that burn cleaner and improve air quality through reduced pollution of the environment is possible by biotechnological means. Micro-organisms are used to decompose wastes and clean up contaminated sites by the technology of bioremediation. The use of disease-resistant cultivars can make crop production less environmentally intrusive by reducing the use of agrochemicals (Chatterjee et al., 2008).

Forensics

Since the DNA profile, i.e. the nucleotide sequence of the genome, is unique in every individual, it can be used as a powerful basis of identifying individuals in a population. DNA-based evidence is used in cases involving paternity disputes and family relationships. Furthermore, it is used in health care and judicial systems. In the judicial system, forensic experts use DNA profiling to identify suspects in criminal cases, especially when body fluids and other particles like hair and skin samples can be retrieved (Jobling and Gill, 2004).

Agriculture

Biotechnology can complement conventional breeding for crop and animal improvement. Instead of extensive re-arrangement of genes, as occurs in conventional breeding, biotechnology enables targeted gene transfer to occur. The genome of the recipient individual remains intact, except for the introduced gene (or genes), thus accelerating breeding programmes and the development of organisms with desirable characteristics. Furthermore, biotechnology enables gene transfer across natural breeding boundaries, overcoming mating barriers and creating a "universal gene pool" or "universal breeding population" accessible to all organisms. Likewise, it is possible to specifically introduce novel, desirable traits and characteristics into existing species. This biotechnological application is used to improve the yield of crop and animal species and their product quality such as nutritional value and shelf life (Shewry et al, 2008). In addition to these benefits, this methodology reduces the need for agrochemicals by creating disease and pest-resistant species, thereby reducing environmental pollution from chemical runoff. Increased yields and higher food quality can contribute to reducing world hunger and malnutrition (FAO, 2004).

Several technologies in the field of agricultural biotechnology exist that do not rely on the creation of GMOs. Molecular techniques are being used to monitor breeding populations and to diagnose animals and plants infected with diseases. Micropropagation techniques are being widely used to generate clonal plant materials, allowing rapid large-scale clonal propagation of many plant species including trees. Biofertilizers and biopesticides can be applied in place of conventional fertilizer and pesticides to promote plant growth and health in an environmentally sustainable way (FAO, 2001).

To summarize, the field of biotechnology is very diverse, both in terms of methodologies and techniques applied and the potential applications and outcomes. Biotechnology has the potential to contribute to a worldwide sustainable development and the reduction of world hunger, including the branches of biotechnology concerned with agricultural research and development (FAO, 2004). Importantly, biotechnology is not only based on GMOs, but offers several important and well established techniques that are not dependent on or derived from genetic modifications. However, the focus of this publication is on GMOs and related products. Following an introduction to the molecular background and the scientific basis of GMO development and to the aims and prospects of this research, the main part of this book will introduce biosafety concepts related to the use of GMOs.

STRUCTURE AND FUNCTION OF GENES

Genes and Heredity

The study of genes and heredity is called genetics. Heredity phenomena have been of interest to humans since long before the underlying principles were scientifically investigated and understood. Ancient peoples were improving plant crops and domesticating animals by selecting desirable individuals for breeding. Genetics as a set of scientific principles and analytical procedures emerged in the 1860s when the Augustinian monk Gregor Mendel performed a set of experiments that revealed the existence of biological "factors" responsible for transmitting traits from generation to generation. These factors were later called genes, following the discovery of chromosomes and genetic linkage in the early twentieth century. Up to this point genetics looked at genes as abstract entities that somehow control hereditary traits. Through genetic analyses the inheritance of different genes was studied, but the physical and biochemical nature of the gene remained unknown. Further work revealed that chromosomes consist of DNA and protein, and subsequent studies allowed the conclusion that DNA is, in fact, the hereditary material (Morange and Cobb, 2000).

DNA was thought to be a simple molecule, thus many scientists did not believe that it indeed carried and stored the information for an entire organism. How can such huge amounts of information be contained and passed on from one generation to the next? Clearly, the genetic material must have both the ability to encode specific information and the capacity to duplicate that information precisely during every cell division. What kind of molecular structure could allow such complex functions?

The Structure of DNA

Although the exact DNA structure was not known until 1953, its basic building blocks had been known for many years. It had been shown that DNA is composed of four basic molecules called nucleotides, which are identical except that each contains a different nitrogen-containing base. Each nucleotide is made up of a phosphate group, a sugar (of the deoxyribose type), and one of the four bases. The four bases are adenine (A), guanine (G) (the purines) and cytosine (C) and thymine (T).

In 1953 James Watson and Francis Crick were the first to succeed in putting the building blocks together and came up with a reasonable DNA structure. They used DNA X-ray diffraction patterns produced by Rosalind Franklin and Maurice Wilkins and data from Erwin Chargaff. The X-ray data showed the DNA molecule to be long, thin and helical (spiral-like) in shape.

Chargaff had established certain empirical rules about the amounts of each component of DNA:

- The total amount of pyrimidine nucleotides (T + C) always equals the total number of purine nucleotides (A + G).
- The amount of T always equals the amount of A, and the amount of C always equals the amount of G. But the amount of A + T is not necessarily equal to the amount of G + C.

The structure that Watson and Crick derived from these clues is a double helix (Figure 1). Each helix is a chain of nucleotides held together by phosphodiester bonds, in which a phosphate group forms a bridge between -OH groups on two adjacent sugar residues. The two DNA chains (helices) are running in an antiparallel direction and are held together by hydrogen bonds between opposing bases, thus forming a double helix. Each base pair (bp) consists of one purine and one pyrimidine base, paired according to the following rule: G pairs with C, and A pairs with T (Watson et al., 2008).

Molecular Biology and Genetic Engineering 149

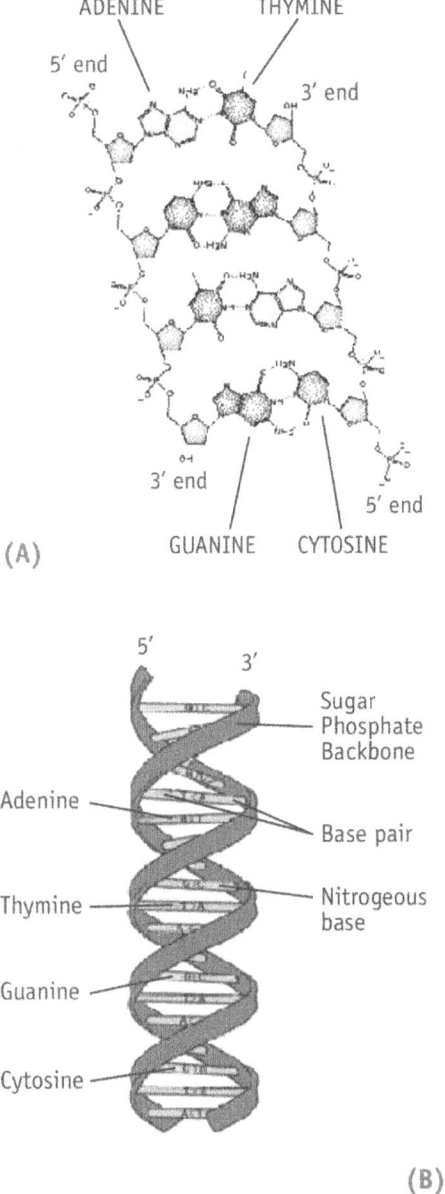

Figure 1. The structure of DNA In part (A), the four bases, the pairing of the bases and the connection of the bases through the sugarphosphate-backbone are depicted. Note that the two DNA strands are held together through base pairing and are running in opposite direction, labelled 3' and 5' end respectively (read: three prime and five prime). In part (B), a schematic drawing of the actual DNA double helix structure is depicted, containing the same elements in simplified form and labelling as in (A).

Elucidation of the structure of DNA caused a lot of excitement in the scientific community for two major reasons. First, the structure suggests an obvious way in which the molecule can be duplicated, or replicated, since each base can specify its complementary base by hydrogen bonding. Thus each strand can serve as a template for the synthesis of a complementary strand. Second, the structure suggests that the sequence of nucleotide pairs in DNA is dictating the sequence of amino acids in a protein encoded by a gene. In other words, some sort of genetic code may comprise information in DNA as a sequence of nucleotide pairs, which can be translated into the different language of amino acid sequence in protein.

The Flow of Genetic Information: The Central Dogma

In the early 1950s, Francis Crick suggested that there is a unidirectional flow of genetic information from DNA through ribonucleic acid (RNA) to protein, i.e. "DNA makes RNA makes protein". This is known as the central dogma of molecular biology, since it was proposed without much evidence for the individual steps. Now these steps are known in detail: DNA is transcribed to an RNA molecule (messenger RNA [mRNA]), that contains the same sequence information as the template DNA, and subsequently this RNA message is translated into a protein sequence according to the genetic code (Miller et al., 2009).

The Genetic Code

The basic building blocks of DNA are the four nucleotides; the basic building blocks of proteins are the amino acids, of which there are 22 that naturally occur in proteins (the so-called proteinogenic amino acids). The genetic code is the correspondence between the sequence of the four bases in nucleic acids and the sequence of the 22 amino acids in proteins. It has been shown that the code is a triplet code, where three nucleotides (one codon) encode one amino acid. Since there are only 22 amino acids to be specified and 64 different codons (4^3 =64), most amino acids are specified by more than one codon and the genetic code is said to be degenerate, or to have redundancy. The genetic code has colinearity, which means that the order of the bases in the DNA corresponds directly to the order of amino acids in the protein (Watson et al., 2008).

Clearly, if the genetic code is to be read as we would read a sentence in a book, we need to know where to start and stop. The codon AUG serves as a start signal, encoding the amino acid methionine, which is therefore the first amino acid incorporated into all proteins. However, methionine is also found elsewhere, not only at the beginning. Therefore, the translational machinery has to find the correct methionine codon to start and not just any given AUG codon anywhere in the gene sequence. This process is facilitated by sequences surrounding the initiation AUG codon. These sequences are therefore highly important for the translation process. The end of the translated region is specified by one of three codons which encode "stop". These are UAA, UAG and UGA. If mutations, i.e. unintended changes in the DNA sequence, take place that create one of the stop codons instead of an amino acid encoding codon, the results may be severe as the resultant protein will be shorter than intended. Such proteins are referred to as being truncated, and are very likely non-functional. Other mutations alter one codon to another, resulting in the replacement of the original amino acid by a different one, which can have severe or negligible effects, depending on the importance of the amino acid, for the entire protein. The addition or deletion of a single nucleotide can also have a severe effect, since all following codons will be shifted by one nucleotide, resulting in a very different message – a so-called frameshift mutation. The region between the start-methionine and the first stop codon is referred to as the open reading frame (ORF).

Finally, the genetic code is virtually universal, i.e. it is the same in all organisms living on this planet. Genes taken from plants can be decoded by animal cells, while genes from prokaryotes can be decoded by eukaryotic systems, and vice versa. Without such a universal nature of the code, genetic manipulation and genetic engineering would be much more difficult (Voet and Voet, 2004).

The Gene

Historically, a gene is defined as a heritable unit of phenotypic variation. From a molecular standpoint, a gene is the linear DNA sequence required to produce a functional RNA molecule, or a single transcriptional unit (Pearson, 2006). Genes can be assigned to one of two broad functional categories: structural genes and regulatory genes. It is the function of the end product of a gene that distinguishes structural and regulatory genes.

- Structural genes code for polypeptides or RNAs needed for the normal metabolic activities of the cell, e.g. enzymes, structural proteins, transporters, and receptors, among others.
- Regulatory genes code for proteins whose function is to control the expression of structural genes. With regard to molecular composition both classes of genes are similar.

A gene usually occupies a defined location on a chromosome, of which there are 46 in every human cell and which contain the entire human genome. The exact chromosomal gene location is defined by specific sequences for the start and termination of its transcription. Each gene has a specific effect and function in the organism's morphology or physiology, can be mutated (i.e. changed), and can recombine with other genes. It is a store of information (in the form of nucleotide base sequence); consequently it does not initiate any action, but is acted upon, e.g. during the process of gene expression. The complete set of genes of an organism, its genetic constitution, is called the genotype. The human genome, for example, contains an approximate number of 25 000 protein-coding genes. The physical manifestation, or expression, of the genotype is the phenotype (i.e. the organism's morphology and physiology). If a particular characteristic, such as brown eye color, is part of an organism's phenotype, one can conclude that the individual carries the gene(s) for that characteristic. If, however, a particular characteristic is not expressed, one cannot implicitly conclude that the particular gene is absent because expression of that gene might be repressed. Different varieties of the same gene, resulting in different phenotypic characteristics, are called alleles (Griffiths et al., 2007).

Genes may be located on either strand of the double-stranded DNA. But, regardless of which strand contains a particular gene, all genes are read in a 5' to 3' direction, and the strand containing the particular gene is referred to as the sense or coding strand.

As stated above, every cell of a human body, except germ line cells, contains 46 chromosomes. From each parent, we inherit 23 chromosomes, representing the complete genome. Thus, each body cell is diploid, i.e. contains two copies of the human genome and likewise two copies (alleles) of each gene.

The haploid set of the human genome (23 chromosomes), consists of approximately 3 200 megabases (Mb; 1 Mb = 10^6 bp) and contains an estimated number of 20 000 to 25 000 protein-coding genes (International Human Genome Sequencing Consortium, 2004). In fact, protein-coding DNA sequences only represent approximately 1.5 percent of the total genome; the remaining majority of DNA represent regulatory sequences, RNA encoding genes, or simply DNA sequences that have not yet been assigned to a certain function (sometimes inappropriately referred to as "junk DNA"). Interestingly, the estimated number of proteins is somewhat higher than the number of genes, due to alternative splicing and other variations in gene expression.

In comparison, the genome of E. coli, a widely used model bacterium, consists of one chromosome of 4.6 Mb in size, encoding approximately 4 400 genes in total. The genome of Arabidopsis thaliana, probably the most important model plant, consists of five chromosomes, of 157 Mb in size and encodes approximately 27 000 genes. Importantly, there is no straight connection between genome size, number of genes and organism complexity; some plants, vertebrates and even protozoans (single-cell organisms) have significantly larger genomes than the human genome (Patrushev and Minkevich, 2008).

The Arrangement and Layout of Genes

In eukaryotic organisms each cell contains more than one DNA molecule packaged into individual chromosomes; a diploid human cell, as stated, contains 46 chromosomes. Along the length of each DNA molecule/chromosome one can find thousands of genes, with more or less random spacing. In bacteria, one can frequently find clusters of genes that are related, in the sense that the proteins encoded by these genes are required in the same metabolic pathway. Therefore, as the cell needs all the gene products more or less simultaneously in order to keep that pathway running, it is appropriate for the cell to arrange these genes in clusters and employ a mechanism to express them together. These clusters of genes are known as operons; the most studied operon is the lactose operon in E. coli. This operon contains three genes which are adjacent on the DNA and are required for the utilization of lactose as a metabolic energy source in the cell. The operon also contains all the control sequences needed to ensure efficient expression of the genes as an ensemble (Reznikoff, 1992). Operons do not occur in higher organisms but related genes are sometimes found in clusters as well, and comparable regulatory mechanisms are found.

Many genes in eukaryotes have a distinctive structural feature: the nucleotide sequence contains one or more intervening segments of DNA that do not code for the amino acid sequence of the protein. These non-translated sequences interrupt the otherwise co-linear relationship between the nucleotide sequence of the gene and the amino acid sequence of the protein it encodes. Such non-translated DNA segments in genes are called introns. The pieces that constitute mature mRNA, and therefore ultimately for protein, are referred to as exons. During and after transcription the exons are spliced together from a larger precursor mRNA that contains, in addition to the exons, the interspersed introns. The number of exons that constitute a final mRNA molecule depends on the gene and the organism, but can range from as few as one to as many as fifty or more. The origin of intron/ exon structure is a matter of scientific debate. To date it is not clear whether it predated the divergence of eukaryotes and prokaryotes with the subsequent loss of introns in prokaryotes, or if introns and the splicing mechanism evolved in eukaryotes after their evolutionary separation from prokaryotes (Mattick, 1994).

In addition to introns and exons, the structural features of the eukaryotic gene include regulatory elements, a promoter region, a transcription start site and a transcription termination site (Figure 2). Specific proteins in the cell nucleus, the cellular compartment where DNA is stored, can bind to regulatory element sequences of a gene, thus controlling the expression of that gene. The promoter region is the sequence of the gene where the transcription machinery (the assembly of proteins required for transcription) binds to the DNA in order to start transcription to RNA. The start site indicates to the transcription machinery where to start and the termination site indicates were to stop transcription of the gene.

Gene Expression

Genes exert their function through a process called gene expression, a process by which heritable information from a gene, encoded on DNA, is transformed into a functional gene product, such as protein or RNA (some genes code for functional RNA molecules, such as tRNA and rRNA). Genes are expressed by being first transcribed into RNA, and may then subsequently be translated into protein. A cell employs many different mechanisms to regulate gene expression.

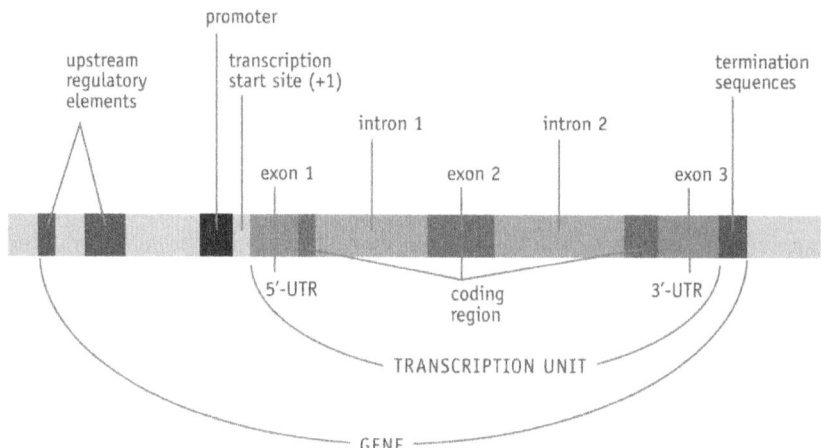

Figure 2. A general structural arrangement of the different components making up a eukaryotic gene Upstream regulatory elements (enhancers) and the promoter are required for regulation and initiation of transcription. Exons, which constitute the actual protein-coding regions, and interspersed introns are indicated. The 5' and 3' untranslated regions (UTRs) are mRNA sequences that do not encode protein, but are required for a correct translation process. Transcription start and termination sites are also indicated.

Gene expression can be regulated at many different levels, from DNA transcription, pre-mRNA processing, mRNA stability and efficiency of translation up to protein modification and stability. Thus, a cell can precisely influence the expression level of every gene, and studying and predicting gene expression levels is a difficult task. Nevertheless, this is especially important for biotechnological applications, since it is desirable to precisely define the expression levels of introduced genes in transgenic organisms. In the following section, the major processes of gene expression will be introduced.

Transcription and Translation

The first step in gene expression is transcription, namely the production of a singlestranded RNA molecule known as mRNA in the case of protein-coding genes. The nucleotide sequence of the mRNA is complementary to the DNA from which it was transcribed. In other words, the genetic messages encoded in DNA are copied precisely into RNA. The DNA strand whose sequence matches that of the RNA is known as the coding strand and the complementary strand on which the RNA was synthesized is the template strand.

Transcription is performed by an enzyme called RNA polymerase, which reads the template strand in 3' to 5' direction and synthesizes the RNA from 5' to 3' direction. To initiate transcription, the polymerase first recognizes and binds a promoter region of the gene. Thus a major regulatory mechanism of gene expression is the blocking or sequestering of the promoter region. This can be achieved either by tight binding of repressor molecules that physically block the RNA polymerase, or by spatially arranging the DNA so that the promoter region is not accessible (Thomas and Chiang, 2006).

In eukaryotes, transcription occurs in the nucleus, where the cell's DNA is sequestered. The initial RNA molecule produced by RNA polymerase is known as the primary transcript and must undergo post-transcriptional modification before being exported to the cytoplasm for translation. The splicing of introns present within the transcribed region is a modification unique to eukaryotes. The splicing reaction offers various possibilities for regulating and modulating gene expression in eukaryotic cells.

Following transcription and post-transcriptional mRNA processing, the mRNA molecule is ready for translation. In eukaryotes, the mRNA must first be transported from the nucleus to the cytoplasm, whereas in prokaryotes no nucleus exists and transcription and translation take place in the same compartment.

Translation is the process by which a mature mRNA molecule is used as a template for synthesizing a protein. Translation is carried out by the ribosome, a large macromolecular complex of several rRNA and protein molecules. Ribosomes are responsible for decoding the genetic code on the mRNA and translating it into the amino acid sequence of proteins. Likewise, they are catalysing the chemical reactions that add new amino acids to a growing polypeptide chain by the formation of peptide bonds (Ramakrishnan, 2002).

The genetic code on the mRNA is read three nucleotides at a time, in units called codons, via interactions of the mRNA with specialized RNA molecules called transfer RNA (tRNA). Each tRNA has three unpaired bases, known as the anticodon, that are complementary to the codon it reads. The tRNA is also covalently attached to the amino acid specified by its anticodon. When the tRNA binds to its complementary codon in an mRNA strand, the ribosome ligates its amino acid cargo to the growing polypeptide chain. When the synthesis of the protein is finished, as encoded by a stop-codon on the mRNA, it is released from the ribosome. During and after its synthesis, the new protein must fold to its active three-dimensional structure before it can carry out its cellular function (Voet and Voet, 2002).

A single mRNA molecule can be translated several times and thus produce many identical proteins, depending on its half-life in the cell, i.e. the average time it remains within the cell before it is degraded.

Regulation of Gene Transcription

Promoters

The promoter region of a gene is usually several hundred nucleotides long and immediately upstream from the transcription initiation site. The promoter constitutes the binding site for the enzyme machinery that is responsible for the transcription of DNA to RNA, the RNA polymerase.

In eukaryotic cells several RNA polymerases are present, the most prominent one that is responsible for the transcription of protein-coding genes being RNA polymerase II. There are different types of promoters for different RNA polymerases. Promoters for RNA polymerase II, the polymerase that transcribes protein-coding genes into mRNA, often contain the consensus sequence 5'-TATA-3', 30 to 50 bp upstream of the site at which transcription begins. Many eukaryotic promoters also have a so-called CAAT box with a GGNCAATCT consensus sequence centred about 75 bp upstream of the initiation start site (with N representing any of the four bases). RNA polymerases I and III are mainly responsible for the transcription of RNA molecules that possess an intrinsic function as catalytic or structural molecules, such as tRNA and rRNA, and that are not translated into proteins (Okkema and Krause, 2005).

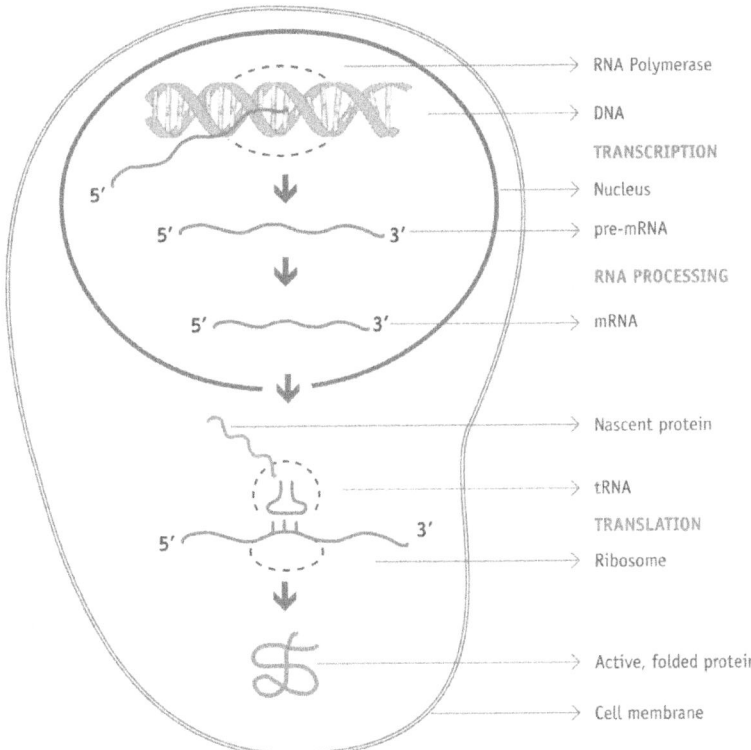

Figure 3. Transcription and Translation In the nucleus, DNA is transcribed to a pre-mRNA molecule by RNA polymerase. The pre-mRNA is processed, e.g. by intron excision, to the mature mRNA. The mRNA is exported to the cytoplasm and translated into protein, which is accomplished by ribosomes and tRNA that together decode the genetic code into amino acid sequence. Following translation, the synthesized protein adopts its correct 3-dimensional shape and is ready to perform its cellular function.

In general, the promoter region has a high importance for the regulation of expression of any gene. This concept will come up again later on in this module when the production of transgenic animals is introduced. Careful choice of promoters to drive gene expression in transgenic organisms is very important to ensure the transgenic organism possesses the desired characteristics.

Enhancers

Enhancers were first described as sequences that increase transcription initiation but, unlike promoters, were not dependent on their orientation or the distance from the transcription start site. It is now apparent that enhancers are generally short sequences (less than 20 to 30 bp) that bind specific transcription factors, which then facilitate the assembly of an activated transcriptional complex (i.e. the RNA polymerase) at the promoter. Most enhancers function both on the coding and non-coding strand of the DNA (i.e. in either orientation), can act up to several thousand bps distant from their target promoter, and are a rather unspecific form of regulatory element (Visel et al., 2007). This implies that an enhancer element may influence several, possibly very distant, promoters. Most enhancers are only active in specific cell types and therefore play a central role in regulating tissue specificity of gene expression. Some regulatory elements bind transcription factors that act to reduce the efficiency of transcriptional initiation, and many genes contain a combination of both positive and negative upstream regulatory elements, which then act in concert on a single promoter. This allows gene expression to be controlled very precisely in a temporal and spatial manner with regard to cell type, developmental stage and environmental conditions. Mutations of promoters or enhancers can significantly alter the expression pattern, but not the structure of a particular gene product.

Operators

Operators are nucleotide sequences that are positioned between the promoter and the structural gene. They constitute the region of DNA to which repressor proteins bind and thereby prevent transcription. Repressor proteins have a very high affinity for operator sequences. Repression of transcription is accomplished by the repressor protein attaching to the operator sequence downstream of the promoter sequence (the point of attachment of the RNA polymerase). The enzyme must pass the operator sequence to reach the structural genes start site. The repressor protein bound to the operator physically prevents this passage and, as a result, transcription by the polymerase cannot occur (Reznikoff, 1992). Repressor proteins themselves can be affected by a variety of other proteins or small molecules, e.g. metabolites that affect their affinity for the operator sequence. This allows a further level of gene expression regulation to be accomplished.

Attenuators

The attenuator sequences are found in bacterial gene clusters that code for enzymes involved in amino acid biosynthesis. Attenuators are located within so-called leader sequences, a unit of about 162 bp situated between the promoter-operator region and the start site of the first structural gene of the cluster. Attenuation decreases the level of transcription approximately 10-fold. As the concentration of an amino acid in the cell rises and falls, attenuation adjusts the level of transcription to accommodate the changing levels of the amino acid. High concentrations of the amino acid result in low levels of transcription of the structural genes, and low concentrations of the amino acid result in high levels of transcription. Thus the biosynthesis of an amino acid can be linked to the actual concentration of that amino acid within the cell. Attenuation proceeds independently of repression, the two phenomena are not dependent on each other. Attenuation results in the premature termination of transcription of the structural genes (Yanofsky et al., 1996).

Several other regulatory elements have been described that regulate gene expression at the level of transcription. In general, the interplay of all involved factors and sequences is, in most cases and especially in eukaryotes, very complex and not entirely understood. The expression level of a gene is therefore the net result of all stimulating and repressing factors acting on it (Watson, 2008). This combinatorial system of positive and negative influences allows the fine-tuning of gene expression and needs to be carefully considered when designing transgenic organisms.

Regulatory mRNA Sequences

In the preceding paragraph, DNA sequences were described that regulate transcription of DNA to an mRNA transcript. This transcript, sometimes referred to as pre-mRNA, contains a variety of sequences in addition to the protein-coding sequences. This includes 5' and 3' untranslated sequences which are important in the regulation of translation, and introns (in the case of eukaryotes) which need to be excised before the process of translation can take place. In eukaryotes, processing of a pre-mRNA to a mature mRNA that is ready for translation takes place in the same compartment as transcription, the nucleus.

Introns and Splice Junctions

In eukaryotic pre-mRNA processing, intervening sequences (introns) that interrupt the coding regions are removed (spliced out), and the two flanking protein-coding exons are joined. This splicing reaction occurs in the nucleus and requires the intron to have a GU-dinucleotide sequence at its 5'-end, an AG-dinucleotide at its 3'-end, and a specific branch point sequence. In a two-step reaction, the intron is removed as a tailed circular molecule, or lariat, and is subsequently degraded. This splicing reaction is performed by RNA-protein complexes known as snRNPs (small nuclear ribonucleoproteins). The snRNPs bind to the conserved intron sequences to form a machinery called spliceosome, in which the cleavage and ligation reactions take place (Matthew et al., 2008).

5' Untranslated Sequences

During the processing of precursor mRNA in the nucleus, the 3' terminus as well as introns are removed. In addition, shortly after initiation of mRNA transcription, a methylguanylate residue is added to the 5' end of the primary transcript. This 5' "cap" is a characteristic feature of every mRNA molecule, and the transcriptional start or initiation site is also referred to as the capping site. The 5' UTR extends from the capping site to the beginning of the protein coding sequence and can be up to several hundred bps in length. The 5' UTRs of most mRNAs contain the consensus sequence 5' –CGAGCCAUC-3 involved in the intiation of protein synthesis (i.e. translation). In addition, some 5' UTRs contain "upstream AUGs" that may affect the initiation of protein synthesis and thus could serve to control expression of selected genes at the translational level (Hughes, 2006).

3' Untranslated Sequences and Transcriptional Termination

The 3' end of a mature mRNA molecule is created by cleavage of the primary precursor mRNA and the addition of a several hundred nucleotide long polyadenylic acid (poly-A) tail. The site for cleavage is marked by the sequence 5' AAUAAA 3' some 15 to 20 nucleotides upstream and by additional uncharacterized sequences 10 to 30 nucleotides downstream of the cleavage site. The region from the last protein codon to the poly-A addition site may contain up to several hundred nucleotides of a 3' UTR, which includes signals that affect mRNA processing and stability. Many mRNAs that are known to have a short half life contain a 50 nucleotide long AU-rich sequence in the 3' UTR. Removal or alteration of this sequence prolongs the half life of mRNA, suggesting that the presence of AU-rich sequences in the 3' UTR may be a general feature of genes that rapidly alter the level of their expression. In general, the half-life of an mRNA indicates the average time that an mRNA molecule persists in the cell and thus can be translated before it is degraded. The mRNA half-life is therefore an important variable for the level of gene expression (Gray and Wickens, 1998).

Regulation of Gene Expression

Regulation of gene expression refers to the all processes that cells employ to convert the information carried by genes into gene products in a highly controlled manner. Although a functional gene product may be RNA or protein, the majority of known regulatory mechanisms affect the expression level of protein coding genes. As mentioned above, any step in the process of gene expression may be modulated, from transcription, to RNA processing, to translation, to posttranslational modification of the protein. Highly sophisticated gene expression regulatory systems allow the cell to fine-tune its requirements in response to environmental stimuli, developmental stages, stress, nutrient availability etc. (Nestler and Hyman, 2002; Watson et al., 2008).

To conclude, this section has provided an overview of genes, gene expression and hereditary phenomena. Although this text offers only a brief introduction to the topic, it should have become clear that correct gene expression is based on a highly complicated network and interplay of numerous factors, and a complete comprehension of these networks is only beginning to emerge. However, a good understanding of the basic principles is required to follow and understand biotechnological applications and developments, as well as the associated current limits and difficulties of this technology.

VECTORS AND PROMOTERS

Recombinant DNA Technology – An Overview

Following the elucidation of the DNA structure and the genetic code, it became clear that many biological secrets were hidden in the sequence of bases in DNA. Technical and biological discoveries in the 1970s led to a new era of DNA analysis and manipulation. Key among these was the discovery of two types of enzymes that made DNA cloning possible: cloning, in this sense, refers to the isolation and amplification of defined pieces of DNA. One enzyme type, called restriction enzymes, cut the DNA from any organism at specific sequences of a few nucleotides, generating a reproducible set of fragments. Restriction enzymes occur naturally in many bacteria, where they serve as defense mechanisms against bacteriophage (viruses infecting bacteria) infection by cutting the bacteriophages genome upon its entry into the cell. The other enzyme type, called DNA ligases, can covalently join DNA fragments at their termini that have been created by restriction enzymes. Thus, ligases can insert DNA restriction fragments into replicating DNA molecules such as plasmids (bacterial, circular DNA molecules), resulting in recombinant DNA molecules. The recombinant DNA molecules can then be introduced into appropriate host cells, most often bacterial cells. All descendants from such a single cell, called a clone, carry the same recombinant DNA molecule (Figure 4). Once a clone of

cells bearing a desired segment of DNA has been isolated, unlimited quantities of this DNA sequence can be prepared (Allison, 2007). Furthermore, in case the DNA fragment contains protein-coding genes, the recombinant DNA molecule introduced into a suitable host can direct the expression of these genes, resulting in the production of the proteins within the host. These developments, DNA cloning and the production of recombinant proteins, were major breakthroughs in molecular biology and set the stage for modern biological research.

Vectors

A vector is a DNA molecule which can replicate in a suitable host organism, and into which a fragment of foreign DNA can be introduced. Most vectors used in molecular biology are based on bacterial plasmids and bacteriophages (bacteria-infecting viruses).

Vectors need to have the following characteristics:

- Possess an origin of replication (ori), which renders the vector capable of autonomous replication independent of the host genome.
- Have a site (or sites) which can be cleaved by a restriction enzyme, where the foreign DNA fragment can be introduced.
- Contain convenient markers for identifying the host cell that contains the vector with the inserted DNA of interest. A common selection marker is an antibiotic resistance gene. If the host bacteria cells contain the vector then the bacteria will grow in the presence of that antibiotic, whereas growth of bacteria without the plasmid is restricted.

In addition to the above-listed features, the vector should be easily introducible into the host organism where it has to replicate and produce copies of itself and the foreign DNA. Furthermore, it should be feasible to easily isolate the vector from the host cell (Watson, 2008).

Types of Cloning Vectors

Plasmids

Plasmids are circular, double-stranded DNA molecules that are independent from a cell's chromosomal DNA. These extra chromosomal DNAs occur naturally in bacteria and in the nuclei of yeast and some higher eukaryotic cells, existing in a parasitic or symbiotic relationship with their host cell. Most naturally occurring plasmids contain genes that provide some benefit to the host cell, fulfilling the plasmid's portion of a symbiotic relationship. Some bacterial plasmids, for example, encode enzymes that inactivate antibiotics. Therefore, a bacterial cell containing such a plasmid is resistant to the antibiotic, whereas the same type of bacterium lacking the plasmid is killed. Plasmids range in size from a few thousand bps to more than 100 kilobases (kb).

The plasmids most frequently used in recombinant DNA technology are derived from and replicate in E. coli (Jana and Deb, 2005). In general, these plasmids have been modified to optimize their use as vectors in DNA cloning. One such modification, for example, is the reduction in size to approximately 3 kb, which is much smaller than that of naturally occurring E. coli plasmids. In addition, most plasmids contain a multiple cloning site (MCS), a short sequence of DNA containing many restriction enzyme sites close together. Thus, many different restriction enzymes can be used for the insertion of foreign DNA fragments. In addition to antibiotic resistance genes, many modern plasmid vectors also contain a system for detecting the presence of a recombinant insert, such as the blue/white β-galactosidase system that allows simple visual screening of bacterial clones.

Bacteriophages

Bacteriophages, or phages, are viruses that infect bacteria. They can display either lytic life cycles, leading to the death of the host bacterium and release of new phage particles, or more complex lysogenic cycles during which the phage genome is integrated into the bacterial genome.

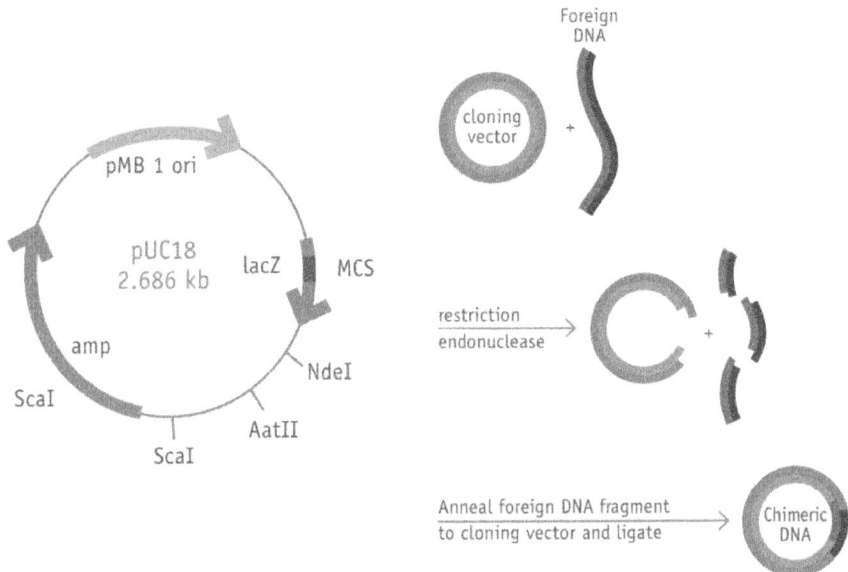

Figure 4. A typical plasmid cloning vector and the principle of DNA cloning. The pUC18 plasmid is a frequently used plasmid for DNA cloning. The plasmid size, ampicillin resistance gene (amp), origin of replication (ori) and multiple cloning site (MCS) are indicated. On the right hand side, the overall principle of DNA cloning is depicted.

One of the best studied phages is bacteriophage λ (Lambda) whose derivatives are commonly used as cloning vectors (Chauthaiwale et al., 1992). The λ phage particle consists of a head containing the 48.5 kb double-stranded DNA genome, and a long flexible tail. During infection, the phage binds to certain receptors on the outer membrane of E. coli and subsequently injects its genome into the host cell through its tail. The phage genome is linear and contains single-stranded ends that are complementary to each other (the socalled cos ends). Due to the complementarity, the cos ends rapidly bind to each other upon entry into the host cell, resulting in a nicked circular genome. The nicks are subsequently repaired by the cellular enzyme DNA ligase. A large part of the central region of the phage genome is dispensable for lytic infection, and can be replaced by unrelated DNA sequence. The limit to the size of DNA fragments which can be incorporated into a λ particle is 20 kb, which is significantly larger than fragments suitable for plasmids (around 10 kb maximum). A further advantage of λ-based vectors is that each phage particle containing recombinant DNA will infect a single cell. The infection process is about a thousand times more efficient than transformation of bacterial cells with plasmid vectors.

Cosmids

Both λ phage and E. coli plasmid vectors are useful for cloning only relatively small DNA fragments. However, several other vectors have been developed for cloning larger fragments of DNA. One common method for cloning large fragments makes use of elements of both plasmid and λ-phage cloning. In this method, called cosmid cloning, recombinant plasmids containing inserted fragments up to a length of 45 kb can be efficiently introduced into E. coli cells. A cosmid vector is produced by inserting the cos sequence from λ-phage DNA into a small E. coli plasmid vector about 5 kb long. Cosmid vectors contain all the essential components found in plasmids. The cosmid can incorporate foreign DNA inserts that are between 35 and 45 kb in length. Such recombinant molecules can be packaged and used to transform E. coli. Since the injected DNA does not encode any λ-phage proteins, no viral particles form in infected cells and likewise the cells are not killed. Rather, the injected DNA circularizes, forming in each host cell a large plasmid containing the cosmid vector and the inserted DNA fragment. Cells containing cosmid molecules can be selected using antibiotics as described for ordinary plasmid cloning.

A recently developed approach similar to cosmid cloning makes use of larger E. coli viruses such as bacteriophage P1. Recombinant plasmids containing DNA fragments of up to ≈100 kb can be packaged in vitro with the P1 system.

Yeast Artificial Chromosomes (YAC)

YACs are constructed by ligating the components required for replication and segregation of natural yeast chromosomes to very large fragments of target DNA, which may be more than 1 Mb in length (Ramsay, 1994). YAC vectors contain two telomeric sequences (TEL), one centromere (CEN), one autonomously replicating sequence (ARS) and genes which act as selectable markers in yeast. YAC selectable markers usually do not confer resistance to antibiotic substances, as in E. coli plasmids, but instead enable growth of yeast on selective media lacking specific nutrients.

Bacterial Artificial Chromosomes (BAC)

BAC vectors were developed to avoid problems that were encountered with YACs to clone large genomic DNA fragments. Although YACs can accommodate very large DNA fragments they may be unstable, i.e. they often lose parts of the fragments during propagation in yeast.

In general, BACs can contain up to 300–350 kb of insert sequence. In addition, they are stably propagated and replicated in E. coli, are easily introduced into their host cell by transformation, large amounts can be produced in a short time due to the fast growth of E. coli, and they are simple to purify (Giraldo and Montoliu, 2001). The vectors are based on the naturally occurring plasmid F factor of E. coli, which encodes its own DNA polymerase and is maintained in the cell at a level of one or two copies. A BAC vector consists of the genes essential for replication and maintenance of the F factor, a selectable marker gene (SMG) and a cloning site for the insertion of target fragment DNA.

To summarize, cloning vectors are DNA molecules that can incorporate foreign DNA fragments and replicate in a suitable host, producing large quantities of the desired DNA fragment. Such methods are highly important for a variety of molecular biology applications and are the basis of recombinant DNA technology. However, for the production of transgenic organisms and related biotechnological applications, such vectors need to possess additional sequence elements and properties that allow targeted transfer of specific genes and controlled expression of these genes in a host organism. The necessary features to accomplish these tasks will be discussed in the following paragraphs.

Promoters

The promoter sequence is the key regulatory region of a gene that controls and regulates gene expression. More specifically, the promoter has a major importance in the regulation of transcription, i.e. the transfer of the information contained in a DNA coding region into an mRNA transcript. Promoters play an important role in the regulation of gene expression at different locations and times during the life cycle of an organism or in response to internal and external stimuli (Juven-Gershon et al., 2008). Investigating and unravelling the precise function of promoter components and the additional factors associated with their performance revealed new possibilities of genetic engineering. Nowadays, it is feasible to modulate the expression of defined genes in an organism by combining them with a promoter of choice, resulting in the desired gene expression profile.

This approach can be used to modulate the expression of endogenous genes (i.e. genes that the organism possesses already) or to introduce foreign genes in combination with a foreign or endogenous promoter to create an organism with defined novel traits. Thus, promoters have a huge influence in follow-on research and development in biotechnology, and a more detailed understanding will certainly further influence the development of GMOs.

Types of Promoters

In general, promoters can be divided into different classes according to their function:

- ***Constitutive Promoters***: Constitutive promoters direct the expression of a gene in virtually all cells or tissues of an organism. The genes controlled by such promoters are often "housekeeping genes", i.e. genes whose products are constantly needed by the cell to survive and maintain its function. Constitutive promoters are to a large extent, or even entirely, insensitive to environmental or internal influences, thus the level of gene expression is always kept constant. Due to the insensitivity to external or internal stimuli and the high sequence conservation of such promoters between different species, constitutive promoters are in many cases active across species and even across kingdoms. An important example is the Cauliflower mosaic virus (CaMV) 35S promoter, which is frequently used to drive transgene expression in transgenic plants.

- ***Tissue-Specific Promoters***: Tissue-specific promoters direct the expression of a gene in a specific tissue or cell type of an organism or during certain stages of development. Thus, the gene product is only found in those cells or tissues and is absent in others, where the promoter is inactive. In plants, promoter elements that specifically regulate the expression of genes in tubers, roots, vascular bundles, other vegetative organs or seeds and reproductive organs have been used for genetic engineering, both within a certain species and across different species. Frequently, such promoters rely on the presence or absence of endogenous factors to function, so in fact it is the presence or absence of these factors that defines the tissue-specificity of gene expression.

- ***Inducible Promoters***: Inducible promoters are of high interest to genetic engineering because their performance is dependent on certain endogenous or external factors or stimuli. In the ideal case, gene expression by an inducible promoter can be controlled by the experimenter by simply adding a certain substance to the cell culture/ the organism. This will result in expression of all genes controlled by this promoter – in the case of transgenic organisms, usually only the genes that have been specifically introduced (Padidam, 2003). Within the class of inducible promoters, one can find promoters controlled by abiotic factors such as light, oxygen level, heat, cold and wounding, while others are controlled by certain chemicals or metabolites. As it may be difficult to control some of these factors in the field, promoters that respond to chemical compounds, which are not found naturally in the organism of interest, are of particular interest. Substances that have been found to control certain promoters include rare metabolites, antibiotics, some metals, alcohol, steroids and herbicides, among other compounds. Once a promoter that responds to a certain compound has been identified it can be further engineered and adapted to induce gene expression in GMOs at will, independent of other factors encountered by the organism (Gurr and Rushton, 2005).

Expression Vectors

Cloning a gene encoding a particular protein is only the first of many steps needed to produce a recombinant protein for agricultural, medical or industrial use. The next step is to transfer the DNA sequence containing the gene into the desired host cell for its expression and the production of the protein of interest. In order to allow expression of the gene of interest in the host cell or organism, it must be transferred into a vector that has several distinct sequence features. These features include all sequences that are required to drive and regulate expression of the gene, i.e. all components that are associated with a functional gene. Thus, in addition to the characteristics described for cloning vectors, an expression vector must carry a promoter, a polyadenylation site, and a transcription termination sequence. These sequences should have a correct orientation with regard to the multiple cloning site, where the foreign DNA is integrated. Inserting a coding sequence in proper orientation in between these expression control sequences will result in the expression of the gene in an appropriate host. A simplified version of an expression vector is depicted in Figure 5.

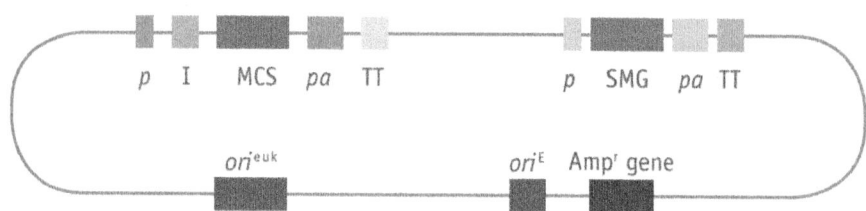

Figure 5. Generalized mammalian expression vector The multiple cloning site (MCS), where the foreign DNA can be inserted, and selectable marker gene (SMG) are under control of a eukaryotic promoter (p), polyadenylation (pa), and termination of transcription (TT) sequences. An intron (I) enhances the production of heterologous protein. Propagation of the vector in E. coli and mammalian cells depends on the origins of replication oriE and orieuk, respectively. The ampicillin gene (Ampr) is used for selecting transformed E. coli cells.

In some cases, it is necessary and helpful to fuse some translation control and protein purification elements to the gene of interest (Figure 6) or to add them to the expression vector MCS. This is especially important if a recombinant protein is purified after its expression in a certain host cell or organism. For this purpose, short specific amino acid sequences, commonly referred to as tags, can be added to the protein by adding the sequence encoding them to the coding sequence of the protein. These tags can greatly facilitate protein purification, due to certain properties they possess and that are specific for each tag. If necessary, such tags can be removed from the final, purified protein by introducing a further specific amino acid sequence, which is recognized by a protease that cleaves the protein at this position and thus removes the tag. An example of a gene with such added sequences is given in Figure 6.

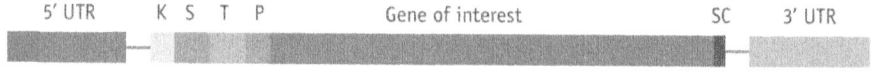

Figure 6. A gene of interest fitted with sequences that enhance translation and facilitate both secretion and purification of the produced protein These include the Kozak sequence (K) [5'-ACCAUGG-3', its presence near the initiating AUG greatly increases the effectiveness of initiation], signal sequence (S) required for secretion, protein affinity tag (T), proteolytic cleavage site (P), and stop codon (SC). The 5' and 3' UTRs increase the efficiency of translation and contribute to mRNA stability.

In the case of transgenic plant and animal production, the general layout of an engineered gene as depicted in Figure 6 also holds true in most cases. However, other types of vectors to deliver the transgene to the plant or animal cells are frequently employed. Whereas cells in cell culture can be easily monitored for the presence of the desired expression vector and the expression vector is stably maintained within the cells, this is not necessarily the case for complex organisms. Therefore, the genes of interest are usually integrated in a vector that mediates integration of the transgene into the host organism's genome (i.e. into a chromosome). Thus, the transgene becomes an integral part of the organism's genome, and as such is present in all cells of an organism and is stably passed on to subsequent generations (Somers and Makarevich, 2004). This is usually not the case for plasmid vectors, which are maintained as extra-chromosomal entities and are frequently lost during cell divison and propagation.

To summarize, this section has provided an introduction to the field of recombinant DNA technology. Specific DNA fragments can be cloned, by means of cloning vectors, and subsequently be isolated, investigated and further modified with great ease. Furthermore, specific DNA fragments containing protein-coding genes can be transferred to expression vectors, which will result in expression of the encoded proteins upon introduction of the vector into an appropriate host cell or organism. Thus, desired proteins can be produced in large quantities. Careful choice of the vector, the production host and promoter and other regulatory sequences is of high importance for the success of such approaches. Modern biotechnology offers the possibility to freely combine genes with promoters and other desired sequences, regardless of the original source of the genes and DNA sequences.

This technology also sets the basis for the creation of transgenic plants and animals, which are engineered to express new traits and properties by the specific introduction or modulation of genes and regulatory sequences.

PLANT TRANSFORMATION AND SELECTION TECHNIQUES

Plant Transformation

Genetic transformation is the (sometimes heritable) change in the genome of a cell or organism brought about by the uptake and incorporation of introduced, foreign DNA. Transformation encompasses a variety of gene transfer events, which can be characterized by the stability of transformation, the subcellular compartment transformed (nuclear, mitochondrial or plastid) and whether the transferred DNA is stably integrated into the host genome (Shewry et al., 2008).

Table 3. documents the generally accepted definitions of these alternative transformation events.

Table 3. Definitions of transformation

Term	Definitions
Stable transformation	The transgene and novel genetic characteristics are stably maintained during the life of the cell culture or organism. The transgene is usually, but not necessarily, integrated into the host genome.
Transient expression	Expression of the transgene is detected in the first few days after its introduction into host cells. A subsequent decline in expression indicates that expression was based on non-integrated, extra-chromosomal DNA.
Integrative transformation	The transgene is covalently integrated into the genome of the host cell. In fertile plants (or animals) the transgene is inherited by the next generation (a form of stable integration).
Nuclear transformation	Gene transfer into the nuclear genome of the host cell, as confirmed by cellular fractionation, eukaryotic-type expression or mendelian inheritance.
Organellar transformation	Gene transfer into the plastid or mitochondrial genome of the host cell, as confirmed by cellular fractionation, prokaryotic-type expression or maternal inheritance.
Episomal transformation	Viral genomes or "mini-chromosomes" are introduced which replicate independently from the host genome. Stable over several generations in some cases.

Plant Tissue Culture

An important phenomenon that is a key determinant to plant transformation, and thus the generation of transgenic plants, is the finding that whole plants can be regenerated from single cells. Plant transformation thus depends on two events: successful introduction of foreign DNA into target plant cells, and subsequent development of a complete plant derived from the transformed cells.

In vitro regeneration is the technique of developing plant organs or plantlets from plant cells, tissues or organs isolated from the mother plant and cultivated on artificial media under laboratory conditions (Thorpe, 2007). Depending on different physical and physiological factors, in combination with various growth regulators, regeneration occurs via organogenesis (initiation of adventitious roots or shoots from plant cells or tissues) or embryogenesis (formation of plants from somatic cells through a pathway resembling normal embryogenesis from the zygote). Both organogenesis and embryogenesis can be initiated either directly (from meristematic cells) or after formation of a callus (mass of undifferentiated parenchymatic cells induced by wounding or hormone treatment). Transformed plants can thus be regenerated from calli or wounded plant tissues, such as leaf disks, into which foreign DNA has previously been introduced (Figure 7).

Plant Transformation Techniques

There is an expanding repertoire of plant transformation techniques available, ranging from established techniques to highly experimental methodologies (Newell, 2000). In Table 4 these alternative approaches to gene delivery are listed with brief comments on their application, efficiency and limitations. The most widely used techniques are the Agrobacterium tumefaciens-mediated transfer, microprojectile bombardment ("gene gun" or biolistic method) and direct gene transfer to protoplasts. The biolistic technique has proven especially useful in transforming monocotyledonous species like maize and rice, whereas transformation via Agrobacterium has been successfully practised in dicotyledonous species. Only recently has it also been effectively employed in monocotyledons. In general, the Agrobacterium-mediated method is considered preferable to the gene gun due to the higher frequency of single-site insertions of the foreign DNA into the host genome, making the transformation process easier to monitor. All available and currently employed transformation techniques are briefly described in the following sections.

Microprojectile Bombardment

This technique uses high velocity particles, or microprojectiles, that are coated with DNA and deliver exogenous genetic material into the target cell or tissue. Transformed cells are selected, cultured in vitro and regenerated to produce mature transformed plants (Kikkert et al., 2005).

The particles, either tungsten or gold, are small (0.5-5 µm) but big enough to have the necessary mass to be sufficiently accelerated and penetrate the cell wall carrying the coated DNA on their surface. Once the foreign DNA is integrated into the plant genome in the cell nucleus, which is a somewhat spontaneous process, it can be expressed. Gold particles are chemically inert, although rather costly, and show a high uniformity. Tungsten particles, despite showing mild phytotoxicity and being more variable in size, are adequate for most studies. Furthermore, the chosen microprojectile should have good DNA binding affinity but, at the same time, be able to release the DNA once it has hit the target. DNA coating of surface-sterilized particles can be accomplished by defined DNA treatments using, for instance, the calcium chloride method, with the addition of certain chemicals to protect the DNA. However, a recent report describes the novel use of Agrobacterium as coating material for the microprojectiles, which are then shot into the target tissue. Once coated the particles are ready for shooting; the particles are accelerated and ultimatively collide with the target, usually plant cells or calli grown on a Petri dish. The DNA, delivered with this strategy, is expressed after reaching the nucleus and integrating randomly into the plant genome.

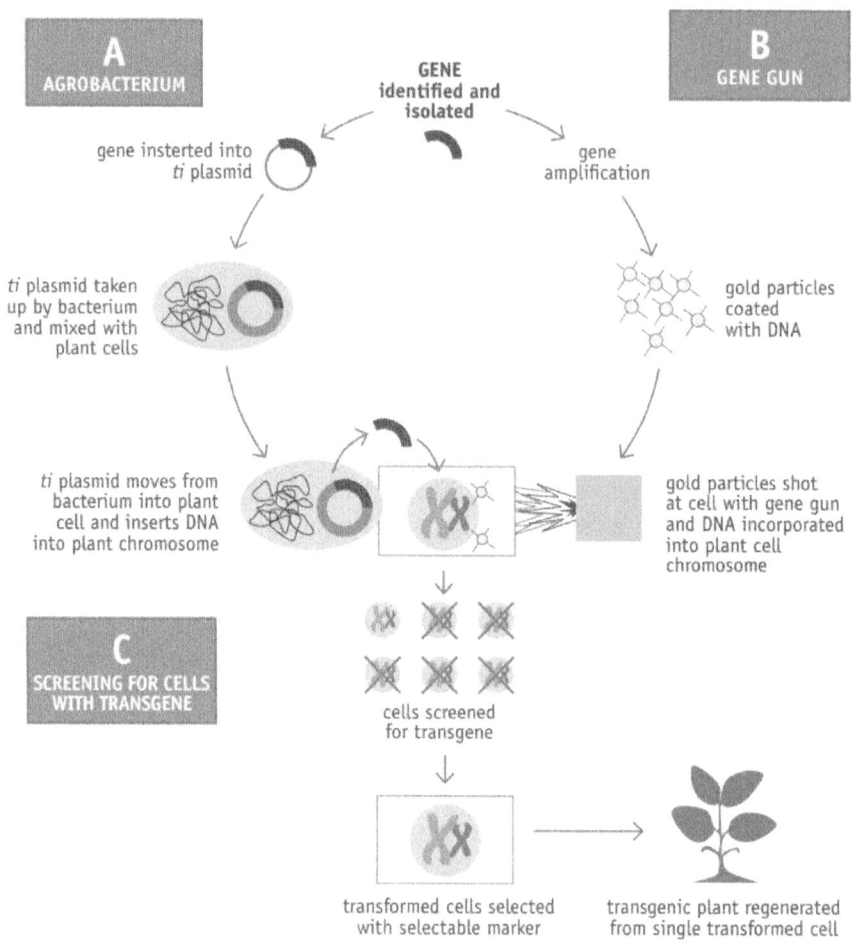

Figure 7. Steps involved in the generation of genetically transformed plants using either the Agrobacterium tumefaciens (A. tumefaciens) or microprojectile bombardment approaches Following introduction of foreign DNA into the plant cell, successfully transformed cells are selected and used to regenerate a transgenic plant.

Agrobacterium-Mediated Plant Transformation

A. tumefaciens are soil bacteria that have the ability to infect plant cells and transfer a defined sequence of their DNA to the plant cell in the infection process. Upon integration of the bacterial DNA into a plant chromosome, it directs the synthesis of several proteins, using the plant cellular machinery, that ensure the proliferation of the bacterial population within the infected plant. Agrobacterium infections result in crown gall disease (Gelvin, 2003).

In addition to its chromosomal genomic DNA, an A. tumefaciens cell contains a plasmid known as the Ti (tumour-inducing) plasmid. The Ti plasmid contains a series of vir (virulence) genes that direct the infection process, and a stretch of DNA termed T-DNA (transfer DNA), approximately 20 kb in length, that is transferred to the plant cell in the infection process. The T-DNA encodes proteins required for the maintenance of infection. These proteins include certain plant hormones that stimulate cell growth, resulting in the formation of galls, and proteins required for a certain metabolic pathway that secures the availability of nutrients for the bacteria (Figure 8).

Agrobacterium can only infect plants through wounds. When a plant root or stem is wounded it gives off certain chemical signals. In response to these signals, agrobacterial vir genes become activated and direct a series of events necessary for the transfer of the T-DNA from the Ti plasmid to the plant cell through the wound.

To harness A. tumefaciens and the Ti-plasmid as a transgene vector, the tumorinducing section of T-DNA is removed, while the T-DNA border regions and the vir genes are retained. The desired transgene is inserted between the T-DNA border regions, applying recombinant DNA technology. Thus, in the infection process, the transgene DNA is transferred to the plant cell and integrated into the plant's chromosomes (Lacroix et al., 2006). To achieve transformation, Agrobacterium cells carrying an appropriately constituted Ti plasmid vector containing the desired transgene can be inoculated into plant stems, leaf disks etc., to allow infection and T-DNA transfer to the plant cells. The explants that have been co-cultivated with Agrobacterium are subsequently processed through various tissue culture steps resulting in the selection and production of transformed cells and plants.

Protoplast Transformation Techniques

One of the characteristic features of plant cells is that they are surrounded by a rigid, cellulose-based cell wall. Protoplasts are plant cells in which the cell wall has been removed (Davey et al., 2005). Therefore protoplasts behave like animal cells, which naturally have no cell wall barrier. Plant regeneration from single protoplasts is possible due to the totipotency of plant cells, i.e. the potential of a single cell to reconstitute a complete plant.

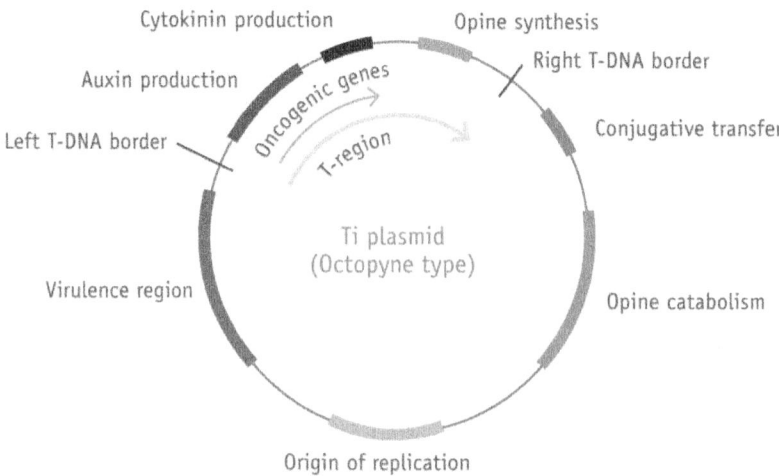

Figure 8. Wild type Ti plasmid of Agrobacterium tumefaciens (A. tumefaciens) The T region, i.e. the region of the plasmid that can be exchanged and replaced with the transgene of interest, is highlighted in light blue.

Removal of the cell wall is achieved by treating the plant material (leaves, tissue cultures, suspended cells, etc.) with a cocktail of cell wall-destroying enzymes, including pectinases, cellulases, and/ or hemicellulases in an appropriate incubation medium of the correct osmolality (i.e. the concentration of solutes in the medium). After removal of the cell wall, the protoplasts must be kept immersed in a solution of the appropriate solute concentration to prevent them from bursting. Thus, monitoring the correct osmolality of the culture medium until a new cell wall has formed is of high importance.

Different approaches exist for the delivery of transgene DNA into protoplasts through the plasma membrane. These include chemical treatments, electroporation and micro-injection techniques (Davey et al., 2005).

Chemical Techniques

The most commonly applied chemical protoplast transformation methods include polyethylene glycol (PEG) treatment, Ca^{2+} -DNA co-precipitation and liposomal DNA delivery. PEG treatment is the most widely used technique, employing solutions of 10-15 percent PEG in combination with high calcium content and a high pH. After mixing the isolated DNA and the protoplasts, followed by different washes, the DNA may be taken up by the protoplast. The role of PEG is to alter the plasma membrane properties, causing a reversible membrane permeabilization, thus enabling exogenous macromolecules to enter the cell cytoplasm.

Ca^{2+}-DNA co-precipitation depends on the formation of a co-precipitate of plasmid DNA and calcium phosphate. On contact with protoplasts under high pH conditions, the co-precipitate trespasses the cell's plasma membrane.

Liposomes, which are negatively-charged spheres of lipids, are also employed for DNA transfer and uptake into cells. DNA is first encapsulated into the liposomes which are subsequently fused with protoplasts, employing PEG as a fusogen.

Electroporation

Electrical pulses are applied to the DNA-protoplast mixture, provoking an increase in the protoplast membrane permeability to DNA. This technique is much simpler than the chemical method, providing satisfying results. However, the electrical pulses must be carefully controlled as cell death can occur above a certain threshold. The pulses induce the transient formation of micropores in the membrane lipid bilayer which persist for a few minutes, allowing DNA uptake to occur.

Micro-Injection

This technique was originally designed to transform animal cells, and was later adapted for and gained importance in transforming plant cells. However, in plant cells the existence of a rigid cell wall, a natural barrier, prevents micro-injection. Furthermore, the presence of vacuoles that contain hydrolases and toxic metabolites that may lead to cell death after vacuole breakage presents a severe restriction to micro-injection. Therefore, protoplasts, rather than intact plant cells, are more suitable for micro-injection. This method is labor-intensive and requires special micro-equipment for the manipulation of host protoplasts and DNA. However, some success in transforming both monocotyledonous and dicotyledonous species has been achieved employing this technique.

Virus-Mediated Plant Transformation/Transduction

Virus-based vectors have been shown to be efficient tools for the transient, highlevel expression of foreign proteins in plants (Chung et al., 2006). These vectors are derived from plant viruses, e.g. Tobacco Mosaic Virus (TMV), and are manipulated to encode a protein of interest. Initial delivery of the virus-based vector to the plant can be achieved by Agrobacterium - the vector is encoded in the T-DNA, which is transferred to the plant. This method is applicable to whole plants, by the process of agroinfiltration, circumventing the need for labor-intensive tissue-culture.

Within a plant cell, the virus-based vectors are autonomously replicated, can spread from cell to cell and direct the synthesis of the encoded protein of interest. The advantages of this method are the applicability to whole plants and thus a much faster outcome than the establishment of a transgenic plant, and the high-level expression of the desired protein within a short time. The major disadvantage is that the process is transient: the expression level decreases over time, and the genetic change is not passed on to subsequent generations, i.e. it is not heritable.

The process of virus-mediated DNA transfer is referred to as transduction. Several other plant transformation techniques, which have been reported but could not be reproduced or did not gain significant importance, are listed below in Table 4.

Table 4. Summary of plant transformation techniques

Gene delivery method	Characteristics
Agrobacterium	Well-established transformation vector for many dicots and several monocots and a promising vector for gymnosperms. A wide range of disarmed Ti- or Ri-derived plasmid vectors are available. Additional value for the delivery of viral genomes to suitable hosts by agroinfiltration.
Direct DNA transfer to protoplasts	Well-established transformation technique with wide host range. Permeabilization of the plasma membrane to DNA by chemical agents or electroporation. Alternatively, genes can be delivered to protoplasts by injection or fusion with DNA in encapsulated liposomes.
Microprojectile bombardment	A widely used technique for introducing DNA via coated particles into plant cells. No host range limitation. Gene transfer to *in situ* chloroplasts has been documented.
Micro-injection	Effective gene delivery technique allowing visual DNA targeting to cell type and intracellular compartment. Labour-intensive and requiring specialist skills and equipment.
Macroinjection	Technically simple approach to deliver DNA to developing floral tissue by a hypodermic needle. Germline transformation not reproducibly reported.
Impregnation by whiskers	Suspensions of plant cells mixed with DNA and micron-sized whiskers. Both transient expression and stable transformation observed.
Laser perforations	Transient expression observed from cells targeted with a laser microbeam in DNA solution.
Impregnation of tissues	Transient and stable expression from tissue bathed in DNA solution or infiltrated under vacuum.
Floral dip	Stable DNA integration and expression following dipping of floral buds into DNA solution.
Pollen tube pathway	Claims of germ line transformation by treating pollen or carpels with DNA; remains controversial.
Ultrasonication	Stable transformation by ultrasonication of explants in the presence of DNA reported. Confirmation required.

Selection of Successfully Transformed Tissues

Following the transformation procedure, plant tissues are transferred to a selective medium containing a certain selective agent, depending on which SMG was used in the transgene expression cassette. Selectable markers are genes which allow the selection of transformed cells, or tissue explants, by enabling transformed cells to grow in the presence of a certain agent added to the medium (Miki and McHugh, 2004). One can differentiate between negative and positive selection: in positive selection, transformed cells possess a growth advantage over non-transformed cells, while in negative selection transformed cells survive whereas non-transformed cells are killed. Negative selection is the method of choice for most approaches. Thus, only cells/plants expressing the SMG will survive and it is assumed that these plants will also possess the transgene of interest. All subsequent steps in the plant regeneration process will only use the surviving cells/plants. In addition to selecting for transformants, marker genes can be used to follow the inheritance of a foreign gene in a segregating population of plants.

In some instances, transformation cassettes also include marker/reporter genes that encode gene products whose enzymatic activity can be easily assayed, allowing not only the detection of transformants but also an estimation of the level of foreign gene expression in the transgenic tissue. Markers such as β-glucuronidase (GUS), green fluorescent protein (GFP) and luciferase allow screening for enzymatic activity by histochemical staining or fluorimetric assays of individual cells and can be used to study cell-specific as well as developmentally regulated gene expression. These types of transgene constructs are usually used for optimizing transformation protocols and not for the development of commercial GM crops.

In some cases, it may be desirable to produce a transgenic plant that does not contain the SMG used for the initial selection of transformed cells. Concerns have been raised about the release of transgenic plants containing antibiotic resistance or herbicide resistance genes, since the possibility of gene transfer to other species cannot be ruled out. Therefore, techniques for producing marker-free plants have been developed, by either using markers not based on herbicide/antibiotic tolerance or by specifically deleting the SMG after selection of transformed cells (Darbani et al., 2007).

Selectable Marker Genes (SMG)

The selectable portions on most transformation vectors are prokaryotic antibiotic resistance enzymes, which will also confer resistancy when they are expressed in plant cells. In some experiments, enzymes providing protection against specific herbicides have also been used successfully as marker genes (Miki and McHugh, 2004). The selective agent employed, i.e. the antibiotic or herbicide, must be able to exert stringent selection pressure on the plant tissue concerned, to ensure that only transformed cells survive. Below, some commonly used marker genes are briefly presented.

Neomycin Phosphotransferase (NPT-II) Gene and Hygromycin Phosphotransferase (HPT) Gene

Neomycin phosphotransferase-II (npt-II) is a small bacterial enzyme which catalyses the phosphorylation of a number of aminoglycoside antibiotics including neomycin and kanamycin. The reaction involves transfer of the γ-phosphate group of adenosine triphosphate (ATP) to the antibiotic molecule, which detoxifies the antibiotic by preventing its interaction with its target molecule - the ribosome. The hygromycin phosphotransferase (hpt) gene, conferring resistance to the antibiotic hygromycin, is also commonly used as selection marker.

Chloramphenicol Acetyltransferase (CAT) Gene

The chloramphenicol resistance (cat) gene encodes the enzyme chloramphenicol acetyltransferase (CAT) and was the first bacterial gene to be expressed in plants.

The enzyme specifically acetylates chloramphenicol antibiotics, resulting in the formation of the 1-, 3-, and 1,3-acetylated derivatives, which are inactive. Although not used as a selection system in plants, the gene is used frequently as a reporter gene in plant promoter studies.

Phosphinothricin Acetyltransferase Genes (Bar and Pat Genes)

A commonly used herbicide is phosphinothricin (PPT, also known as Glufosinate). This compound binds to and inhibits glutamine synthethase, which is an important enzyme in the nitrogen metabolism and ammonium fixation pathways. PPT-induced glutamine synthetase inhibition results in elevated cellular ammonium levels and cell death. The enzyme phosphinothricin acetyltransferase (PAT), first identified in Streptomyces hygroscopicus, acetylates and thus detoxifies PPT. This allows transformed cells, or complete transgenic plants, to survive and grow in the presence of PPT.

β-Glucuronidase Gene (GUS)

The E. coli β-glucoronidase gene has been adapted as a reporter gene for the transformation of plants. β-glucuronidase, encoded by the uidA locus, is a hydrolase that catalyses the cleavage of a wide variety of β-glucuronides, many of which are available commercially as spectrophotometric, fluorometric and histochemical substrates.

There are several features of the GUS gene which make it a useful reporter gene for plant studies. Firstly, many plants assayed to date lack detectable intrinsic glucuronidase activity, providing a null background in plants. Secondly, glucuronidase is easily, sensitively and cheaply assayed both in vitro and in situ and is sufficiently robust to withstand fixation, enabling histochemical localization in cells and tissue sections. The preferred histochemical substrate for tissue localization of GUS is 5-bromo-4chloro-3-indolyl-β-D-glucuronide (X-gluc). The advantage of these substrates is that the indoxyl group produced upon enzymatic cleavage dimerizes to indigo which is virtually insoluble in an aqueous environment. The histochemical assay for GUS consists of soaking tissue in substrate solution and analysing the appearance of blue colour.

Luciferase Gene

The luciferase (luc) gene isolated from Photinus pyralis (firefly) encodes the enzyme catalysing the ATP/oxygen-dependent oxidation of the substrate luciferin, resulting in the emission of light (bioluminescence). As a reporter, the gene is the basis of highly sensitive assays for promoter activity and for protein targeting sequences, involving the measurement of light emission using liquid scintillation counter photomultipliers, luminometers, X-ray film exposure or sensitive camera film.

Green Fluorescent Protein (GFP)

GFP is a widely used marker protein in modern biological research. The protein shows green fluorescence upon exposure to blue light. Originally, the protein was isolated from the jellyfish Aequorea victoria, but nowadays several other varieties from other marine organisms, as well as engineered versions (with different colour fluorescence), are available. GFP is widely applied for studies addressing gene expression or promoter efficiency as well as protein localization, stability and degradation.

Molecular Analysis of Transgenic Plants

After the successful transformation and selection of plant cells and the subsequent regeneration of a transgenic plant, it is desirable to monitor the presence of the transgene in the plant and to investigate the expression levels of the introduced genes encoding the protein(s) of interest (Stewart, 2005).

Analysis of transgenic plants at the molecular level is mainly performed by PCR and Southern blot analysis. PCR indicates the presence of the desired transgene within the plant, whereas stable integration of the transgene into the cellular genome is confirmed by Southern blot analysis. If plants are analysed that have been transformed using A. tumefaciens, it is important to prepare plant DNA from sterile tissue, as contamination with A. tumefaciens DNA will interfere with the interpretation of the results. Southern blot analysis using genomic DNA also yields information on the copy number of the integrated DNA sequences, whether any multiple inserts are tandemly linked or dispersed throughout the genome, and on the stability of the integrated DNA in the F1 progeny of the transformed plants.

Molecular analysis of the protein expression levels, including tissue-specific expression, developmental stage-specific expression, expression upon certain stimuli and so on, can be assayed by enzyme-linked immunosorbent assay ELISA or immunostaining of plant tissue. Expression of a gene of interest can also be assayed by determining the presence and quantity of the corresponding RNA transcript, e.g. by applying a modified PCR protocol (reverse transcriptase PCR) [RT-PCR]).

Application of Transgenic Plants

Numerous applications of transgenic plants are already reality or are envisaged and under investigation for the future; the main transgene targets being pest resistance and herbicide tolerance. In addition, resistances to abiotic stresses, such as drought, or improved nutrient profiles are increasingly investigated. Further possible applications that are under development are the production of medically valuable proteins or chemicals in plants (biopharmaceuticals), or the production of edible plants containing vaccines. In recent years, the technique of gene stacking, i.e. the introduction and targeting of several traits within one plant species, has also gained significant importance. Since the applications of transgenic plants are diverse and numerous, no complete coverage of the field will be provided at this point.

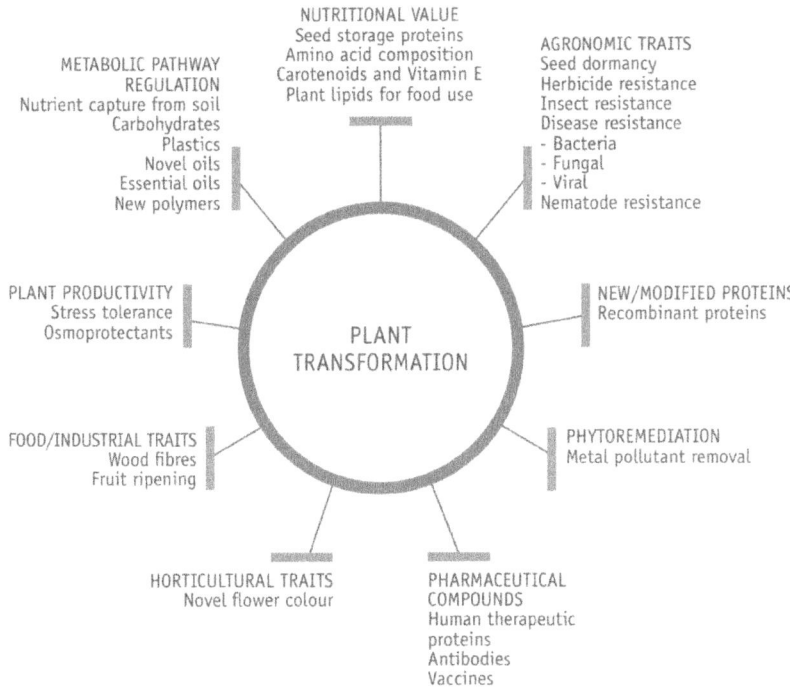

Figure 9. Applications of transgenic plants that are already available or envisaged for the future

BIOTECHNOLOGY IN ANIMAL PRODUCTION

Modern biotechnology provides a number of possible applications in animal and livestock production. Research and development in the field focuses on improving animal growth, enhancing reproduction rates, enhancing breeding capacity and outcomes, improving animal health and developing new animal products (Basrur and King, 2005). The major techniques used to achieve these goals are the creation of transgenic animals, manipulation of animal reproduction, marker-assisted selection (MAS), molecular disease diagnostic and application of biotechnology to modify animal feed.

Biotechnology in Animal Breeding and Reproduction

Animal breeding, nowadays, is a field that is influenced by a whole range of biotechnological applications and developments (Bazer and Spencer, 2005). The common goal of all efforts undertaken in this field is genetic progress within a population, i.e. the improvement of the genetic resources and, ultimately, the phenotypic outcome. Genetic progress is influenced by several factors, namely the accuracy of choosing candidates for breeding, the additive genetic variation within the population, the selection intensity (i.e. the proportion of the population selected for further breeding), and the generation interval (the age of breeding). Note that the first three factors need to be increased in order to increase genetic progress, whereas the last factor, being generation interval, needs to be decreased. All factors can be influenced, to a varying extent, by modern biotechnology.

The techniques that are currently available to reach this end can be divided into two different groups. The first group includes all technologies that interfere with reproduction efficiency: artificial insemination, embryo transfer (ET), embryo sexing, multiple ovulation, ova pick-up and cloning, amongst others. The outcome of these technologies is an increased breeding accuracy, selection intensity and, in some cases, a shortened generation interval.

The second group of applications is based on the molecular determination of genetic variability and the identification of genetically valuable traits and characteristics. This includes the identification and characterization of quantitative trait loci (QTL) and the use of molecular markers for improved selection procedures. Quantitative traits are phenotypic characteristics that show a distribution of expression degree within a population (usually represented by a normal distribution), and that are based on the interaction of at least two genes (also known as polygenic inheritance). A typical example in humans is skin colour, which is based on the interaction of several genes resulting in a large variety of phenotypes. A QTL is a DNA sequence that is associated with a certain quantitative trait – not even necessarily a gene that contributes to the trait, but possibly a sequence that is close in space to involved gene(s). Knowledge of the loci responsible for a certain quantitative trait and the underlying genes can help to select individuals for further breeding, or to start genetic engineering of the trait in question. Below, the most frequently applied techniques in animal breeding and reproduction will be summarized and explained in more detail.

Artificial Insemination (AI)

Artificial insemination (AI) is the process of collecting semen from a particular male (e.g. a bull) that is subsequently used for the fertilization of many females (e.g. cows) (Galli et al., 2003). The semen can be diluted and preserved by freezing (cryopreservation). This technique can enable a single bull to be used for fertilization simultaneously in several countries for up to 100 000 inseminations a year. The high intensity and accuracy of selection arising from AI can lead to a four-fold increase in the rate of genetic improvement in dairy cattle relative to that from natural mating. Since its establishment in the 1950s, AI has proven to be a very successful biotechnology, greatly enhancing the efficiency of breeding programmes (Rege, 1994). Use of AI can reduce the transmission of venereal diseases in a population and the need for farmers to maintain their own breeding males. Furthermore, it facilitates more accurate recording of pedigree and minimizes the cost of introducing improved stock. AI has significant importance for the breeding of cattle, swine and poultry.

Embryo Transfer (ET)

Although not economically feasible for commercial use on small farms at present, embryo technology can greatly contribute to research and genetic improvement in local breeds. There are two procedures presently available for the production of embryos from donor females (McEvoy et al., 2006). One consists of superovulation using a range of hormone implants and treatments, followed by AI and then flushing of the uterus to gather the embryos. The other, called in vitro fertilization (IVF) consists of recovery of eggs from the ovaries with the aid of the ultrasound-guided transvaginal oocyte pick-up (OPU) technique. When heifers reach puberty at 11-12 months of age, their oocytes may be retrieved weekly or even twice a week. These are matured and fertilized in vitro and kept until they are ready for implantation into foster females. In this way, high-value female calves can be used for breeding long before they reach their normal breeding age. IVF facilitates recovery of a large number of embryos from a single female at a reduced cost, thus making ET techniques economically feasible on a large scale. Additionally, IVF produces embryos suitable for cloning experiments. However, ET is still not widely used despite its potential benefits.

Embryo Sexing

Technologies for rapid and reliable sexing of embryos allow the generation of the desired sex at specific points in a genetic improvement program, markedly reducing the number of animals required and enabling increased breeding progress. A number of approaches to the sexing of semen have been attempted; however, the only method of semen sexing that has shown any promise has been the sorting of spermatozoa according to the DNA content by means of flow cytometry (Rath and Johnson, 2008). Embryo sexing has been attempted by a variety of methods, including cytogenetic analysis, assays for X-linked enzyme activity, analysis of differential development rates, detection of male-specific antigens, and the use of Y-chromosome specific DNA sequences.

Animal Cloning

Animal cloning is defined as the process of producing organisms that are genetically identical. The cloning of animals can be achieved by two strategies: embryo splitting and somatic cell nuclear transfer (SCNT) (somatic cell cloning). Both techniques offer the possibility for creating clone families from selected superior genotypes and to produce commercial clone lines (Vajta and Gjerris, 2006).

Somatic cell cloning is based on the procedure of removing the DNA from an unfertilized oocyte and replacing it with the DNA obtained from a somatic cell. The somatic cell DNA can be obtained from any individual, preferably an individual with desirable traits. Once introduced to the oocyte, the somatic cell's DNA is reprogrammed by the oocyte and the unfertilized oocyte can develop as an embryo. The resulting animal will be genetically identical as the somatic cell donor. In theory it is possible to obtain practically unlimited numbers of somatic donor cells from an individual, which allows cloning technology to be applied for the production of many genetically identical individuals. In addition, this technique offers another advantage: the somatic cells genome can be subjected to genetic manipulation prior to the introduction into the oocyte, resulting in a transgenic organism.

Embryo splitting, the second cloning technology, is the process of dividing a developing embryo, typically at the 8-cell stage, into two equal parts that continue to develop. The procedure can be repeated several times, but usually only four viable embryos can be obtained from a founder embryo. The technology has no significant importance in research and development nowadays.

Genetic Markers and Marker-Assisted Selection (MAS)

A genetic marker is defined as a DNA sequence that is associated with a particular trait, in terms of spatial proximity of sequence, and thus segregates in an almost identical and predictable pattern as the trait. This marker can include the gene (or a part thereof) which is responsible for the trait, or DNA sequences that are sufficiently close to the gene(s) so that co-segregation is ensured.

Genetic markers facilitate the "tagging" of individual genes or small chromosome segments containing genes which influence the trait of interest. Availability of large numbers of such markers has raised the likelihood of detection of major genes influencing quantitative traits. The process of selection for a particular trait using genetic markers is called marker assisted selection (MAS). MAS can accelerate the rate of breeding progress by increasing the accuracy of selection and by reducing the generation interval. Marker identification and use should enhance future prospects for breeding for such traits as tolerance or resistance to environmental stresses, including diseases (Dekkers, 2004; FAO, 2007).

Two types of marker can be considered. First, markers that are sufficiently close to the trait gene on the chromosome so that, in most cases, alleles of the marker and the trait gene are inherited together. This type of marker is called a linked marker. At the population level, alleles at linked markers cannot be used to predict the phenotype until the association between alleles at the marker and alleles at the trait gene is known (called "phase"). To determine phase, inheritance of the marker and trait gene has to be studied in a family. However, information on phase is only valid within that family and may change in subsequent generations through recombination (Ron and Weller, 2007).

The second type of marker is a functional trait. These markers are called "direct" markers. Once the functional polymorphism is known it is possible to predict the effect of particular alleles in all animals in a population, without first having to determine the phase. Therefore, "direct" markers are more useful than "linked" markers for predicting the phenotypic variation of target traits within a population. A further complication is that the mechanisms of genetic control differ between traits. The variation seen in some traits is directly controlled by a single gene (monogenic traits), which may have a limited number of alleles. In the simplest situation a gene will have two alleles: one allele will be associated with one phenotype and another allele with a different phenotype. An example is black versus brown coat colour in cattle: the brown coat colour occurs as a result of a mutation in the melanocyte hormone receptor gene, which results in the creation of a different allele with a different function.

However, the traits that are important in livestock production are generally more complex and have a very large range of variation in the observed phenotype, caused by the interaction of multiple genes (polygenic traits). Growth rate and milk yield are examples of two traits that exhibit a continuous phenotypic variation. Such traits are called quantitative traits. The variation in quantitative traits is controlled by several genetic loci (called quantitative trait loci [QTL]), each of which is responsible for a small amount of the overall variation (Rocha et al., 2002). The behaviour of genes (including major genes) that control a trait is likely to be dependent on the genetic background.

The myostatin allele responsible for double muscling in Belgian Blue cattle is also found in other breeds; however, the phenotype associated with the allele is variable between the breeds. This suggests that there are genes at other loci in the genome that act to modify the phenotypic expression of the major gene. Thus, information is required not only on the major genes that control a trait, but also on the interactions between genes. It is therefore premature to start using DNA-based selection widely, without further knowledge of gene interaction networks. However, some DNA tests for specific polymorphisms are being offered commercially, e.g. the GeneSTAR test for tenderness (based on variations in the calpastation gene, Pfizer Animal Genetics) and marbling (based on variations in the thyfoglobulin gene), and the Igenity test for fat deposition (based on variations in the leptin gene, Merial). These tests can be used by breeders and evaluated in their populations.

MAS and gene mapping are also considered as important tools to investigate, maintain and conserve the genetic diversity and the genetic resources of agricultural species. During the last decades an increasing portion of breeds became extinct, mainly local breeds that are not used in a sustainable manner and are not covered by breeding programmes. However, these local breeds are of high importance since they are adapted to local conditions, contribute to local food security and represent a unique source of genes that can be used for the improvement of industrial breeds. Molecular marker techniques can play an important role in the characterization and protection of agricultural genetic resources (FAO, 2006).

Transgenic Animals

A transgenic animal is an animal that carries a specific and deliberate modification of its genome – analogous to a transgenic plant. To establish a transgenic animal, foreign DNA constructs need to be introduced into the animal's genome, using recombinant DNA technology, so that the construct is stably maintained, expressed and passed on to subsequent generations. The last point, heritability of the genetic modification, can be achieved by creating an animal that carries the modification in the genome of its germ line: all offspring derived from this animal will be completely transgenic, as they will carry this modification in all their somatic and germ line cells.

Transgenic animals can be created for a variety of different purposes: to gain knowledge of gene function and further decipher the genetic code, study gene control in complex organisms, build genetic disease models, improve animal production traits, and produce new animal products (Melo et al., 2007).

In 1982, the first transgenic animal was produced: a mouse, obtained by microinjection of a DNA construct into a fertilized, single-cell stage oocyte (Palmiter et al., 1982). The transgene construct used was composed of the rat growth hormone gene, fused to the mouse metallothionein-I promoter. The study was published in Nature magazine, and the impressive outcome of the study was chosen as the cover photo: the produced transgenic mice were unnaturally large, approximately twice the size as non-transgenic control mice. The impact of this study on both the scientific and public community was huge, and raised speculations about the potential applications of this technology for animals of agricultural importance. Since the insertion of a single growth hormone gene was sufficient to have tremendous effects on mice, it was anticipated that this procedure would also be applicable for agricultural animals, resulting in highly increased growth rate, feed efficiency and reduced fat deposition. Many other possible applications were also subject of speculation, such as a manipulation of milk production or production of milk with novel ingredients, increased wool production or increased resistance of farm animals to diseases and parasites.

By 1985, transgenic pigs and sheep had been obtained, with cattle and chicken following somewhat later (Melo et al., 2007). Since that time the development of transgenic animals and the exploration of agricultural applications has been a steady process, although at a slower rate than what was initially expected. Engineering a specific trait proved to be much more difficult than simply introducing the responsible gene, and technical limitations, the high costs of the process and insufficient knowledge about gene function and regulation of gene expression severely restricted progress. This is particularly true for agricultural species such as cattle, which proved to be much more complicated than mice.

Nevertheless, research in the field continues and several agricultural applications are envisaged, and the approval and market release of the first transgenic animals is expected to take place in the next few years. The knowledge of gene expression and regulation is constantly extending, facilitating genetic engineering in complex animals, such as mammals. Likewise, the repertoire of available techniques to manipulate DNA and animals is constantly increasing. Therefore, it is likely that genetic engineering techniques applied to animals of agricultural importance will play an increasingly important role in the years to come. The techniques that are currently applied to produce transgenic animals are listed below.

Micro-Injection

Micro-injection, the first successful approach for the creation of transgenic animals, has already been described in the preceding paragraphs. Briefly, it is based on the injection of a foreign DNA construct into a fertilized oocyte (Figure 10). The construct integrates randomly into the host oocyte genome, subsequently the zygote continues embryonic development, the embryo is transferred to a foster mother and eventually develops to a transgenic animal. However, this method has strong limitations: on average, less than 1 percent of embryos injected and 10 percent of animals born are transgenic, genes can only be added, not replaced or deleted, and multiple copies of the transgene are inserted at random, hindering the correct regulation of gene expression and possibly interfering with endogenous gene function (Robl et al., 2007). This requires large amounts of oocytes to be injected, as the overall efficiency of the process is very low and basically a trialand-error process, whose outcome can only be influenced to a small extent.

ES Cell Based Cloning and Transgenesis

To overcome the problems associated with micro-injection techniques, embryonic stem cell (ES cell) technology has been developed (Denning and Priddle, 2003). Embryonic stem cells, as the name suggests, are derived from embryos at a very early stage (the blastula), and possess the important characteristic of pluripotency. Pluripotency is the ability of these cells to differentiate to any of the cell types and tissues found in the adult organism. ES cells can be grown in culture for many passages and can be subjected to transformation with transgene constructs, resulting in modifications of their genome. The constructs used not only permit the selection of successfully transformed cells, but also allow gene targeting to be accomplished Thus, genes can be specifically introduced, replaced or deleted (so-called knock-ins and knock-outs). Transformed ES cells are re-introduced into the blastocoel cavity of an embryo, where they integrate and produce a mosaic (chimaeric) animal, i.e. an animal that is made up of transformed and non-transformed cells. Possibly, the chimaeric animal carries the transgene in the germ line; in this case, it is possible to obtain completely homozygous transgenic animals through selective breeding (Figure 10).

This technique, mainly through the feature of gene targeting, allows a broad variety of genetic modifications to be introduced. For many years, several laboratories worldwide have tried to produce ES cells from farm animals, and although some success has been claimed, no robust and reproducible method has been published. Indeed, even in mice the production of ES cells is a costly and labour-intensive technology (Melo et al., 2007).

Somatic Cell Nuclear Transfer (SCNT)

The method of choice nowadays for the production of transgenic animals is somatic cell nuclear transfer (SCNT). This method, also known as somatic cell cloning, initially gained importance for the possibility to clone animals in theoretically unlimited numbers. However, it can also be adapted to produce transgenic animals, with the additional benefit of targeted genetic manipulation (Heyman, 2005).

The insertion of a transgene construct into a specific, pre-determined DNA site of the host genome is called gene targeting. The process and the construction of the transgene is more complex than random gene insertion, as is the case during microinjection. Nevertheless, gene targeting is a powerful and widely used technique due to the ability to insert the transgene into a specific site (knock-in), inactivate specific genes (knock-out) or replace the endogenous version of a gene with a modified version. This helps to overcome many of the problems and limitations that are associated with random transgene insertion.

The usual procedure is to produce a series of transgenic founder animals of both sexes, which are subjected to breeding, with the aim of producing homozygous offspring. The entire method is based on the following protocol: oocytes from a donor animal are enucleated, i.e. their nucleus containing the genome is removed. Subsequently a donor nucleus is injected into the enucleated oocyte, and the cells are fused by electrofusion. Following fusion, the oocyte is activated by chemical or mechanical means to initiate embryonic development, and the resulting embryo is transferred to a foster mother (Hodges and Stice, 2003). The donor nuclei can be derived from either somatic cells or ES cells that have been subjected to targeted genetic manipulation prior to injection into the oocyst. Thus, a large number of identical animals with targeted genetic modifications can be obtained (Figure 10).

Homozygous transgenic animals may also be obtained by a slightly modified approach: by targeting a transgene to one member of a pair of chromosomes, and subsequently target the same site on the other chromosome with the same transgene (Robl et al., 2007).

Another approach based on the techniques outlined above makes use of a rejuvenation system for bovine fibroblast (connective tissue) cells (Kuroiwa et al., 2004). A bovine fibroblast cell line is derived from a bovine foetus, and subjected to genetic manipulation. Such primary cell lines grow for only a limited number of cell divisions in culture, allowing only a limited number of genetic manipulations to be introduced (usually only one) before the cells stop dividing and eventually die (Robl et al., 2007).

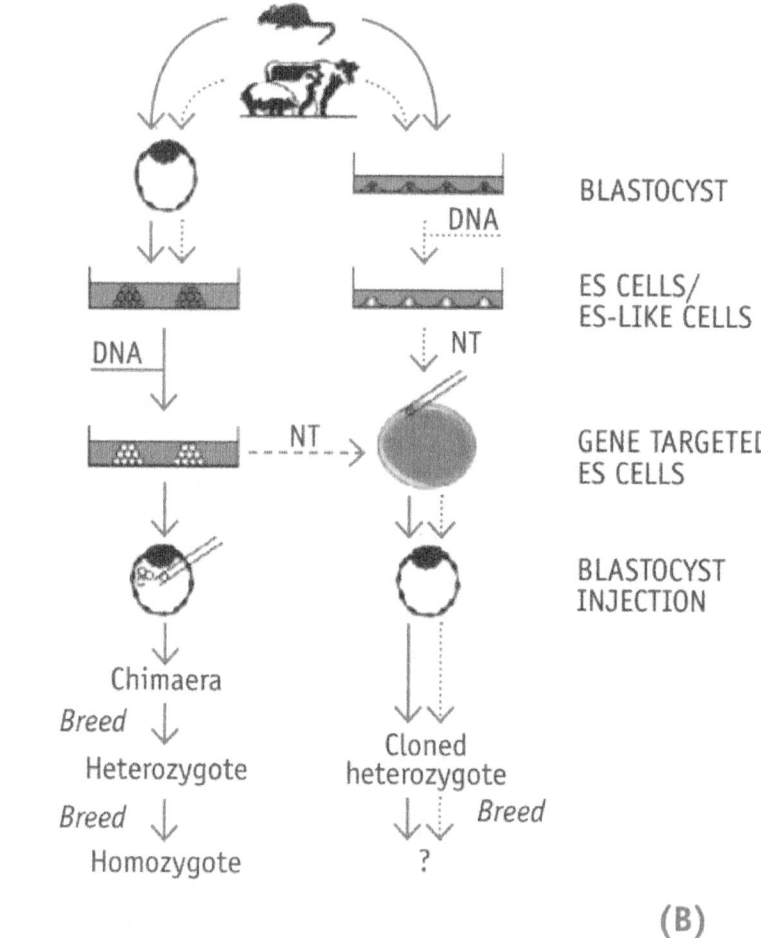

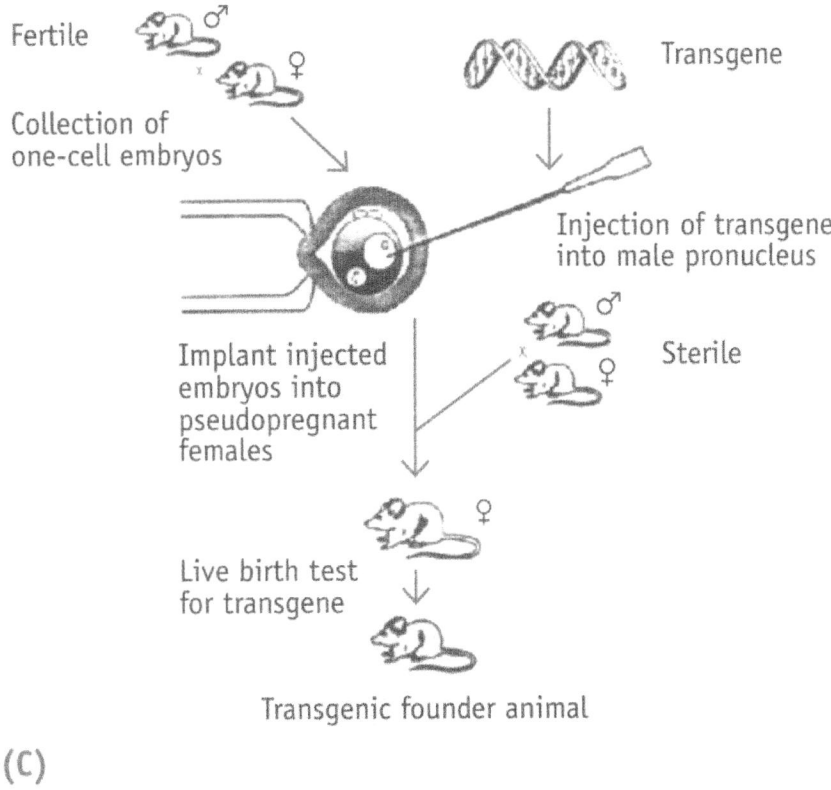

(C)

Figure 10. Comparison of micro-injection, ES cell techniques and somatic cell nuclear transfer (SCNT) for the creation of transgenic animals. (A) ES cells are obtained from an early embryo, can be subjected to genetic modification in culture and are subsequently re-injected into an embryo or are used for nuclear transfer (NT) to an enucleated oocyte. Transgenic animals can be obtained by breeding in both cases. (B) In SCNT, the nucleus from a somatic donor cell is removed and injected into an enucleated oocyte, resulting in a cloned animal. (C) During micro-injection, the DNA construct is injected into a fertilized oocyte, resulting in random DNA integration and the production of a transgenic founder animal.

After the genetic manipulation, the cells are used in a cloning procedure to obtain cloned foetuses. These foetuses can be subjected to a second round of fibroblast isolation, manipulation and cloning. Once the genetic manipulations are completed, the final cell line can be used in a cloning procedure to produce transgenic offspring.

Artificial Chromosome Transfer

Artificial chromosomes are a relatively recent development in animal transgenics (Robl et al., 2003). One outstanding characteristic is their ability to carry very large fragments of DNA, up to several Mb (compared with 5-30 kb on a typical plasmid vector). Artificial chromosomes possess a centromere, telomeres and origins of replication, sequences that are responsible for their stable maintenance within the cell as autonomous, self-replicating chromosomes. This eliminates the need for integration into the host genome. Due to these properties, artificial chromosomes can be used to transfer either very large, complex genes or many small genes and regulatory elements to a target animal. The actual process of chromosome transfer and subsequent cloning of animals is similar to the SCNT approach.

The feasibility of this technique has been proven by the transfer of a human artificial chromosome, encoding the human antibody genes of 10 Mb in size to cattle (Kuroiwa et al., 2002). The transferred chromosome was stable in the adult transgenic animals, and the encoded antibody genes were expressed to a certain extent.

Sperm-Mediated DNA Transfer

Several reports describe the use of sperm as a vector to deliver transgene DNA to the oocyte during the process of fertilization. The efficiency of this process varies considerably between species, and several approaches are under investigation to improve uptake and incorporation of foreign DNA. These include, among others, intracytoplasmic injection of DNA-coated sperm into the oocyte, or liposome treatment of sperm to facilitate DNA uptake (Robl et al., 2007). Nevertheless, the entire process is not completely understood and far from being used routinely.

Viral-Vector Mediated DNA Transfer

Transgenesis may also be accomplished by employing virus-derived vectors, namely vectors based on the retrovirus-class of lentiviruses (Whitelaw et al., 2008). Genes that are essential for viral replication are deleted from the viral genome, maintaining only the capacity for integration of the viral genome into the host genome. Parts of the vector that were occupied by viral genes can then be replaced by the transgene of interest – an approach analogous to the modification of the A. tumefaciens Ti-plasmid. Viruses carrying the modified vector are then produced in vitro and subsequently injected into the perivitelline space of the zygote (or an unfertilized oocyte), resulting in infection of the zygote and integration of the viral genome into the host genome. Transgenesis rates reaching up to 100 percent of injected embryos have been described (Park, 2007).

Major drawbacks of this method are a limited transgene size and random transgene integration. The maximal transgene size is 8 kb, which is rather low compared with other techniques. Random and possibly multiple transgene integration may lead to position effects, disturbance of the host genome and dose effects, as is the case with pronuclear injection. Solving these problems holds great promise for the further development and application of lentiviral vectors.

Other, less frequently used, methods include biolistics, liposome-mediated DNA transfer to cells and embryos, or DNA transfer to cells and embryos by electroporation. As mentioned in the introduction, constant progress is being made in the field of animal transgenesis, although no approval for commercial release has been obtained so far. Some of the envisaged applications of GM animals are given in the following paragraph.

Applications for Transgenic Animals

Since the production of the first transgenic mice almost three decades ago much work has been performed on the development of technologies for efficient transgenes is. Many initial problems, such as low efficiencies, random transgene insertions and unexpected and undesirable behavior of transgenic animals, have been overcome or are at least understood in more detail. Furthermore, an ever-increasing knowledge of genes, gene function and regulation of gene expression facilitates the planning and creation of transgenic animals with desired traits.

The main interest of modern agricultural research with regard to transgenic animals can be divided into two broad categories: production of animals with improved intrinsic traits, such as higher growth rates, improved milk production, disease resistance etc. The other is the production of animals that produce novel products, such as pharmaceuticals, proteins of medical relevance, vaccines etc. (Wheeler et al., 2003; Niemann et al., 2005). Examples of both categories will be given in the following sections.

Transgenic Animals for Food Production

Engineering transgenic animals with an application in food production focuses mainly on improved meat production, improved carcass quality and enhanced milk production. Milk is a complex biological fluid and has a high importance for contributing to the nutrition of many societies (Melo et al., 2007). The major goals for transgenic animal development concerning milk production are increased milk production, higher nutrient content or milk containing novel substances. Most milk proteins (circa 80 percent) belong to the caseins, and transgenic cattle were created that contain extra copies of casein genes. This resulted in elevated casein protein levels in milk (Brophy et al. 2003). Another milk application that is being investigated is the production of milk with no lactose (milk sugar) present, since approximately 70 percent of the world population cannot metabolize lactose and thus cannot consume dairy products. Engineering milk with novel properties, e.g. milk containing the immune-stimulating human protein lactoferrin, is a further approach. Many other additives, e.g. different growth hormones or substances that stimulate health and development, have been proposed for overexpression in milk and thus possibly contribute to growth and health of developing offspring. In pigs, the transfer of the bovine a-lactalbumin gene led to increased milk production, resulting in faster piglet growth and survival rate.

One of the first reports with relevance for enhanced meat production was the article about the first transgenic mice, expressing rat growth hormone and showing increased body size and mass. However, transferring this approach to pigs initially did not yield promising results. Nevertheless, pigs showing increased muscle weight gain and feed efficiency by introducing porcine growth hormone or human insulin-like growth factor have been created (Niemann et al., 2005). Furthermore, pigs expressing the enzyme phytase in their salivary glands have been created: these animals can metabolise the phosphor present as phytic acid in corn and soy products, thus needing less phosphor as feed additives and releasing less phosphor with their manure, reducing the environmental impact of pig farming (Haefner et al., 2005).

Experiments in cattle are focusing on the myostatin gene, a negative regulator of muscle mass, resulting in a high increase in muscle mass in animals with a myostatin mutation or deletion.

Transgenesis is also employed for fish; injection of embryos with constructs containing either the bovine or Chinook salmon growth hormone has been reported, with the aim of improving fish growth in general and especially under adverse conditions, e.g. low water temperatures. This has resulted in an up to 5-11 fold increase in weight after one year of growth for transgenic salmon and 30-40 percent increased growth of transgenic catfish (Wheeler, 2007).

All these studies demonstrate the fundamental feasibility of applying transgenesis to agricultural animals for improved food production, but so far no transgenic food producing animal has been released for commercial use. In addition to the research and development necessary for the establishment of a transgenic animal, there are several other factors that strongly influence the use of transgenic animals for food production. Among these are considerations concerning the economic practicability, social acceptance of transgenic food and, possibly most important, regulations concerning the approval of GMOs and derived products.

Regulatory authorities need to consider three factors:
- Safety of the food product for human consumption;
- Environmental impact of the genetically modified animals;
- Welfare of the animals.

These factors need to be considered on a case-to-case approach for every new transgenic animal or product that has been obtained using GMOs. In principle, this safety investigation is identical to the safety regulations and procedures that apply for transgenic plants. A detailed description of the safety evaluation procedures and the underlying regulatory documents and treaties is provided in the accompanying modules of this compendium.

Transgenic Animals for Production of Human Therapeutics

One major application of animal transgenesis nowadays is the production of pharmaceutical products, also known as animal pharming. The costs for producing transgenic animals are high, but since the pharmaceutical industry is a billion-dollar market the input is likely to be a feasible and economically worthwhile investment (Sullivan et al., 2008). Since many human proteins cannot be produced in microorganisms and production in cell culture is often labor-intensive with low yields, the production of biopharmaceuticals in transgenic animal bioreactors is an attractive alternative (Kind and Schnieke, 2008). Furthermore, many human proteins cannot be produced in microorganisms, since they lack post-translational modification mechanisms that are essential for the correct function of many human proteins.

Pharmaceutical proteins or other compounds can be produced in a variety of body fluids, including milk, urine, blood, saliva, chicken egg white and seminal fluid, depending on the use of tissue-specific promoters (Houdebine, 2009). Nevertheless, milk is the preferred medium due to its large production volume. Furthermore, it has been shown that the mammary glands can produce up to 2 g of recombinant protein per liter of milk; assuming average protein expression and purification levels, only relatively small herds of transgenic animals would be required to supply the world market with a specific recombinant protein (e.g. 100 transgenic goats for the production of 100 kg monoclonal antibodies required per year [Melo et al., 2007]). In Table 5, biomolecules expressed in mammary glands and their anticipated applications are listed:

Table 5. Pharmaceuticals produced by transgenic animals

Pharmaceutical	Bioreactorspecies	Application/treatment	Company
Antithrombin III	goat	thrombosis, pulmonary embolism	GTC Biotherapeutics (USA)
tPA	goat	thrombosis	PPL Therapeutics (UK)
α-antitrypsin	sheep	emphysema and cirrhosis	PPL Therapeutics (UK)
Factor IX	sheep	hemophilia b	PPL Therapeutics (UK)
Factor VIII	sheep	hemophilia a	PPL Therapeutics (UK)
Polyclonal antibodies	cattle	vaccines	Hematech (USA)
Lactoferrin	cattle	bactericide	Pharming Group (NED)
C1 inhibitor	rabbit	hereditary angioedema	Pharming Group (NED)
Calcitonin	rabbit	osteoporosis and hypercalcemia	PPL Therapeutics (UK)

Adapted from: Melo *et al.*, 2007.

Another advantage of biopharmaceutical production in transgenic animals is the reduced risk of transmitting diseases, compared with human-derived material. Several cases are known where hundreds of patients were infected with HIV, Hepatitis C or Creutzfeld-Jakob-Disease following treatment with human-derived pharmaceuticals. Of course, animal-derived material needs to be subjected to a thorough purification procedure to exclude transmission of animal diseases (zoonoses) or contamination with animal DNA or protein that might induce an immune reaction.

Nevertheless, the development of transgenic animals that secrete high contents of the desired product in their milk, and the subsequent development of an effective and high-yield purification protocol to get rid of contaminating proteins, requires a lot of knowledge and financial and intellectual input. So far, only GTC Biotherapeutics Antithrombin III has been approved for the United States market and is sold under the name of ATryn (FDA, 2009). Furthermore, many potential target proteins as well as the technologies to develop a transgenic animal are covered by patents and intellectual property rights, thus only a small number of proteins are being investigated by a small number of pharmaceutical companies at the moment (Kind and Schnieke, 2007).

A particularly promising approach is the development of transgenic animals that express human polyclonal antibodies. Antibodies are the fastest growing set of new biopharmaceuticals, for therapeutic use in cancer, autoimmune diseases, infections, transplantations, biodefence and immune deficiencies. Currently all approved therapeutic antibodies are produced by cell culture techniques.

The possibilities for the production of polyclonal human antibodies in transgenic cattle are currently being investigated; such antibodies would mimic the natural human immune response to a pathogen. Cattle would be especially suited for this purpose, since the total amount of antibodies in an adult animal is approximately 1 kg. One approach towards this end is the use of artificial chromosomes to transfer the human antibody genes to the target animal (Kuroiwa et al., 2002). Concomitantly, the endogenous antibody genes of the animal are knocked out to prevent their expression and thus allow purification of human antibodies without contaminating bovine antibodies. To obtain human polyclonal antibody sera from the animal, the animal would need to be immunized with a vaccine containing the pathogen of interest, e.g. a bacterium or a virus. Subsequently, the animal would build up an immune response and express the human antibodies directed against that pathogen. These antibodies could subsequently be extracted and purified from the animal's blood plasma and used to treat humans suffering from an infection with that particular pathogen. This perspective for a quick availability of large amounts of human antibody sera targeted against a certain pathogen or disease agent has raised speculations about a transformation of medicine similar to the introduction of antibiotics in the 1940s and 50s (Kind and Schnieke, 2007). Similar approaches, based on the same methodology, are being pursued for the use of plants as bioreactors for the production of medically valuable proteins and small-molecule drugs (Twyman et al., 2005).

Transgenic Animals for Improved Disease Resistance

Resistance or susceptibility to diseases and the immune response typically depend on a variety of genes, but identification of some key genes has brought up the possibility of gene transfer to target important and specific aspects of the immune system (Niemann et al., 2005). Diseases that are under investigation, by either introducing resistance genes or removing susceptibility genes, include bovine spongiform encephalopathy (BSE), brucellosis, other viral or bacterial infections, parasitic organisms, and intrinsic genetic disorders.

One often-cited example is resistance against mastitis: mastitis is a bacterial infection of the bovine mammary gland, leading to decreased productivity and milk contamination. Transgenic cattle have been produced that secrete the small protein lyostaphin in their milk, which is a potent inhibitor of Staphylococcus aureus (S. aureus), the bacterium responsible for the majority of mastitis cases. According to first trials, the transgenic cows are resistant to S. aureus – mediated mastitis (Donovan et al., 2005).

Further approaches of animal transgenics target animal reproductive performance and prolificacy, development of organs for transplantations (xenotransplantation) that do not evoke a rejection response, or improvement of animal fibre and wool.

Biotechnology in Animal Health

Apart from the aforementioned possibilities to generate transgenic animals with enhanced resistances to diseases, biotechnology offers a variety of other techniques that contribute to improved animal health. These include the production of vaccines to immunize animals against diseases, and the development of improved disease diagnostic tools.

Vaccines

Vaccines are substances, derived from a pathogen, that are used to stimulate an animal's immune system to produce the antibodies needed to prevent infection from that particular pathogen. Vaccination is therefore the main approach to protect animals from infectious diseases. The majority of vaccines are based on material directly derived from inactivated bacteria or viruses, which potentially revert to their virulent (disease-causing) form. Modern biotechnology offers possibilities to engineer specific vaccines that are free from pathogen-derived material and are more effective and safe in stimulating the immune response (Rogan and Babiuk, 2005).

One approach is based on recombinant protein technology: once a protein from a pathogen that serves as antigen (i.e. a molecule that stimulates an immune response) has been identified, this protein can be safely expressed in cell culture, e.g. in E. coli or mammalian cells, using recombinant DNA technology. Subsequently, this protein can be harvested, purified and used as a vaccine (also known as subunit vaccines). In addition, it has also become possible to create fusions of several pathogen proteins, so that one final protein stimulates a variety of immune responses (Meeusen et al., 2007).

A second approach consists of using DNA-based vaccines. This methodology is based on the delivery of plasmid DNA to the cells of a host animal that encodes pathogenic proteins. Once expressed within the cell, the proteins stimulate the animal's immune response in the same way as if the proteins were delivered from outside; thus the animal serves as its own bioreactor for vaccine production (Rogan and Babiuk, 2005). The efficiency of this method is largely dependent on effective plasmid delivery to the animal cells; methods for delivery include chemical transformation, electroporation, injection and the gene gun.

A third approach is the delivery of pathogen-derived antigens by live recombinant vectors. Bacteria, viruses or even parasites can be engineered to express foreign proteins from the pathogen of interest that act as antigens. The engineered organism is then delivered to the animal, where it induces a limited infection and presents the foreign pathogenic protein, thus stimulating an immune response against that pathogen.

Recently, a very interesting combination of transgenic plant technology and animal vaccination has emerged: plants are engineered to express an antigenic protein from a pathogen at high levels in their tissues or storage organs. Subsequently these plants can be fed to animals and the vaccine is presented to and taken up by the mucosal surfaces in the intestine, thus providing a direct feed-vaccination (Floss et al., 2007).

In addition to the vaccine itself, substances that stimulate vaccine uptake and activity (so-called adjuvants) and the route of vaccine delivery (injection, inhalation, feed, etc.) are factors that are strongly investigated and further developed by biotechnological methods.

Diagnosis of Disease and Genetic Defects

Successful control of a disease requires accurate diagnosis. Modern biotechnology offers many applications to diagnose diseases caused by pathogens as well as diseases caused by intrinsic genetic disorders of an organism. The currently available and deployed techniques are outlined below.

The ability to generate highly specific antigens by recombinant DNA techniques has significantly raised the number of ELISAs that have the capacity to differentiate between immune responses generated by vaccination from those due to infection. This has made it possible to overcome one of the major drawbacks of antibody detection tests: the fact that, because antibodies can persist in animals for long periods, their presence may not indicate a current infection (Rege, 1996).

The advent of PCR has enhanced the sensitivity of DNA detection tests considerably. For example, PCR used in combination with DNA hybridization analysis has been shown to provide a sensitive diagnostic assay to detect bovine leukosis virus. This holds true for many other pathogenic organisms that are difficult to detect by serological methods (Schmitt and Henderson, 2005).

Other diagnostic techniques include nucleic acid hybridization assays and restriction endonuclease mapping. A good example of the specificity of nucleic acid hybridization is its application in distinguishing infections caused by peste des petits ruminants (PPR) virus from rinderpest, diseases whose symptoms are clinically identical and which cannot be distinguished with available serological reagents. This technique also allows comparison of virus isolates from different geographical locations.

Molecular epidemiology is a fast growing discipline that enables characterization of pathogen isolates (virus, bacteria, parasites) by nucleotide sequencing, allowing the tracing of their origin. This is particularly important for epidemic diseases, where the possibility of pinpointing the source of infection can significantly contribute to improved disease control. Furthermore, the development of genetic probes, which allow the detection of pathogen DNA/RNA (rather than host antibodies) in livestock, and the advances in accurate, pen-side diagnostic kits can considerably enhance animal health programs (FAO, 2001).

DNA testing is also being used to diagnose hereditary weaknesses of livestock. One available test identifies the gene which is responsible for Porcine Stress Syndrome in pigs. Animals that carry this gene tend to produce pale, low-quality meat when subjected to the stress of transport or slaughter. The identification of pigs that carry this gene excludes them from breeding programs, resulting in an overall decrease in the frequency of that gene within a population (Madan, 2005).

Another example of DNA analysis is the diagnosis of a mutation of Holstein cattle that causes leucoyte adhesion deficiency. Cattle with this condition suffer diseases of the gum, tooth loss and stunted growth. The disease is fatal, and animals usually die before reaching one year of age. The available test identifies carriers of the defective gene, allowing the elimination of such animals from breeding herds. Ideally, all animals used for breeding should be tested to exclude any carriers of the gene (Madan, 2005).

DNA Technologies in Animal Nutrition and Growth

Nutritional Physiology

Applications are being developed for improving the performance of animals through better nutrition. Specific enzymes can chemically modify feedstuffs and thus improve the nutrient availability and uptake by the animal. This lowers feed costs and reduces output of waste into the environment. Prebiotics (substances that stimulate microbial growth) and probiotics (live micro-organisms) as feed additives or immune supplements can either stimulate growth of beneficial microorganisms in the digestive system, or inhibit pathogenic gut micro-organisms and render the animal more resistant to them. Administration of the recombinantly produced growth hormone somatotropin (ST) results in accelerated growth and leaner carcasses in meat animals and increased milk production in dairy cows. Immunomodulation, i.e. administration of substances that stimulate or repress immune system function, can be used for enhancing the activity of endogenous anabolic hormones (FAO, 2001).

In poultry nutrition, possibilities for improvement include the use of feed enzymes, probiotics and antibiotic feed additives. The production of tailor-made plant products for use as feeds that are free from anti-nutritional factors through recombinant DNA technology is also a possibility.

Plant biotechnology may produce forages with improved nutritional value or incorporate vaccines or antibodies into feeds that may protect the animals against diseases.

Rumen Biology

Rumen biology has the potential to improve the nutritive value of ruminant feedstuffs that are fibrous, low in nitrogen and of limited value for other animal species. Biotechnology can alter the amount and availability of carbohydrate and protein in plants as well as the rate and extent of fermentation and metabolism of these nutrients in the rumen (FAO, 2001).

Methods for improving rumen digestion in ruminants include the use of probiotics, which is the supplementation of animal feed with beneficial live micro-organisms, to improve the intestinal microbial balance for better utilization of feed and for good health (Weimer, 1998). The added bacteria may improve digestion of feed and absorption of nutrients, stimulate immunity to diseases, or inhibit growth of harmful micro-organisms. Transgenic rumen micro-organisms could also play a role in the detoxification of plant poisons or inactivation of antinutritional factors. Successful introduction of a caprine rumen inoculum into the bovine rumen to detoxify 3-hydroxy 4(IH) pyridine (3,4 DHP), a breakdown product of the non-protein amino acid mimosine found in Leucaena forage is an example (Rege, 1996).

To conclude this section, it should be noted that many biotechnological applications are already available in the field of animal production and utilization. However, all techniques that have been successfully adopted so far are based on conventional biological methodologies, such as assisted reproduction and MAS. On the contrary, the approval and commercialization of techniques based on the creation of GM animals is only beginning to emerge. This is in sharp contrast with the field of transgenic plants, which have been in commercial use since the mid-1990s. Nevertheless, research in the field of GM animals is actively searching for solutions to the problems that are still linked to the production and application of GM animals. The approval of the first drug that is produced in a transgenic organism is a positive sign in this respect, and many other applications of GM animals, both in agriculture and medicine, are envisaged to follow in the near future.

GENETIC ENGINEERING OF MICRO-ORGANISMS OF INTEREST TO AGRICULTURE

Micro-organism is a term employed to cover all organisms that are not visible to the naked eye; this includes bacteria, archae, fungi, protists, green algae and small animals, such as plankton. The development of genetically modified micro-organisms of interest to agriculture is of significant importance. These micro-organisms may be used as gene transfer systems or donors and recipients of desirable genes. Micro-organisms functioning as gene transfer systems and as donors of genes have already been discussed.

Micro-organisms play important roles in different sectors of agriculture, food processing, pharmaceutical industries and environmental management. This development already started early in the history of humankind with the use of micro-organisms for the fermentation process. In the early 1970s, micro-organisms, notably E. coli, were used at the forefront of molecular biology research, resulting in the advent of recombinant DNA technology. The first recombinant protein, produced in a micro-organism and approved as a drug by the FDA in 1982, was human insulin. Since then hundreds of recombinant proteins have been engineered and expressed in micro-organisms and approved for use as pharmaceuticals. Nowadays, many microbial processes and pathways are understood and deciphered at the genetic level and can thus be subjected to specific and targeted genetic manipulation (Bull et al., 2000). Traditionally this approach largely depended on the identification and selection of random mutants with desirable characteristics; recombinant DNA technology presents a significant advance in this respect, since specific metabolic pathways can be manipulated with high precision and completely new functions can be introduced into an organism. The following sections give some examples of micro-organisms of economic importance that have been genetically modified through recombinant DNA technology.

Genetically Modified Micro-Organisms as Biopesticides and Biofertilizers

Biopesticides are defined as all substances derived from natural materials, including plants, animals and micro-organisms, that exhibit pesticidal activity. Such biological control agents are increasingly targeted for genetic enhancement due to a rising recognition of their potential benefits to modern agriculture (Rizvi et al., 2009). Biological control represents an alternative to chemical pesticides which have been subjected to much criticism due to their adverse impacts on the environment and human health. Therefore, there is a strong requirement to develop safer and environmentally amenable pest control using existing organisms in their natural habitats. Several such organisms, referred to as biological control agents, are available that offer protection against a wide range of plant pests and pathogenic microbial agents without damaging the ecosystem.

If biological control agents are to be effective in plant disease management, they must be efficacious, reliable and economical (Fravel, 2005). To meet these conditions superior strains are often required that are not found in nature. In this case the existing attributes of the biocontrol agents can be genetically manipulated to enhance their biocontrol activity and expand their impact spectrum.

The foreign genes used for transforming biological control agents can be integrated into the host genome or a plasmid. To express a heterologous gene in fungi or bacteria, the regulatory region of this gene must be modulated in its promoter and terminator regions in order to optimize the expression of the inserted gene in the new host. The addition of specific genes that are known to confer biocontrol activity may enhance or improve biocontrol capacity of organisms that do not naturally possess these genes.

Free-living bacteria associated with plants have been targeted to enhance their capacity either as soil inoculants or as biocontrol agents of plant pathogens. Studies on micro-organisms capable of enhancing plant growth have concentrated on the rhizosphere (root zone) whereas those on biocontrol target both the rhizosphere and phylloplane (leaf zone). Several important rhizobacteria including Sinorhizobium meliloti and Pseudomonas putidrii, both of which are excellent root colonizers, lack the ability to synthesize chitinases. Chitinases are enzymes that destroy chitin, a major component of fungi cells (Dahiya et al., 2006). Introducing genes encoding chitinases into their genome have enabled them to provide protection against plant pathogenic fungi. These two bacteria are good targets because of the unique beneficial characteristics they confer. Sinorhizobium is a symbiotic bacterium which stimulates formation of root nodules in legumes involved in fixing atmospheric nitrogen. Many Pseudomonas species in the rhizosphere environment produce siderophores which chelate iron ions, thereby increasing iron uptake by plants. The genetically modified commercial strain (RMBPC-2) of Sinorhizobium meliloti has added genes that regulate the nitrogenase enzyme involved in nitrogen fixation (Scupham et al., 1996).

The Trichoderma species are widely present in soils and are antagonistic to other fungi. T. harzianum, in particular, is a strong rhizosphere colonizer which is also able to parasitize plant pathogenic fungi. It establishes tight physical contact with hyphae of target fungi with the aid of binding lectins. Several extracellular enzymes, including chitinases, glucanases, lipases and proteases, are produced by the Trichoderma species, which has been improved further with the transfer of chitinase genes, notably from Serratia marcescens (Benitez et al., 2004).

The Agrobacterium radiobacter strain k84 protects plants against crown galls caused by A. tumefaciens strains carrying Ti-plasmids of the nopaline type. Protection conferred by A. radiobacter strain k84 is due to agrocin 84, an A nucleotide derivative. When taken up by A. tumefaciens, it inhibits DNA synthesis, resulting in cell death (Vicedo et al., 1993). A. radiobacter has an additional negative effect on soil pathogens by being a very effective rhizosphere colonizer. Although A. radiobacter strain k84 has been widely used commercially for a long time, there was concern about its long-term effectiveness as a biocontrol agent. This is because the gene encoding agrocin is carried on a transmissible plasmid, which can be transferred by conjugation to A. tumefaciens. In the event of agrocin-encoding plasmid transfer, recipient A. tumefaciens strains would no longer be subjected to biocontrol by A. radiobacter strain k84. This concern was addressed by modification of the agrocin-encoding plasmid to prevent its transfer to A. tumefaciens. The ensuing genetically engineered strain, known as A. radiobacter strain K1026, is a transgenic organism approved for use as a pesticide (EPA).

Bacillus thuringiensis (Bt) has been used as a biopesticide for many years. The insecticidal activity of B. thuringiensis is based on the production of crystalline protein inclusions during sporulation. The crystal proteins are encoded by different cry genes and are also known as delta-endotoxins. The protein crystals are highly toxic to a variety of important agricultural insect pests; when the proteins are taken up by susceptible insect larvae they induce lysis of gut cells, resulting in death of the larvae by starvation and sepsis (Roh et al., 2007). The toxin can be applied to plants as a spray consisting of a mixture of spores and protein crystals. However, the toxin has the disadvantage of fast degradation in sunlight. To overcome this limitation, different cry genes encoding the Bt toxin have been cloned and introduced into another bacterium, Pseudomonas flourescens. The transgenic P. flourescens strains are killed and used as a more stable and persistent biopesticide compared to the B. thuringiensis sprays (Herrera et al., 1994). Furthermore, cry genes are widely used to create transgenic plants that directly express the toxin and are thus protected from susceptible insect pests.

Baculoviruses (although, per definition, viruses are not micro-organisms) are also being manipulated to be effective biopesticides against insect pests such as corn borer, potato beetle and aphids (Szewczyk et al., 2006).

Micro-Organisms for Enhancing the Use of Animal Feeds

Animal digestive tracts harbor beneficial microflora that aid in the digestibility of various feeds. However, the function of these micro-organisms is easily affected by the unfavourable conditions within the gut, such as acidity and antibiotics used to treat pathogenic micro-organisms. Examples of gut micro-organisms that have been genetically modified include Prevotella ruminicola with a tetracycline resistance gene, cellulolytic rumen bacteria with acid tolerance, hind gut bacteria with cellulose activity, rumen bacteria transformed with genes to improve protein yield and yeast (Saccharomyces cerevisiae) containing a transgene from the closely related Saccharomyces diastaticus, allowing it to increase the digestibility of low-quality roughage in conventional feeds (Weimer, 1998). The major limitation to the use of these engineered organisms has been their establishment in the appropriate regions of the gut. Some organisms are being used as beneficial supplements in animal feeds. These are called probiotics and their use aims at improving digestion of feed and absorption of nutrients, stimulate immunity to diseases and inhibit growth of harmful micro-organisms (Gomez-Gil et al., 1998). For the improvement of silage, strains of the bacterium Lactobillus planetarium are being developed with the aim of increasing the lactate content and reduce the pH and ammonia content.

Micro-organisms are being extensively used as bioreactors for the production of hormones and other substances that enhance animal size, productivity and growth rates. The recombinantly produced hormone bST (bovine somatotropin) was among the first recombinant hormones commercially available. It can increase milk yield by as much as 10 to 15 percent when administered to lactating cows (Etherton and Bauman, 1998). Current development efforts are looking at a wide spectrum of genes that affect growth and productivity within the animal and which could be expressed in recombinant micro-organisms to obtain the respective protein in large quantities.

Genetically Modified Micro-Organisms in Food Processing

Many micro-organisms are being manipulated with the objective of improving process control, yields and efficiency as well as the quality, safety and consistency of bioprocessed products. Modifications target food enzymes, amino acids, peptides (sweeteners and pharmaceuticals), flavours, organic acids, polysaccharides and vitamins. A classical example is the production of the recombinant cheese making enzyme, chymosin, in bacteria. Its use was approved in 1990 in the United States, and nowadays 80 percent of US cheese is produced using this product (Law and Mulholland, 1991).

Genetically Modified Micro-Organisms in Bioremediation

Micro-organisms are widely used in cleaning up pollution such as oil spills or agricultural and industrial wastes by degrading them into less toxic compounds (Chatterjee et al., 2008). Some bacteria are being used as "bioluminescensors" that give luminescence in response to chemical pollutants. An example is the mercury resistance gene mer that is expressed in some bacteria and can result in bioluminescence upon encountering the presence of even very low levels of mercury in the environment.

A modified bacterium, Rhodopseudomonas capsulate, has the ability to grow rapidly in simple synthetic media. It is being used in advanced swine waste treatment plants in both Japan and Republic of Korea. The concentration of short chain fatty acids, one of the main sources of the bad odour of swine wastes, decreased dramatically after treatment. The residue after treatment can be used as a safe organic fertilizer. Several other applications of micro-organisms or plants for the purpose of bioremediation are being investigated.

To conclude this section, micro-organisms have always been at the forefront of research and development in the field of recombinant DNA methodology and biotechnology. This can be largely attributed to the comparative ease of culturing, analyzing and manipulating many micro-organisms. Nevertheless, many micro-organisms and their potential benefits remain unexplored and new species are being discovered regularly; therefore, research and development of biotechnological applications for micro-organisms in the field of agriculture and nutrition holds great promise for the future (Bull et al., 2000).

GMO DETECTION, IDENTIFICATION AND QUANTIFICATION METHODS

The precise and accurate detection of GMOs with high sensitivity in a given biological sample is of significant importance. This need for exact GMO detection methods will become increasingly clear in the following modules, when concepts for GMO surveillance, monitoring, biosafety measures and the implementation of relevant regulations are introduced.

Different stakeholders involved in the development, use and regulation of GMOs do at some point need to monitor and verify the presence and the amount of GMO material in agricultural products. Furthermore, comprehensive GMO monitoring also includes the analysis of biological samples, such as material derived from plant species that are related to an introduced GMO, to check for horizontal transfer of the transgene. This need has generated a demand for analytical methods capable of detecting, identifying and quantifying either the unique DNA sequences introduced or the protein(s) expressed in transgenic plants and animals. Thus, comprehensive GMO analysis techniques consist of three steps: detection, identification and quantification of GMO material (Anklam et al., 2002).

- Detection (screening for GMOs). The objective of this first step is to determine if a product contains GMO material or not. For this purpose, a screening method can be used. The result is a qualitative positive/negative statement. Analytical methods for detection must be sensitive and reliable enough to obtain accurate and precise results and reliably identify small amounts of GMO material within a sample.
- Identification. The purpose of the identification step is to reveal how many different GMOs are present in a sample, to precisely identify each single one and determine if they are authorized or not. Specific information (i.e. details on the molecular make-up of the GMOs) has to be available for the identification of GMOs.
- Quantification. If a food product has been shown to contain one or more authorized GMOs, it becomes necessary to assess compliance of the set threshold level regulations for the product in question. This is achieved by determining the exact amount of each GMO that has been found in the sample.

This testing framework is depicted in Figure 11, with labelling regulation thresholds of the European Union (adapted from Anklam et al., 2002).

In general, the range of sample types that need to be tested for GMO content is extensive and covers raw commodities as well as highly processed food. Furthermore, the number and variety of worldwide commercially grown GMOs is constantly increasing. Therefore, it is necessary to carefully approach each sample on a case-by-case basis and thus determine the most appropriate testing method (Jasbeer et al., 2008)

Every method developed for the detection of GMOs that is considered for routine use by official testing authorities and laboratories has to undergo several testing procedures to verify the analytical performance of the method (Michelini et al., 2008). The performance requirements of each method include applicability (if it is suited for the detection purpose), practicability (costs, material and machine requirements), specifity, dynamic range (range of different concentrations that can be detected), accuracy, limits of detection and quantification, and robustness (reproducibility of results) (Lipp et al., 2005).

Sampling Procedures

Irrespective of the analytical method selected for GMO detection, correct sampling procedures are critically important for reliable and reproducible

GMO analysis. An insufficient sampling plan can have strong effects on the reliability of the detected GMO level. In fact, the variance associated with the sampling procedure likely represents the major contribution to the overall variance of the detection procedure (Michelini et al., 2008). Furthermore, GMO material usually shows a heterogeneous distribution within the bulk of a product, additionally contributing to sampling-dependent variance. Raw materials, in particular, may show a significant heterogeneity, whereas processed materials and food usually display a more uniform distribution. The influence of the sampling strategy is more relevant when the overall GMO concentration is low.

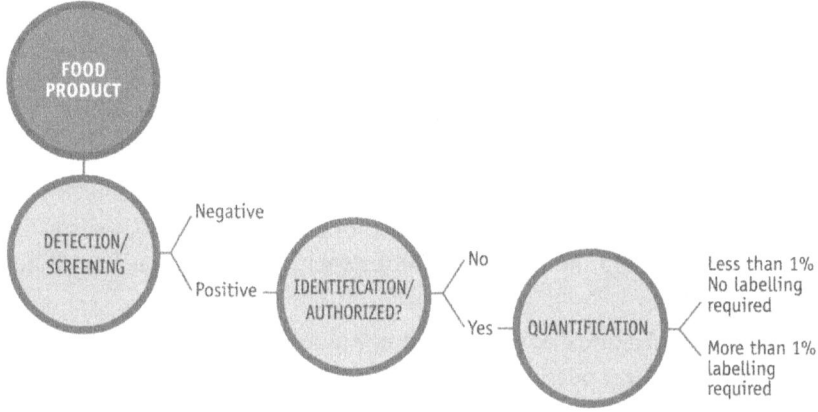

Figure 11. GMO detection framework. A comprehensive testing scheme consists of GMO detection/screening, GMO identification and GMO quantification.

Samples must therefore be taken in a manner that ensures that they are statistically representative of the larger lot volume or quantum of material. The sample size has to be adjusted to the required sensitivity and allow reliable GMO detection; the smaller the sample, the weaker the statistic significance of the testing procedure. So far, no generally accepted sampling guidelines have been established, and different control authorities employ different sampling schemes (Anklam et al., 2002). The major parameters that influence the sampling plan are lot size, lot heterogeneity, the defined tolerance level and the applied testing methods. Furthermore, parameters that are specific for each event, i.e. the size of the host genome, the copy number of the transgene event involved, and the amount of material that can be analysed in a single test, need to be taken into consideration (Lipp et al., 2005). Efforts are underway to define and internationally harmonize sampling plans, based on sound statistical requirements and analyses (Miraglia et al., 2004).

An example of a sampling plan, based on kernels, is calculated by Grothaus et al., 2006:

- To detect a lot concentration of 0.01 percent GMO material with 99 percent probability, 46 050 particles are required.
- To detect a lot concentration of 0.1 percent GMO material with the same confidence of 99 percent, 4 603 particles are required.
- If the confidence level for the detection of 0.1 percent GMO material is decreased to 95 percent, 2 995 articles are required.

Other calculations based on kernels state that at least 3 500 particles are required to detect a 1 percent contamination with a confidence level of 95 percent (Ovesna et al., 2008). The International Organization for Standardization (ISO) has issued a brochure on sampling procedures for GMO testing (ISO, 2006); a handbook from the International Seed Testing Agency (ISTA) on this topic is also available (ISTA, 2004). However, it should be noted that in any case the sampling strategy is highly dependent on the material analysed (raw, processed ingredients and processed food) and the required sensitivity, and it should be revised on a case-by-case basis. The establishment of a sampling plan that takes into account all relevant parameters and factors is a complex statistical procedure (refer to Remund et al., 2001 for further information). The reduction of sampling errors and thus more reliable test results are important for all involved parties: for consumers, the probability of consuming food that has been accepted although containing GMO above set threshold limits is reduced, and for producers the probability of lot rejection although the GMO content is below the set threshold limit is reduced as well. Therefore, the adoption and implementation of standardized sampling procedures should be of interest to all parties involved in GMO production, trading and consummation (Miraglia et al., 2004).

The actual sampling procedure consists of various steps: (1) sampling the lot of seed, grain or other material to obtain the bulk sample; (2) sampling the bulk sample to obtain the laboratory sample; (3) subsampling the laboratory sample to obtain the test sample; (4) homogenization (grinding etc.) of the test sample and sampling of the resulting meal to obtain the analytical sample; (5) extracting the analyte of interest (DNA, protein) from the analytical sample and using subsamples of it as final test portions (Lipp et al., 2005). The final test portion, for example in the case of PCR analysis of DNA, is typically around 100-200 ng of DNA which can be used in a single PCR.

Sample Preparation Procedures

The next step in GMO detection and quantification analyses, following the sampling procedure, is sample preparation for subsequent analytical procedures. Since all officially approved detection techniques rely on either DNA or protein-based assays, this section will focus on sample preparation and extraction techniques for these two compounds.

The ultimate aim of sample preparation is the isolation of DNA or protein with sufficient integrity, purity and quantity to allow reliable detection and quantification analyses. The choice of extraction procedure depends on the sample matrix, the target analyte and the type of analysis to be performed (GMO screening, identification or quantification). Different sample matrixes in combination with different extraction procedures have been shown to strongly influence the outcome of subsequent analyses (Cankar et al., 2006), therefore the appropriate extraction method needs to be determined for each individual sample (Jasbeer et al., 2008).

A further complication is the fact that samples often consist of highly processed food, i.e. the original plant or animal material has undergone several manufacturing steps. This might include simple mechanical procedures, such as milling, or complex chemical or enzyme-catalyzed modifications. Since proteins and DNA are likely to be degraded during such processing steps, the detection of these compounds in highly processed food requires sensitive and reliable detection methods (Michelini et al., 2004).

DNA Extraction Procedures

Compared with protein, DNA is a relatively stable molecule that can still be identified when it is partially degraded or denatured, contributing to its prime importance for GMO detection. It is possible to obtain DNA suitable for subsequent analyses from highly processed and refined food matrices; examples of failures to isolate DNA, to date, include refined soybean oil, soybean sauce and refined sugar (Jasbeer et al., 2008). DNA can be isolated as intact, high molecular weight DNA from fresh material, or as fragmented DNA from processed, old material (Ovesna et al., 2008).

Three parameters are characteristic for DNA extraction procedures:
- The DNA quantity: the overall amount of extracted DNA.
- The DNA quality: as mentioned, food processing has a negative effect on DNA quality. Heat exposure, enzymatic degradation or unfavourable chemical conditions contribute to DNA fragmentation or damage. Target sequences for subsequent analyses, therefore, often do not exceed 100-400 bp in length.

- DNA purity: DNA in food matrices might be severely contaminated, by substances such as polysaccharides, lipids or polyphenols. Obtaining DNA of high purity is important to avoid complications or misleading results during subsequent analyses.

The key steps in sample preparation include homogenization of the material, chemical or enzymatic pretreatment, extraction and purification (Jasbeer et al., 2008). Concerning plant material, small aliquots of 100-350 mg are sufficient for DNA isolation, given that this laboratory sample is representative of the field sample and has been correctly homogenized (Anklam et al., 2002).

Five DNA extraction methods are commonly used, depending on the food matrix to be analysed. These are the DNeasy Plant Mini Kit (Qiagen), Wizard extraction (Promega), GENESpin Kit (GeneScan), cetyl trimethylammonium bromide (CTAB) based extraction, or a combination of CTAB-extraction with DNA-binding silica columns (Michelini et al., 2008). It is important to carefully determine the extraction method that is most suited for the food matrix in question in order to obtain reliable and reproducible extraction and analysis results.

Protein Extraction Procedures

In contrast to DNA, proteins are very heat-labile molecules. Furthermore, they are easily affected by chemical treatments or enzymatic degradation. The detection of a specific protein depends on the recognition of this protein by an antibody directed against that protein. If the target protein is degraded or denatured (i.e. loses its specific 3-dimensional shape), this antibody-mediated detection can no longer be performed. Therefore, it is not possible to reliably and reproducibly detect and quantify proteins in complex food matrices, such as processed agricultural material and food products, that have been subjected to mechanical, thermal, enzymatic or chemical processing (Anklam et al., 2002).

Due to these limitations, protein analysis is only applicable for materials in their raw state (Jasbeer et al., 2008). However, the basic steps in sample preparation are the same as in DNA extraction: material homogenization, pretreatment, extraction and purification.

GMO Detection by Phenotypic Characterization

Phenotypic characterization is possible if the inroduced transgene(s) result in the absence or presence of a specific trait that can be screened by analysing the phenotype of the organism. Detection methods using this approach are referred to as bioassays. This approach can be used, for example, to test for the presence or absence of herbicide resistance transgenes. One such test is based on the germination of seeds in the presence of the herbicide of interest and subsequent analysis of germination capacity. Herbicide assays are considered to be accurate and inexpensive. Controls, including seeds with or without the trait targeted, should be included in all samples tested. Typically, a test sample consists of 400 seeds. The test accuracy is dependent on the overall germination efficiency of the seeds: the higher the germination efficiency, the higher the confidence level of the test. Obviously, only viable seed or grain can be tested (no processed products), and each test requires several days to complete. Furthermore, bioassays require separate tests for each trait in question and at present such tests will not detect non-herbicide tolerance traits. Therefore, the tests are only of limited value for inspection authorities.

Molecular Detection and Quantification of GMOs – DNA-based Methods

As stated above, the methods of choice for detecting and quantifying GMO material on a molecular level are based on detecting either the inserted, foreign DNA fragments or the novel proteins that are expressed from this DNA. Methods for the detection of foreign DNA rely mainly on PCR that allows amplification and detection of specific DNA fragments from the entire genome. Another advantage of DNA-based detection is the finding that there is usually a linear relationship between quantity of GMO present in a sample and quantity of transgenic DNA, thus it can be used to accurately quantify the amount of GMO material present in a sample. Finally, the stability of DNA and the extractability of suitable DNA even from highly processed food matrices contribute to its prime importance for GMO analysis.

PCR-based GMO Detection

As evident from the name, GMOs are the result of genetic modification. Therefore, the most suitable GMO detection methods are those that directly target the modification itself – the modified DNA.

Polymerase chain reaction (PCR), including variants of the technique such as competitive PCR and real-time-PCR, is the method of choice for DNA-based GMO detection, identification and quantification (Lipp et al., 2005). Due to its very high sensitivity, PCR is well suited for the analysis of processed food matrices containing degraded DNA or material that has only low GMO content.

PCR-based GMO detection is dependent on detailed knowledge of the molecular makeup of a GMO, i.e. the sequence of the transgene and, optimally, the transgene integration site in the host genome. For authorized and commercially released GMOs, such information is available in public databases such as AGBIOS (Ovesna et al., 2008). In general, a typical gene construct for the production of a GMO consists of at least three elements: a promoter to drive expression of the inserted gene(s), the inserted/altered gene(s), and a terminator as a stop signal behind these genes. Such sequences can be specifically detected in a PCR analysis.

If no detailed sequence information about a GMO is available, PCR-based methods rely on the detection of commonly used genetic elements. Such frequently used elements are, for example, the CaMV 35S promoter, the A. tumefaciens nopaline synthase terminator (nos3'), or the kanamycin resistance marker gene (nptII) (Michelini et al., 2008). Focusing on such sequences for routine GMO screening purposes is promising, since many commercially available GMOs contain these elements, or varieties thereof, and can thus be detected in standard screening procedures.

GMO detection is frequently based on the detection of the P-35S and nos3' genetic elements; however, several approved GMOs do not contain the P-35S or nos3' sequences and additional target sequences are needed to detect their presence. Furthermore, to detect as many variants of a GMO marker as possible (there are at least eight variants of P-35S used in GM crops), a careful choice of primers is required. In addition, it should be noted that the detection of a common GMO marker solely indicates the presence of material derived from a GMO within a sample, but does not provide any information about the species or the engineered trait (Jasbeer et al., 2008).

Most PCR-based GMO detection methods include a positive control primer set for the amplification of a reference gene. This is often a so-called housekeeping gene, which is present in (and unique to) all varieties of the investigated species (Miraglia et al., 2004). Examples include the lectin gene in soybean or the invertase gene in maize. If a strong signal cannot be obtained with the positive control primer set, then there may be problems with the integrity or purity of the extracted DNA. Negative controls, for example samples with all necessary PCR ingredients but without template DNA, should also be included routinely to test for contamination with undesired DNA.

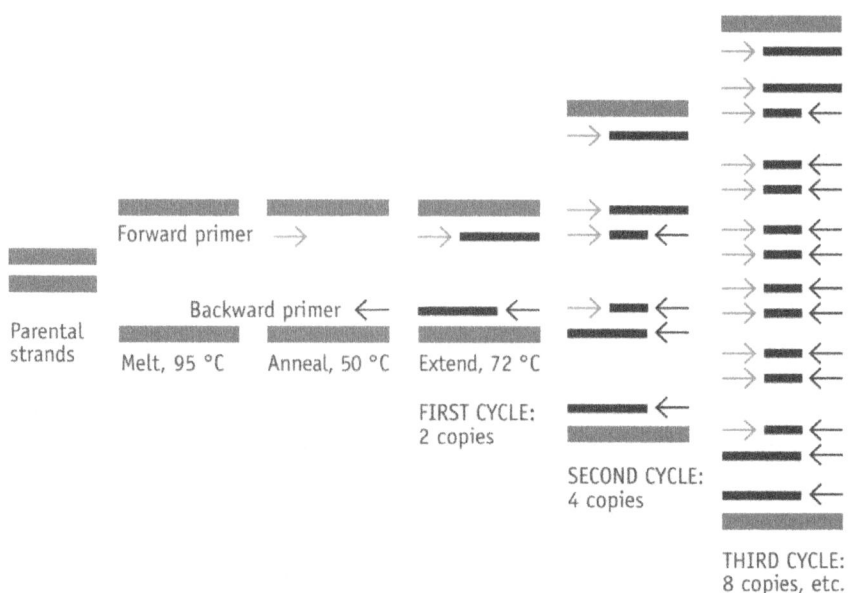

Figure 12. The polymerase chain reaction.

The outcome of a PCR can be evaluated by a variety of methods. Most frequently, amplified DNA fragments are subjected to agarose gel electrophoresis, a method to separate and visualize DNA fragments according to size. Since the expected size of a given target sequence is known, the presence of a fragment of that size indicates the presence of that target sequence in the original sample. If no fragment of the expected size is obtained, the sample did not contain the target sequence (given that the PCR worked well). To further verify the identity of an amplified fragment, it can be subjected to hybridization experiments with a complementary sequence, to analytical restriction enzyme digest, or to sequencing (Michelini et al., 2008).

PCR-based GMO Identification

Following a positive result from a GMO screening procedure, the next step is the unequivocal identification of the GMO(s) contained in a sample and the genetic modification event(s) involved. This can be achieved by PCR as well; however, compared with GMO detection, GMO identification is even more dependent on detailed information about the exact genetic modification of a GMO. In fact, this is a major limitation of PCR-based GMO detection and identification: if no such information is available, the GMO will not be detected or identified. Several approaches for GMO identification by PCR exist, and they are summarized below:

- Gene-specific PCR: In a gene-specific PCR, primers are used that lead to the amplification of a fragment from one gene of the transgenic element. This is rather unspecific, since many GMOs are engineered to contain the same, favourable genes. Thus, this method will fail to distinguish between these GMOs. This approach is therefore only useful if the target gene is present in only one GMO within a sample.
- Construct-specific PCR: This approach is more specific than gene-specific PCR. It is based on primers that target the junctions between different elements of the transgene insert, e.g. between the promoter and the gene or between different genes of the insert. Many GMOs contain identical genes, but the exact layout of their transgenes may differ, for example by a different arrangement of the genes or by the use of different promoters and terminators. By using construct-specific PCR, these different constructs, and thus GMOs, can be distinguished and identified.
- Event-specific PCR: Event-specific PCR is the most specific GMO identification strategy. Event, in this case, refers to the insertion of a transgene cassette into the host genome. The integration site is usually specific for each GMO. PCR primers, in this case, target the junction between the transgenic insert and the adjacent host genomic DNA. In most cases, this allows GMO identification with high certainty.

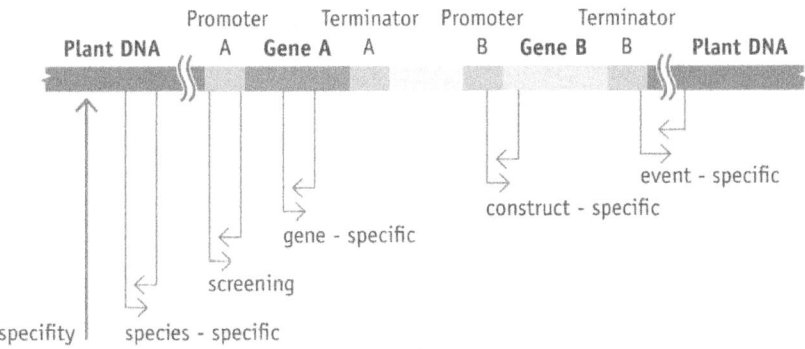

Figure 13. Different PCR strategies with increasing specificity

Due to the dependency of PCR on detailed genetic information about GMOs there is a strong need for a continuous survey of all data available on GMOs – especially the introduced genetic elements and their integration sites. This applies not only for GM products approved for market release but also for any other GMO released for field trials worldwide. Only complete and accessible GMO information can guarantee comprehensive monitoring, detection and identification of GMOs.

PCR-based GMO Quantification

The third step in GMO analysis, following detection and identification, is GMO quantification. Quantifying the GMO content in a sample is important to assess compliance with specific threshold levels for GMOs established by biosafety regulations. The typical approach to quantification utilizes one or more of the broad-spectrum primer sets that target common transgenic elements in GMOs. However, since different GMOs possibly contain these common elements in different numbers, accurate determination of GMO content cannot rely on the use of these common sequence elements alone. Quantification based on event-specific primers is therefore the most accurate means of obtaining quantitative results on GM content.

In general, two quantification approaches can be distinguished: absolute quantification and relative quantification. Absolute quantification, as the name suggests, yields absolute values of an analyte within a sample, e.g. how many milligrams of DNA could be extracted from a sample? This quantification is dependent on the sample size. The second approach is relative quantification: this is a measure of the amount of a substance compared to another substance, e.g. how many copies of transgene DNA per total DNA, or how many copy numbers of a gene per genome? Importantly, the final value obtained is a percentage, and the measurement is independent of the analysed sample size. Relative quantification is required for all GMO-related questions, such as compliance with labelling regulations (Jasbeer et al., 2008).

Use of Conventional PCR Quantification

One possibility for DNA quantification based on conventional PCR is double competitive PCR (DC-PCR). In competitive PCR, one primer pair is used to amplify both the target GMO template DNA and a synthetic template DNA fragment that is added to the same reaction mixture. The second fragment, which has a different size from the GMO target DNA (≤ 40 bp), is called the competitor. By conducting a series of experiments with varying amounts of the added synthetic DNA, it is possible to determine the amount of target GMO DNA in the sample. The competitor DNA serves as internal standard, and is added in different concentrations to the reaction mixture (an experimental setup known as titration). Following PCR amplification, the amplified fragments are visualized by agarose gel electrophoresis. The ratio of the two amplification products then represents the ratio of the initial two template sequences in the PCR mix. In other words, when the two products show equal amplification intensities, the amounts of initial template DNAs were the same. Since the amounts of added competitor DNA are known, this allows quantification of the target DNA in the sample.

Competitive and double-competitive PCR methods are semi-quantitative as a standard is required for comparison. In these cases the standard is the known amount of synthetic DNA. Consequently, the results will only indicate a value below, equal to or above a defined concentration of the standard.

Real-time PCR for GMO Quantification

Another strategy that improves accuracy, specificity and throughput of quantitative PCR is real-time PCR. This technique was originally developed in 1992 and is rapidly gaining popularity due to the introduction of several complete real-time PCR instruments and easy-to-use PCR assays. A unique feature of this PCR technique is that the amplification of the target DNA sequence can be followed during the entire reaction by indirect monitoring of product formation. To this end, the conventional PCR reaction has been modified in order to generate a constantly measurable signal, whose intensity is directly related to the amount of amplified product. This signal is usually fluorescence, which is produced by an interaction between newly amplified DNA with certain added fluorophores. The increases in fluorescence during the reaction, that correspond to increasing concentrations of target DNA, are automatically measured, displayed on a computer screen, and can be analyzed using suitable software.

DNA quantification by real-time PCR is based on the following principle: the PCR reaction mixture is submitted to several cycles of the reaction, until a fluorescent signal is encountered that is statistically significant above the noise level. The number of PCR cycles necessary to reach this threshold is recorded and referred to as Ct (cycle threshold) value. It is important to measure the Ct value in the exponential phase of the amplification procedure. During this stage, the Ct value is inversely proportional to the initial amount of template DNA molecules. In other words, a sample with many template molecules will reach a certain fluorescence threshold level faster than a sample with fewer molecules. For example, if a sample contains twice as many template molecules as a second sample, it will reach the threshold one cycle before the second sample since the amount of DNA is doubled during each reaction cycle. Thus, a low Ct value corresponds to a high initial concentration of target DNA.

Quantification of GMO DNA in a sample by RT-PCR is based on a combination of two absolute quantification values; one for the GMO target transgenic DNA and one for a species-specific reference gene. The GMO content in a sample can be calculated as a percentage using these two absolute values (Michelini et al., 2008). Careful choice of suitable reference material is therefore of crucial importance for determining exact ratios of GMO to non-GMO material. Furthermore, it is important to know the copy number of the inserted transgenic sequences. The detection limit of real-time PCR is very high; for corn, a detection of 0.01 percent GM corn versus non-GM corn has been demonstrated (Anklam et al., 2002).

Several types of fluorescent probes for quantification of DNA using real-time PCR are currently available. One can discriminate between two classes of fluorophores: general DNA-binding dyes and fluorescent reporter probes. The first ones, a prominent example being SYBR Green, bind to double-stranded DNA in an unspecific manner and the resulting dye-DNA complex shows fluorescence. Since the overall amount of dsDNA in a PCR reaction increases, so does the intensity of fluorescence. The second type of probe consists of an oligonucleotide that is complementary to the target sequence, and a fluorophore and a quencher dye attached to it (e.g. the Taqman system). In the intact probe, the fluorophores' fluorescence is inhibited by the proximity of the quencher dye. During the annealing step of the PCR cycle, the oligonucleotide anneals to the target sequence between the two primers. Upon passage of the DNA polymerase during the elongation step, the oligonucleotide is cleaved and the fluorophore is liberated from the quencher dye. Thus, with increasing PCR cycles, the intensity of fluorescence increases as well. The latter, reporter-probe based method has the advantage that only the amplification of the desired target sequence is measured, while non-specific DNA binding dyes also react with non-specific PCR amplification products or other DNA hybrids (Miraglia et al., 2004).

Confirmatory Assays

Following PCR analysis, the identity of the amplicon needs to be confirmed and verified to ensure that the amplified sequence indeed represents the target sequence and is not an unspecific PCR artifact. Several confirmatory assays are available and commonly applied. Agarose gel electrophoresis, the simplest technique, can be applied to check if the amplicon is the expected size. However, it cannot be excluded that a PCR artifact, by coincidence, has the same size as the target sequence. To further verify amplicon identity, it can therefore be subjected to restriction enzyme digest, since every DNA sequence has specific restriction profile. A further assay is Southern blotting, where the target amplicon is subjected to gel electrophoresis, transferred from the gel to a membrane, and hybridized with a complementary, labelled DNA probe; only the correct target sequence will yield a signal from binding of the complementary probe. A further possibility is nested PCR, where two primer pairs and two rounds of amplification are used: the second primer pair anneals within the target region of the first amplification, thus only the correct first amplification product will yield a second amplification product. The ultimate confirmatory assay is sequencing of the amplicon; however, this is rather expensive and requires special equipment that is not available in standard laboratories.

As stated above, PCR is able to amplify and thus identify very small amounts of initial target DNA. This implies that PCR is very sensitive to contamination with undesired DNA, possibly yielding false results in subsequent analyses. Therefore, high caution must be taken during all steps of PCR sample preparation and reaction setup to avoid cross-contamination. This already begins at sampling and sample preparation: it might already be sufficient to use the same grinding device for homogenization of two samples to produce contamination, even if no visible traces were left. Therefore it is of major importance to thoroughly clean and monitor all devices that come in contact with samples and that could potentially contribute to cross-contamination.

Molecular Detection and Quantification of GMOs – Protein-based Methods

A GMO is typically characterized by the introduction of novel genes, which direct the expression of novel proteins. Therefore, the second approach to detect GMOs is not based on detection of the modified DNA, but on the novel and newly expressed proteins. However, whereas modified DNA can be detected in all parts of a transgenic organism at all times, this may not be the case for proteins: the genetic modification might not be directed at the production of novel proteins, protein expression levels might be too low to be detected, and proteins might only be expressed in certain parts of a plant or during certain stages of development (Jasbeer et al., 2008).

A further limitation for protein-based GMO detection is the susceptibility of proteins to heat denaturation and to chemical, enzymatic or mechanical degradation. Since protein detection requires intact, correctly folded protein molecules, it is only possible to reliably detect proteins in raw, non-processed commodities (Miraglia et al., 2004).

Protein-based methods rely on a specific binding between the protein of interest and an antibody against that protein. The antibody recognizes the protein molecule, binds to it, and the resulting complex can be detected, for example by a chromogenic (color) reaction. This type of assay is referred to as immunoassay, since antibodies are the molecules that are produced during an immune reaction to recognize and eliminate foreign (pathogenic) molecules. The main technique applying this procedure is called ELISA (enzyme-linked immunosorbent assay, Figure 14). The antibody required to detect the protein can only be developed with prior access to the purified protein; the protein can be purified from the GMO itself, or it can be synthesized in a laboratory if the composition of the protein is known in detail. Immunoassays can be applied both for detection and quantification of protein, over a wide range of protein concentration. Such assays are available for many proteins that are expressed in commercially released GMOs (Michelini et al., 2008).

Enzyme-Linked Immunosorbent Assay (ELISA)

In ELISA, a protein-antibody reaction takes place in solution on a solid support (plastic plates) and a protein-antibody complex is formed. This complex is usually visualized by adding a second antibody that binds to the first antibody, and that is linked to a certain enzyme. This enzyme can catalyse the reaction of a specific substrate, which is added to the solution, to a colored product (chromogenic detection). The intensity of the color can be measured photometrically and used for quantitative assessments of protein concentration. ELISAs are available for several frequently engineered proteins in GM plants, including neomycin phosphotransferase (nptII), 5-enolpyruvyl-shikimate 3-phosphate synthase (EPSPS), the Bt insecticide Cry1Ab and phosphinotricin acetyltransferase (PAT) (Jasbeer et al., 2008).

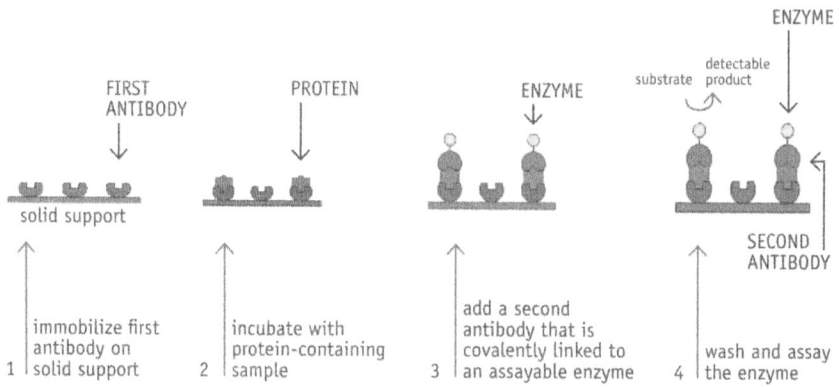

Figure 14. Enzyme-linked Immunosorbent. Assay. In this case a Sandwich-ELISA is depicted. A first antibody is immobilized on a solid support, followed by incubation with the target-protein containing solution. After a washing step (not shown), the second antibody, coupled to an assayable enzyme, is added and binds to the immobilized target protein. Finally the amount of bound secondary antibody, and thus target protein, can be assayed using the attached enzyme, which is usually done colorimetrically. In an easier approach, the target protein can be immobilized directly onto the plate, without a primary antibody.

Some ELISA plates are supplied with a calibration of known concentration of target protein in solution and a negative control defined by the absence of the target. These standards will exhibit distinctively different intensities of a given color at the different concentrations of target molecules provided. By comparing the intensity of color of the sample tested for GMO target molecules with that of the standards, it is possible to work out the concentration range of the target. These immunoassay measurements are semi-quantitative. Quantitative measurements can, however, be obtained by using a microplate reader which measures the absorbance of all samples and standards at the same time. This results in a very high precision of data acquisition and subsequently a precise calculation of target protein concentration in the test samples.

A major advantage of ELISA is the high specificity of the protein-antibody recognition, which allows accurate identification of proteins. Furthermore, they are fast, require only low work input, can be performed automatically to a large extent, and require only small investments in equipment and personnel.

However, ELISA may be around 100 times less sensitive than DNA-based methods, although detection of 0.01 percent of GM material has been described (Grothaus et al., 2006). Furthermore, initial development and validation of a test for a specific protein is more time-consuming, and the supply of antibodies, which are derived from laboratory animals, is a limiting factor (Jasbeer et al., 2008). Furthermore, protein detection and antibody affinity might be affected by the individual matrix under examination (Anklam et al., 2002).

Lateral Flow Devices and Dip Sticks

Lateral flow devices and dip sticks are variations of the technology that ELISAs are based on; paper strips or plastic paddles on which antibody is captured on specific zones are used to detect protein targets derived from GMOs. The strip is dipped into vials containing solutions of the sample to be tested. Each dip is followed by rinsing; the positive reaction is a color change in a specific zone on the stick. Recent improvements of the dip stick have produced lateral flow strips in which reagents are transported through nylon membranes by capillary action. Antibodies specific to the target protein are coupled to a colored reagent and are incorporated into the lateral flow strip. When the strip is brought into contact with a small amount of the sample containing the target protein, an antibody-antigen complex is formed with some of the antibody. The membrane contains two capture zones, one for the bound protein and the other for the colored reagent. A colored band appears in the capture zone corresponding to the bound antibody-protein complex and colored reagent. Appearance of a single colored band in the membrane is a negative test for the presence of the protein targeted. The presence of two bands represents detection of the target (Grothaus et al., 2006).

These tests are available as kits and do not require major equipment or training, and thus represent a rapid GMO testing possibility. Sample preparation only involves homogenization of the sample and mixing with the reagents contained in the kit (Jasbeer et al., 2008).

Molecular Detection and Quantification of GMOs – Other Methods

Several other methods for the detection and quantification of GMOs have been proposed or are in developmental stages. Some of them are presented below – however, the main approved technologies for GMO analysis are PCR-based techniques and ELISA.

Chromatography and Near Infrared Spectroscopy

If the chemical composition of a GMO has been altered, for example fatty acid or triglyceride content, chemical methods based on chromatography or near infrared spectroscopy may be applied to detect these changes. These methods will detect differences in the chemical profile between GM organisms and conventional organisms. The applicability of such approaches has been demonstrated by investigating the triglyceride pattern of oils derived from GM canola by high performance liquid chromatography (HPLC). Triglyceride patterns and content can be compared between GM and non-GM samples. However, it should be noted that such techniques are only applicable when significant changes occur in the biochemical composition of GM plants or derived products. In addition, such methodologies only offer qualitative detection and no quantification (Anklam et al., 2002). In particular, the addition of GM-derived products or raw material in small quantities to a larger lot of conventional material are probably not detectable given the sensitivity of the methods currently used.

Microarrays

Microarray technology (DNA-chip technology) has been developed in recent years for automated rapid screening of gene expression profiles and sequence variation of large numbers of samples. Microarray technology is based on the DNA hybridization principle, with the main difference that many (up to thousands) specific probes are attached to a solid surface and can be simultaneously detected. Different formats have been developed, including macroarrays, microarrays, high-density oligonucleotide arrays (gene chips or DNA chips) and microelectronic arrays.

GMO chip kits are designed to detect species-specific DNA of plants and viruses, frequently used transgene construction elements and specifically introduced genetic modifications, and thus allow the identification of approved and non-approved GMO varieties. One example of a GMO chip version that has been designed and tested for its applicability is capable of detecting species-specific DNA from soybean, maize, oilseed rape, rice, CaMV and several GMOs, including RR-soybean, Maximizer Bt 176 maize, Bt11 maize, Yieldgard Mon810 maize and Bt-Xtra maize. In addition, GMO chips allow the detection of all GMOs that contain the widely used CaMV 35S promoter, Nos-terminator, nptII, bar, and pat genes (Leimanis et al., 2006). Microarrays, in general, thus allow the detection, identification and quantification of a variety of GMOs in a single experimental setup.

Summary of GMO Analysis: Limits and Outlook

As stated in the introduction, the field of GMO detection has a high relevance for all involved parties: research and development, producers, traders, consumers and legislation. Further progress in sampling and detection techniques and in traceability strategies needs to be made to enable adequate implementation and maintenance of GMO-relevant legislation and labelling requirements (Miraglia et al., 2004). Promotion and implementation of reliable, international traceability strategies and agreements may also increase public trust in the transparency of GMOs and related products.

Summary of DNA and Protein-Based Techniques

To summarize the previous sections, DNA and protein-based methods are currently the techniques of choice for GMO analysis. A PCR analysis can take between one to ten days and costs range from 100 to 400 euros. In comparison, an on-site ELISA takes two to eight hours and costs approximately 10 Euros; ELISA-based dipsticks take a few minutes to complete and cost around 3 euros (Miraglia et al., 2004).

DNA-based analysis offers several advantages, including:

- A wide range of applications, from initial GMO screening to event-specific detection;
- The genome is the same in all cells of an organism, i.e. Every part of an organism can be analysed;
- Relative quantification, as required for labelling legislation, is possible;

- DNA is comparatively stable and can be isolated from a wide range of raw and processed matrices;
- A very high sensitivity.

Disadvantages of DNA-based methods include:
- The need for trained staff to operate high-end equipment;
- Expensive, time-consuming and relatively unsuitable for on-site testing;
- DNA may be removed or degraded by certain processing procedures; certain food ingredients possibly interfere with DNA amplification and detection;
- PCR is very susceptible to cross-contamination;
- If no detailed sequence information of a GMO is available, DNA-based analysis is not possible.

Protein-based analysis offers the following advantages:
- Comparatively cheap and less skilled personnel required;
- Cheaper and less sophisticated equipment needed;
- Fast conductance;
- Quantification is possible;
- Comparatively robust and simple assay formats;
- Suitable for batch analysis of samples;
- Possible to conduct on-site tests.

The disadvantages of protein-based analysis include:
- Inferior sensitivity compared to DNA-based methods;
- The development of antibodies is difficult, expensive and requires skilled staff and equipment;
- Only samples containing intact protein, i.e. fresh material, can be analyzed;
- Not possible to distinguish different events that produce the same protein (i.e. less specific than DNA-based methods);
- Protein expression levels in a GM organism may vary significantly in a temporal and spatial manner;
- No relative, but absolute quantification;
- Expression levels of target proteins may be too low to be detectable;

Reactivity of the antibody may be affected by other matrix components.

Thus, a careful evaluation of the most suitable analysis technique for a certain product should be performed to ensure that potential GMO contents are reliably, reproducibly and with high sensitivity detected and quantified. The choice of the technique may depend on a variety of factors, including the purpose (exact quantification for labelling legislation versus a simple yes/no result, GMO monitoring), the need for laboratory or on-site testing, financial background (including availability of personnel and equipment), exact GMO identification or just stating general GMO presence, the speed of analysis, composition of the food matrix to be analyzed, etc. At present, however, PCR-based methods are the most widely applied and validated for GMO analysis purposes.

GENES OF INTEREST TO AGRICULTURE

Transgenic crops with novel agronomic and quality traits are grown in many developed and developing countries. A recent analysis of the current application of transgenic crops and the development over the last decade is provided by the International Service for the Acquisition of Agri-Biotech Applications (James, 2008). For a detailed account on the nature and extent of utilization of the various GM crops, one can consult online databases such as AGBIOS (http://www.agbios.com/dbase.php). The AGBIOS Web site includes details of the transgenes, the scientific background underpinning the traits and information on environmental and food safety issues of a variety of GM plants. A recent publication by the European Commission Joint Research Centre provides information about GM crops that are in the pipeline and expected to be marketed in the short to medium term, i.e. up to 2015 (Stein and Rodriguez-Cerezo, 2009). The database established by the authors is also available online at http://ipts.jrc.ec.europa.eu/publications/pub.cfm?id=2199. By surveying information in these and similar databases it is possible to get information on the genes that have been used for the generation of transgenic crops, how these crops are commercially used and which additional crops are in developmental stages, in field trials or awaiting approval for commercial release. Each GMO is assigned a Unique Identifier, i.c. a code that allows allows direct identification of the GMO (Commission Regulation EC 65/2004).

Herbicide Tolerance Genes

Glyphosate Herbicide Tolerance

The genetically modified glyphosate resistant crops contain a gene encoding the enzyme EPSPS, obtained from a strain of the soil inhabiting bacterium Agrobacterium tumefaciens. The EPSPS enzyme is an important part of the shikimate biochemical pathway which is required to produce aromatic amino acids, which plants need to grow and survive. EPSPS is also constitutively present in plants, but the enzyme is inhibited by binding of glyphosate. Conventional plants treated with glyphosate cannot produce the aromatic amino acids and die, whereas EPSPS from A. tumefaciens does not bind glyphosate and allows plants to survive the otherwise lethal effects of the herbicide (Tan et al., 2006; Gianessi, 2008).

Glufosinate Ammonium Herbicide Tolerance

Glufosinate ammonium is the active ingredient in the PPT herbicides. Glufosinate chemically resembles the amino acid glutamate and functions by inhibiting the enzyme glutamate synthase, which converts glutamate to glutamine. Glutamine synthesis is also involved in the ammonia detoxification of glufosinate resulting in reduced glutamine levels and increases in ammonia concentration. Elevated levels of ammonia damage cell membranes and impair photosynthesis. Glufosinate tolerance is the result of introducing a gene encoding the enzyme phosphinothricin-acetyl transferase (PAT). The gene was originally obtained from the soil actinomycete Streptomyces hygroscopiens. The PAT enzyme catalyses detoxification of phosphinothricin by acetylation (Duke, 2005; Tan et al., 2006).

Sulfonylurea Herbicide Tolerance

Sulfonyl urea herbicides, such as triasulfuron and metsulfuron-methyl, target the enzyme acetolactate synthase (ALS), also called acetohydroxyacid synthase (AHAS), thereby inhibiting the biosynthesis of the branched chain amino acids valine, leucine and isoleucine (Tan et al., 2005). This results in accumulation of toxic levels of the intermediate product alpha-ketoglutarate. In addition to the native ALS gene, herbicide tolerant crops contain the ALS gene from a tolerant line of Arabadopsis thaliana. This variant ALS gene differs from the wild type by one nucleotide and the resulting ALS enzyme differs by one amino acid from the wild type ALS enzyme. Still, this is sufficient to confer resistance to these herbicides, and provides an impressive example for the complexity and sensitivity of genes and proteins and the effects of mutations.

Oxynil Herbicide Tolerance

Oxynil herbicides and bromoxynil are effective against broad leaf weeds. Transgenic herbicide resistant crops contain a copy of the bxn gene isolated from the bacterium Klebsiella pneumoniae. The gene encodes a nitrilase which hydrolyses oxynil herbicides to non-phytotoxic compounds (Duke, 2005).

A recent development in herbicide tolerance is the development of plants containing several tolerance genes, allowing cocktails of different herbicides to be used (Green et al., 2008). This technology is referred to as trait or gene stacking. Ideally, it will become possible to introduce not only herbicide tolerance traits, but also traits conferring insect resistance or quality traits (Halpin, 2005). One possible approach to this end is the development of artificial plant minichromosomes, capable of encoding many different, complex genes and regulatory sequences (Yu et al., 2007).

Resistance to Biotic Stresses

Among insect pests, Lepidoptera (moths and butterflies) represent a diverse and important group. Most insect-resistant transgenic crop varieties developed so far target the control of Lepidoptera, predominantly using transgene cassettes, including toxin-producing cry-type genes obtained from strains of the soil bacterium Bt. The Bt proteins bind to specific sites on the gut lining in susceptible insects (de Maagd et al., 1999). The binding disrupts midgut ion balance which eventually leads to paralysis, bacterial sepsis and death. Important to note is that the original Bt cry-genes have been extensively modified, for example by deleting spurious splicing signals and optimizing the GC content, to improve the expression level in plants. Many cry genes exist that confer resistance to insects other than Lepidoptera. In addition to Bt cry genes, protease inhibitors, neuropeptides and peptide hormones that control and regulate the physiological processes of several insect pests have become candidates for developing insect-resistant crops. Other biocontrol toxins currently studied are chitinases, lectins, alpha-amylase inhibitors, cystatin and cholesterol-oxidase and glucosidase inhibitors (Christou et al., 2006; Ranjekar et al., 2003).

Among disease-causing organisms, viruses have received a lot of attention concerning the development of transgenic crops. This has been possible since the discovery of pathogen-derived resistance, where the expression of a viral protein (e.g. coat protein, replicase, helicase enzyme, etc.) in a transgenic plant renders that plant resistant to the virus (Prins et al., 2008). As a result many viral genes have been cloned and used to transform crops. Genes encoding chitinases and glucanases have been used to generate plants resistant to fungal and bacterial pathogens, respectively. Other strategies for conferring resistance to pathogens in transgenic crops include genes for phytoalexin production pathways which are involved in pathogen-induced infection and defence, and R genes (resistance genes) which have been identified as responsible for additional defence mechanisms in plants (Campbell et al., 2002).

Tolerance to Abiotic Stresses

So far there are no commercialized transgenic crops with resistance to abiotic stresses such as drought, heat, salinity and frost. One possible explanation is that the underlying genetic networks are rather complex, i.e. so far it has not been possible to identify single genes that would confer tolerance to these factors. However, a number of approaches are being developed to tackle these stress factors in crops (Bathnagar-Mathur et al., 2008).

Quality Traits

Modified Flower Color

Many flowers including carnations, roses, lilies, chrysanthemums, roses and gerberas, which are important in the global flower trade, do not produce the blue pigment delphinidin. Transgenic carnation lines with unique violet/mauve color have been developed. The genes of interest here include structural and regulatory genes of the flavanoid biosynthetic pathway.

Delayed Fruit Ripening and Increased Shelf Life

Genes encoding an enzyme which degrades 1-aminocyclopropane 1-carboxylic acid (ACC), an ethylene precursor, and those encoding polygalacturonase (PG) have been suppressed in some transgenic plants. Suppression is accomplished by inserting a truncated or anti-sense version of the gene. Reduced ACC activity results in delayed fruit ripening while decreased activity of PG results in a lower level of cell wall breakdown and hence delays fruit softening and rotting (Prasanna et al., 2007).

Modification of Oil Composition

Oilseed rape and soybean have been modified to increase the content of oleic acid in particular. The modified oils are lower in unsaturated fats and have greater heat stability than oils from the corresponding unmodified crops. In unmodified crops the FAD2 gene encodes a desaturase enzyme that converts C18:1 (oleic acid) to C18:2 and C18:3 acids. In the modified crop a mutant FAD2 gene prevents expression of the active desaturase, resulting in the accumulation of oleic acid (Kinney et al., 2002).

Modified Vitamin and Mineral Profiles

Vitamins and minerals are essential components of the human diet and dietary deficiencies of these nutrients can have severe effects on health and development. In addition to fortification and supplementation strategies for alleviating these deficiencies, transgenic crops with elevated and bio-available vitamins and minerals are being developed (Davies, 2007). Here the strategy is to express the genes responsible for the production or accumulation of the concerned nutrient in the edible parts of the plant. Thus promoters and other control sequences that target the expression of the gene(s) of interest to the correct part of the plant are highly important. In order to improve vitamin A production in rice the genes encoding phytoene synthase and phytoene desaturase have been expressed in the endosperm, resulting in the variety known as "Golden Rice". To improve iron accumulation and bio-availability in rice, genes such as ferritin synthase from soy (Fe storage), metallothionein (cystein-rich storage protein, improves Fe absorption) and a heat stable phytase gene (degrades phytic acid which inhibits Fe absorption) have been expressed in the rice endosperm.

Transgenic Plants as Bioreactors for Biopharmaceuticals and Vaccines

The first trials for the production of human proteins in plants dates back to the early 1990s; however, only in recent years has the use of transgenic plants as bioreactors for the production of small-molecule drugs or pharmaceutical proteins increasingly gained importance (Twyman et al., 2005). The use of transgenic plants as a production platform presents a viable alternative to conventional production of such compounds, such as extraction from natural sources, various cell culture techniques or the use of animal bioreactors. In particular, plant-derived vaccines and antibodies are considered as promising (Tiwari et al., 2009). Trials for the development of plants expressing vaccines in their edible parts, thus allowing cost-effective production and delivery of a vaccine, are a particularly intriguing option (Floss et al., 2007).

REFERENCES

1. Allison, LA. 2007. Fundamental molecular biology. 1st edition. Malden, (MA), Wiley-Blackwell.
2. Anklam, E., Gadani, F., Heinze Hans Pijnenburg, P. & Van Den Eede, G. 2002. Analytical methods for detection and determination of genetically modified organisms in agricultural crops and plant-derived food products. Eur. Food Res. Technol. 214:3–26.
3. Basrur, P.K. & King, W.A. 2005. Genetics then and now: breeding the best and biotechnology. Rev. Sci. Tech. 24(1): 31-49.
4. Bazer, F.W. & Spencer, T.E. 2005. Reproductive biology in the era of

genomics biology. Theriogenology 64(3): 442-56.
5. Benítez, T., Rincón, A.M., Limón, M.C. & Codón, A.C. 2004. Biocontrol mechanisms of Trichoderma strains. Int. Microbiol. 7(4): 249-60.
6. Bhatnagar-Mathur, P., Vadez, V. & Sharma, K.K. 2008. Transgenic approaches for abiotic stress tolerance in plants: retrospect and prospects. Plant Cell. Rep. 27(3): 411-24.
7. Brophy, B., Smolenski, G., Wheeler, T., Wells, D., L'Huillier, P. & Laible, G. 2003. Cloned transgenic cattle produce milk with higher levels of ß-casein and K-casein. Nature Biotechnology 21: 157– 62.
8. Bull, A.T., Ward, A.C. & Goodfellow, M. 2000. Search and discovery strategies for biotechnology: the paradigm shift. Microbiol. Mol. Biol. Rev. 64(3): 573-606.
9. Campbell, M.A., Fitzgerald, H.A. & Ronald, P.C. 2002. Engineering pathogen resistance in crop plants. Transgenic Res. 11(6): 599-613.
10. Cankar, K., Stebih, D., Dreo, T., Zel, J. & Gruden, K. 2006. Critical points of DNA quantification by real-time PCR--effects of DNA extraction method and sample matrix on quantification of genetically modified organisms. BMC Biotechnol. 6:37.
11. CBD (Convention on Biological Diversity). 5 June 1992. Rio de Janeiro. United Nations.
12. Chatterjee S., Chattopadhyay, P., Roy, S. & Sen, S.K. 2008. Bioremediation: a tool for cleaning polluted environments. Journal of Applied Biosciences 11: 594–601.
13. Chatterjee S., Chattopadhyay, P., Roy, S. & Sen, S.K. 2008. Bioremediation: a tool for cleaning polluted environments. Journal of Applied Biosciences 11: 594–601.
14. Chauthaiwale, V.M., Therwath, A. & Deshpandei., V.V. 1992. Bacteriophage lambda as a cloning vector. Microbiological Reviews pp. 577-591.
15. Christon, P., Capell, T., Kohli, A., Gatehouse, J.A. & Gataehouse, A.M. 2006. Recent developments and future prospects in insect pest control in transgenic crops. Trends Plant Sci. 11(6): 302-8.
16. Chung, S.M., Vaidya, M. & Tzafira, T. 2006. Agrobacterium is not alone: gene transfer to plants by viruses and other bacteria. Trends Plant Sci. 11(1): 1-4.
17. Clive, J. 2008. Global Status of Commercialized Biotech/GM Crops: ISAAA Brief No 39. Ithaca, New York, ISAA.
18. Dahiya, N., Tewari, R. & Hoondal, G.S. 2006. Biotechnological aspects

of chitinolytic enzymes: a review. Appl. Microbiol. Biotechnol. 71(6): 773-82.

19. Darbani, B., Eimanifar A., Stewart, C.N. Jr. & Camargo, W.N. 2007. Methods to produce marker-free transgenic plants. Biotechnol. J. 2(1): 83-90.

20. Davey, M.R., Anthony, P., Power, J.B. & Lowe, K.C. 2005. Plant protoplasts: Status and biotechnological perspectives. Biotechnol. Adv. 23(2): 131-71.

21. Davies, K.M. 2007. Genetic modification of plant metabolism for human health benefits. Mutat. Res. 622(1-2): 122-37.

22. de Maagd, R.A., Bosch, D. & Stiekema, W. 1999. Bacillus thuringiensis-toxin-mediated insect resistance in plants. Trends Plant Sci. 4(1): 9-13.

23. Dekkers, J.C. 2004. Commercial application of marker- and gene-assisted selection in livestock: strategies and lessons. J. Anim. Sci. 82 E-Suppl: E313-328.

24. Demain, A.L. & Adrio, J.L. 2008. Contributions of microorganisms to industrial biology. Mol. Biotechnol. 38(1): 41-55.

25. Denning, C. & Priddle, H. 2003. New frontiers in gene targeting and cloning: success, application and challenges in domestic animals and human embryonic stem cells. Reproduction 126(1): 1-11.

26. Donovan, D.M., Kerr, D.E. & Wall, R.J. 2005. Engineering disease-resistant cattle. Transgenic Res. 14(5): 563-7.

27. Duke, S.O. 2005. Taking stock of herbicide-resistant crops ten years after introduction. Pest Manag. Sci. 61(3): 211-8.

28. EPA (Environmental Protection Agency). Factsheet 006474 on Agrabacterium radiobacter strain K1026. Available online at: http://www.epa.gov/pesticides/biopesticides/ingredients/factsheets/factsheet_006474.htm

29. Etherton, T.D. & Bauman, D.E. 1998. Biology of somatotropin in growth and lactation of domestic animals. Physiol. Rev. 78(3): 745-61.

30. FAO. 2001: Agricultural biotechnology for developing countries. Results of an electronic forum.

31. FAO. 2004. The state of food and agriculture. Agricultural biotechnology: meeting the needs of the poor?

32. FAO. 2005. Status of research and application of crop biotechnologies in developing countries. Preliminary assessment

33. FAO. 2006. The role of biotechnology in exploring and protecting agricultural genetic resources.

34. FAO. 2007. Marker-assisted selection: current status and future perspectives in crops, livestock, forestry and fish.
35. FDA (U.S. Food and Drug Administration). Press Release, 6 February 2009. Retrieved on 15.10.2009 at: http://www.fda.gov/NewsEvents/Newsroom/PressAnnouncements/ucm109074.htm
36. Floss, D.M., Falkenburg, D. & Conrad, U. 2007. Production of vaccines and therapeutic antibodies for veterinary applications in transgenic plants: an overview. Transgenic Res. 16(3): 315-32.
37. Fravel, D.R. 2005. Commercialisation and implementation of biocontrol. Annu. Rev. Phytopathol. 43: 337–59.
38. Galli, C., Duchi, R., Crotti, G., Turini, P., Ponderato, N., Colleoni, S., Lagutina, I. & Lazzari, G. 2003. Bovine embryo technologies. Theriogenology 59(2): 599-616.
39. Gavrilescu, M. & Chisti, Y. 2005. Biotechnology – a sustainable alternative for chemical industry. Biotechnology Advances 23: 471–499.
40. Gellisen, G. (Ed). 2005. Production of recombinant proteins. 1st edition. Weinheim, Wiley-VCH.
41. Gelvin, S.B. 2003. Agrobacterium-mediated plant transformation. The biology behind the gene-jockeying tool. Microbiol. Mol. Biol. Rev. 67(1): 16-37.
42. Gianessi, L.P. 2008. Economic impacts of glyphosate-resistant crops. Pest Manag. Sci. 64(4): 346-52
43. Giraldo, P. & Montoliu, L. 2001. Size matters: use of YACs, BACs and PACs in transgenic animals. Transgenic Research 10: 83–103.
44. Gomez-Gil, B., Roque, A., Turnbull, J.F. & Inglis, V. 1998. A review on the use of microorganisms as probiotics. Rev. Latinoam. Microbiol. 40(3-4): 166-72.
45. Gray, N.K. & Wickens, M. 1998. Control of translation initiation in animals. Annu. Rev. Cell. Dev. Biol. 4: 399-458.
46. Green, J.M., Hazel, C.B., Forney, D.R. & Pugh, L.M. 2008. New multiple-herbicide crop resistance and formulation technology to augment the utility of glyphosate. Pest Manag. Sci. 64(4): 332-9.
47. Griffiths, A.J.F., Wessler, S.R., Lewontin, R.C., & Carroll, S.B. 2007. Introduction to genetic analyses. 9th edition. Palgrave Macmillan.
48. Grothaus, G.D., Bandla, M., Currier, T., Giroux, R., Jenkins, G.R., Lipp, M., Shan, G., Stave, J.W. & Pantella, V. 2006. Immunoassay as an analytical tool in agricultural biotechnology. J. AOAC Int. 89(4): 913-28.
49. Gurr, S.J. & Rushton, P.J. 2005. Engineering plants with increased disease

resistance: how are we going to express it? Trends in Biotechnology 23(6): 283-290.

50. Haefner, S., Knietsch, A., Scholten, E., Braun, J., Lohscheidt, M. & Zelder, O. 2005. Biotechnological production and applications of phytases. Appl. Microbiol. Biotechnol. 68(5): 588-97.

51. Halpin, C. 2005. Gene stacking in transgenic plants--the challenge for 21st century plant biotechnology. Plant Biotechnol. J. 3(2): 141-55.

52. Herrera, G., Snyman, S.J. Thomson, J.A. 1994. Construction of a bioinsecticidal strain of pseudomonas fluorescens active against the sugarcane borer, Eldana saccharina. Appl. Environ. Microbiol. 60(2): 682-690.

53. Heyman, Y. 2005. Nuclear transfer: a new tool for reproductive biotechnology in cattle. Reprod. Nutr. Dev. 45(3): 353-61.

54. Hodges, C.A. & Stice, S.L. 2003. Generation of bovine transgenics using somatic cell nuclear transfer. Reprod. Biol. Endocrinol. 1: 81.

55. Houdebine, L.M. 2009. Production of pharmaceutical proteins by transgenic animals. Comp. Immunol. Microbiol. Infect Dis. 32(2): 107-21.

56. Hughes, T.A. 2006. Regulation of gene expression by alternative untranslated region. Trends in Genetics 22(3): 199-122.

57. International human genome sequencing consortium. 2004. Finishing the euchromatic sequence of the human genome. Nature 431: 931-945.

58. ISO (International Organization for Standardization). 2006. EN/TS 21568:2006, Foodstuffs — Methods of analysis for the detection of genetically modified organisms and derived products — Sampling strategies 2006; European Committee for Standardization, Brussels, Belgium

59. ISTA (International Seed Testing Agency). 2004. Handbook on Seed Sampling. 2nd Edition. ISTA Bulking and Sampling Committee. M. Kruse (ed.).

60. Jana, S. & Deb, J.K. 2005. Strategies for efficient production of heterologous proteins in Escherichia coli. Appl. Microbiol. Biotechnol. 67: 289–298.

61. Jasbeer, K., Ghazali, F.M., Cheah, Y.K. & Son, R. 2008. Application of DNA and Immunoassay Analytical Methods for GMO Testing in Agricultural Crops and PlantDerived Products. ASEAN Food Journal 15(1): 1-25.

62. Jobling, M.A. & Gill, P. 2004. Encoded evidence: DNA in forensic

analysis. Nat. Rev. Genet. 10: 739-51.

63. Juven-Gershon, T., Hsu, J-Y., Theisen, J.W.M. & Kadonaga, J.T. 2008. The RNA polymerase II core promoter - the gateway to transcription. Curr. Opin. Cell. Biol. Author manuscript; available in PMC.

64. Kikkert, J.R., Vidal, J.R. & Reisch, B.I. 2005. Stable transformation of plant cells by particle bombardment/biolistics. Methods Mol. Biol. 286: 61-78.

65. Kind, A. & Schnieke, A. 2008. Animal pharming, two decades on. Transgenic Res. 2008 17(6): 1025-33.

66. Kinney, A.J., Cahoon, E.B. & Hitz, W.D. 2002. Manipulating desaturase activities in transgenic crop plants. Biochem. Soc. Trans. 30(Pt 6): 1099-103.

67. Kuroiwa, Y., Kasinathan, P., Choi, Y.J., Naeem, R., Tomizuka, K., Sullivan, E.J., Knott, J.G. Duteau, A., Goldsby, R.A., Osborne, B.A., Ishida, I. & Robl, J.M. 2002. Cloned transchromosomic calves producing human immunoglobulin. Nat. Biotechnol. 20(9): 889-94.

68. Kuroiwa, Y., Kasinathan, P., Matsushita, H., Sathivaselan, J., Sullivan, E.J., Kakitani, M., Tomizuka, K., Ishida, I. & Robl, J.M. 2004. Sequential targeting of the genes encoding immunoglobulin-mu and prion protein in cattle. Nat. Genet. 36(7): 775-80.

69. Lacroix, B., Li, J., Tzifira, T. & Citovsky, V. 2006. Will you let me use your nucleus? How Agrobacterium gets its T-DNA expressed in the host plant cell. Can. J. Physiol. Pharmacol. 84(3-4): 333-45.

70. Law, B.A. & Mulholland, F. 1991. The influence of biotechnological developments on cheese manufacture. Biotechnol. Genet. Eng. Rev. 9: 369-409.

71. Lawrence, S. 2007. Billion dollar babies – biotech drugs as blockbusters. Nature Biotechnology 25: 380-382.

72. Leimanis, S., Hernández, M., Fernández, S., Boyer, F., Burns, M., Bruderer, S., Glouden, T., Harris, N., Kaeppeli, O., Philipp, P., Pia, M., Puigdomènech, P., Vaitilingom, M., Bertheau, Y. & Remacle, J. 2006. A microarray-based detection system for genetically modified (GM) food ingredients. Plant Mol. Biol. 61(1-2): 123-39.

73. Lipp, M., Shillito, R., Giroux, R., Spiegelhalter,F., Charlton, S., Pinero, D. & Song, P. 2005. Polymerase chain reaction technology as analytical tool in agricultural biotechnology. J. AOAC Int. 88(1): 136-55.

74. Madan, M.L. 2005. Animal biotechnology: applications and economic implications in developing countries. Rev. Sci. Tech. 24(1): 127-39.

75. Mattick, J.S. 1994. Introns: evolution and function. Current Opinion in Genetics and Development 4: 823-831.

76. McEvoy, T.G., Alink, F.M., Moreira, V.C., Watt, R.G. & Powell, K.A. 2006. Embryo technologies and animal health - consequences for the animal following ovum pickup, in vitro embryo production and somatic cell nuclear transfer. Theriogenology. 65(5): 926-42.

77. Meeusen, E.N., Walker, J., Peters, A., Pastoret, P.P. & Jungersen, G. 2007. Current status of veterinary vaccines. Clin. Microbiol. Rev. 20(3): 489-510.

78. Melo, E.O., Canavessi, A.M., Franco, M.M. & Rumpf, R. 2007. Animal trangenesis: state of the art and applications. J. Appl. Genet. 48(1): 47-61.

79. Michelini, E., Simoni, P., Cevenini, L., Mezzanotte, L. & Roda, A. 2008. New trends in bioanalytical tools for the detection of genetically modified organisms: an update. Anal. Bioanal. Chem. 392(3): 355-67.

80. Miki, B. & McHugh, S. 2004. Selectable marker genes in transgenic plants: applications, alternatives and biosafety. J. Biotechnol. 107(3): 193-232.

81. Miller, F.P., Vandome, A.F., McBrewster, J. 2009. Central dogma of molecular biology: History of molecular biology, Primary structure, DNA replication, Transcription (genetics). Alphascript Publishing.

82. Miraglia, M., Berdal, K.G., Brera, C., Corbisier, P., Holst-Jensen, A., Kok, E.J., Marvin, H.J., Schimmel, H., Rentsch, J., van Rie, J.P. & Zagon, J. 2004. Detection and traceability of genetically modified organisms in the food production chain. Food Chem. Toxicol. 42(7): 1157-80.

83. Morange. M. & Cobb, M. 2000. A history of molecular biology. 1st edition. Cambridge (MA), Harvard University Press.

84. Nestler, E.J. & Hyman, S.E. 2002. Regulation of gene expression. Neuropsychopharmacology: The Fifth Generation of Progress. Edited by Kenneth L. Davis, Dennis Charney, Joseph T. Coyle & Charles Nemeroff. American College of Neuropsychopharmacology.

85. Newell, C.A. 2000. Plant transformation techniques: Development and Application. Mol. Biotechnol. 16(1): 53-65.

86. Niemann, H., Kues, W. & Carnwath, J.W. 2005. Transgenic farm animals: present and future. Rev. Sci. Tech. 24(1): 285-98.

87. Official Journal of the European Union. Commission Regulation (EC) No 65/2004 of 14 January 2004 establishing a system for the development and assignment of unique identifiers for genetically modified organisms.

88. Okkema, P.G. & Krause, M. 2005. Transcriptional regulation. WormBook,

ed. The C.elegans Research Community, WormBook.

89. Ovesna, J., Demnerova, K. & Pouchova, V. 2008. The detection of genetically modified organisms: an overview. In F. Toldrá (ed.), Meat Biotechnology. Springer Science+Business Media, LLC.

90. Padidam, M. 2003. Chemically regulated gene expression in plants. Curr. Opin. Plant Biol. 6(2): 169-77.

91. Palmiter, R.D., Brinster, R.L., Hammer, R.E., Trumbauer, M.E., Rosenfeld, M.G., Birnberg, N.C. & Evans, R.M. 1982. Dramatic growth of mice that develop from eggs microinjected with metallothionein-growth hormone fusion genes. Nature 300(5893): 611-5.

92. Park, F. 2007. Lentiviral vectors: are they the future of animal transgenesis? Physiol. Genomics 31(2): 159-73.

93. Park, S.J. & Cochran, J.R. 2009. Protein engineering and design. 1st Edition. Boca Raton (FL), CRC Press, Taylor and Francis Group.

94. Patrushev, L.I. & Minkevich, I.G. 2008. The problem of the eukaryotic genome size. Biochemistry (Moscow) 73(13): 1519-1552.

95. Pearson, H. 2006 Genetics: What is a gene? Nature 441: 398-401.

96. Plasmid genome database: http://www.genomics.ceh.ac.uk/plasmiddb/index.html

97. Prasanna, V., Prabha, T.N. & Tharanathan, R.N. 2007. Fruit ripening phenomena—an overview. Crit. Rev. Food Sci. Nutr. 47(1): 1-19.

98. Prins, M., Laimer, M., Noris, E., Schubert, J., Wassenegger, M. & Tepfer, M. 2008. Strategies for antiviral resistance in transgenic plants. Mol. Plant Pathol. 9(1): 73-83.

99. Ramakrishnan, V. 2002. Ribosome structure and the mechanisms of translation. Cell, 108: 557–572.

100. Ramsay, M. 1994. Yeast artificial chromosome cloning. Mol. Biotechnol. 1(2): 181-201.

101. Ranjekar, P.K., Patankar, A., Gupta, V., Bhatnagar, R., Bentur, J. & Ananda Kumar, P. 2003 genetic engineering of crop plants for insect resistance. Current Science 84(3).

102. Rath, D. & Johnston, L.A. 2008. Application and commercialization of flow cytometrically sex-sorted semen. Reprod. Domest. Anim. 43 Suppl 2: 338-46.

103. Rege, J.E.O. 1996. Biotechnology options for improving livestock production in developing countries, with special reference to sub-Saharan Africa. In S.H.B. Lebbie and E. Kagwini. Small Ruminant Research and Development in Africa. Proceedings of the Third Biennial Conference

of the African Small Ruminant Research Network, UICC, Kampala, Uganda, 5-9 December 1994. Nairobi, Kenya, ILRI (International Livestock Research Institute).

104. Remund, K.M., Dixon, D.A., Wright, D.L. & Holden, L.R. 2001. Statistical considerations in seed purity testing for transgenic traits. Seed Science Research 11:101–119. A tool for calculating sample plans, provided by ISTA, is available online at: http://www.seedtest.org/upload/cms/user/seedcalc6.zip.

105. Reznikoff, W.S. 1992. The lactose operon-controlling elements: a complex paradigm. Molecular Microbiology 6(17): 2419-2422.

106. Rizvi, P.Q., Choudhury, R.A. & Ali, A. 2009. Recent advances in biopesticides. In M.S. Khan, A. Zaidi & J. Musarrat (eds.), Microbial Strategies for Crop Improvement. Springer-Verlag Berling.

107. Robl, J.M., Kasinathan, P., Sullivan, E., Kuroiwa, Y., Tomizuka, K. & Ishida, J. 2003. Artificial chromosome vectors and expression of complex proteins in transgenic animals. Theriogenology. 59(1): 107-13.

108. Robl, J.M., Wang, Z., Kasisnathan, P. & Kuroiwa, Y. 2007. Transgenic animal production and animal biotechnology. Theriogenology 67(1): 127-33.

109. Rocha, J.L., Pomp, D. & Van Vleck, L.D. 2002. QTL analysis in livestock. Methods Mol. Biol. 195: 311-46.

110. Rogan, D. & Babiuk, L.A. 2005. Novel vaccines from biotechnology. Rev. Sci. Tech. 24(1): 159-74.

111. Roh, J.Y., Choi, J.Y., Li, M.S., Jin, B.R. & Je, Y.H. 2007. Bacillus thuringiensis as a specific, safe, and effective tool for insect pest control. J. Microbiol. Biotechnol. 17(4): 547-59.

112. Rolinson, G.N. 1998. 40 years of beta-lactam research. Journal of Antimicrobial Chemotherapy 41: 589–603.

113. Ron, M. & Weller, J.I. 2007. From QTL to QTN identification in livestock-winning by points rather than knock-out: a review. Anim. Genet. 38(5): 429-39.

114. Schellenberg, M.J., Ritchie, D.B., & MacMillan., A.M. 2008. Pre-mRNA splicing: a complex picture in higher definition. Trends in Biochemical Sciences 33(6): 243-246.

115. Schmitt, B. & Henderson, L. 2005. Diagnostic tools for animal diseases. Rev. Sci. Tech. 24(1): 243-50.

116. Scupham, A.J., Bosworth, A.H., Ellis, W.R., Wacek, T.J., Albrecht, K.A. & Triplett, E.W. 1996. Inoculation with Sinorhizobium meliloti

RMBPC-2 Increases Alfalfa Yield Compared with Inoculation with a Nonengineered Wild-Type Strain. Appl. Environ/Microbiol. 62(11): 4260-4262.

117. Shewry, P.R., Jones, H.D. & Halford, N.G. 2008. Plant biotechnology: transgenic crops. Adv. Biochem. Eng. Biotechnol. 111: 149–186j.

118. Soccol, C.R., Vandenberghe, L.P., Woiciechowski, A.L., Thomaz-Soccol, V., Correia, C.T. & Pandey, A. 2003. Bioremediation: an important alternative for soil and industrial wastes clean-up. Indian J. Exp. Biol. 41(9): 1030-45.

119. Somers, D.A. & Makarevitch, I. 2004. Transgene integration in plants: poking or patching holes in promiscuous genomes? Curr. Opin. Biotechnol. (2): 126-31.

120. Stein, J.A. & Rodriguez-Cerezo, E. 2009. The global pipeline of new GM crops. Luxembourg, European Commission Joint Research Centre.

121. Stewart, C.N. Jr. 2005. Monitoring the presence and expression of transgenes in living plants. Trends Plant Sci. 10(8): 390-6.

122. Sullivan, E.J., Pommer, J. & Robl, J.M. 2008. Commercialising genetically engineered animal biomedical products. Reprod. Fertil. Dev. 20(1): 61-6.

123. Sullivan, E.J., Pommer. J. & Robl, J.M. 2008. Commercialising genetically engineered animal biomedical products., Reprod, Fertil. Dev. 20(1): 61-6.

124. Szewczyk, B., Hoyos-Carvajal, L., Paluszek, I., Skrzecz, I. & Lobo de Souza, M. 2006. Baculoviruses--re-emerging biopesticides. Biotechnol. Adv. 24(2): 143-60.

125. Tacket, C.O. 2009. Plant-based oral vaccines: results of human trials. Curr. Top. Microbiol. Immunol. 332: 103-17.

126. Tan, S., Evans, R. & Singh, B. 2006. Herbicidal inhibitors of amino acid biosynthesis and herbicide-tolerant crops. Amino Acids 30: 195–204.

127. Tan, S., Evans, R.R., Dahmer, M.L., Singh, B.K. & Shaner, D.L. 2005. Imidazolinonetolerant crops: history, current status and future. Pest Manag. Sci. 61(3): 246-57.

128. Thomas, M.C. & Chiang, C-M. 2006. The general transcription machinery and general cofactors. Crit. Rev. Biochem. Mol. Biol. 41(3): 105-78.

129. Thorpe, T,A. 2007. History of plant tissue culture. Mol. Biotecbnol. 37(2): 169-80.

130. Tiwari, S., Verma, P.C., Singh, P.K. & Tuli, R. 2009. Plants as bioreactors for the production of vaccine antigens. Biotechnol. Adv. 27(4): 449-67.

131. Twyman, R.M., Schillberg, S. & Fischer, R. 2005. Transgenic plants in the biopharmaceutical market. Expert Opin. Emerg. Drugs 10(1): 185-218.

132. Vajta, G. & Gjerris, M. 2006. Science and technology of farm animal cloning: state of the art. Anim. Reprod. Sci. 92(3-4): 211-30.

133. Vicedo, B., Peñalver, R., Asins, M.J. & López, M.M. 1993. Biological Control of Agrobacterium tumefaciens, Colonization, and pAgK84 Transfer with Agrobacterium radiobacter K84 and the Tra Mutant Strain K1026. Appl. Environ. Microbiol. 59(1): 309-315.

134. Visel, A., Bristow, J. & Pennacchio, L.A. 2007. Enhancer identification through comparative genomics. Semin. Cell. Dev. Biol. 18(1): 140–152.

135. Voet, D. & Voet, J.G. 2004. Biochemistry. 3rd edition. Wiley & Sons.

136. Watson, J., Baker, T., Bell, S., Gann, A., Levine, M. & Losick, R. 2008. Molecular biology of the gene. Amsterdam, Addison-Wesley Longman.

137. Weimer, P.J. 1998. Manipulating ruminal fermentation: a microbial ecological perspective. J. Anim. Sci. 76(12): 3114-22.

138. Wheeler, M.B. 2007. Agricultural applications for transgenic livestock. Trends Biotechnol. 25(5): 204-10.

139. Wheeler, M.B., Walters, E.M. & Clark S.G. 2003. Transgenic animals in biomedicine and agriculture: outlook for the future. Anim. Reprod. Sci. 79(3-4): 265-89.

140. Whitelaw, C.B., Lillico, S.G. & King, T. 2008. Production of transgenic farm animals by viral vector-mediated gene transfer. Reprod. Domest. Anim. 43 Suppl 2: 355-8.

141. Williams, J.L 2005. The use of marker-assisted selection in animal breeding and biotechnology. Rev. Sci. Tech. 24(1): 379-91.

142. Yanofsky, C., Konan, K., Salsero, J. 1996. Some novel transcription attenuation mechanisms used by bacteria. Biochemie 78(11): 1017-1024(8).

143. Yu, W., Han, F. & Birchler, J.A. 2007. Engineered minichromosomes in plants. Curr. Opin. Biotechnol. 18(5): 425-31.

Chapter 9

MOLECULAR CLONING, EXPRESSION AND CHARACTERIZATION OF A NOVEL GENEβ-N-ACETYLGLUCOSAMINIDASE FROM BOMBYX MORI

Cheng Chang, Xiaoyong Liu, and Keping Chen

Institute of Life Sciences, Jiangsu University, Zhenjiang, China.

ABSTRACT

Previously, we have reported that a gene encoding Bombyxmoriβ-N-acetylglucosaminidase 2 (BmGlcNA case2) has been identified differentially expressed in the midgut of Bombyxmori strain NB resistant to nucleopolyhedrovirus (BmNPV), strain 306 susceptible to NPV and a near isogenic line BC_9 with similar genetic background to 306 but resistant to NPV by two-dimensional gel electrophoresis (2-DE). To get more knowledge about the relationship between β-N- -acetylglucosaminidase and the resistance of NPV, in this study, the 1542 bp open reading frame of a putative bombyxmoriβ-N-acetylglucosaminidase 2 gene (BmGlcNAcase2) was amplified from a pool of bombyxmoricDNAs and inserted into the prokaryotic expression plasmid pET-30a(+). Western blotting analysis showed that BmGlcNAcase2 was expressed in hemolymph, ovary, testis, fat body, trachea, midgut and silk gland of fifth instar larvae respectively. Immunofluorescence analysis indicated that BmGlcNA case2 was mainly located to the cytoplasm or some structure in cytoplasm.

INTRODUCTION

Sericulture is an important component of ariculture in China, which has a history of over 5,000 years in raising silkworms (BombyxmoriL.). It is also widely practiced in the world, such as in Japan, the former Soviet Union and Brazil. Silkworm viral diseases are major diseases causing great loss in sericulture, and Bombyxmorinucleopolyhedrovirus (BmNPV) is one of the most disastrous.

Therefore, it is a subject of intensive research to control silkworm NPV disease. The key to the sericulture is to develop pathogen-resistant silkworm strains [1]. Recently, comparative gene expression techniques, such as differential display, cDNA microarray assay and two-dimensional gel electrophoresis have become routine to examine changes in gene expression. Such methodologies provide useful approaches to identify differently expressed transcripts, because many genes can be examined simultaneously. To date, although a few BmNPV resistant genes have been reported, such as serine protease [2] and Bmlipase-1 [3], knowledge on the molecular mechanisms of Bombyx mori against this virus remains very limited.

β-N-acetylglucosaminidase(GlcNAcase) are widely distributed in various organisms. These enzymes catalyze the hydrolysis of an O-glycosidic bond in nonreducing terminal N-acetylglucosamine(GlcNAc) residues in an oligosaccharide chain. In insects, GlcNAcase plays an important role in the degradation of various oligosaccharides and glycoconjugates [4-6], a putative bombyxmoriβ-N-acetylglucosaminidase 2 showed broad substrate specificity, and cleaved terminal N-acetylglucosamine residues from the α-3 and α-6 branches of a biantennary N-glycan substrate, and also hydrolyzed chitotriose to chitobiose [7]. In our previous study, by comparison the proteomes of the resistant Bombyx mori strain NB, the susceptible strain 306 and the near isogenic line BC_9 strain by two-dimensional gel electrophoresis (2- DE), many differential protein spots have been obtained, one of which was identified as bombyxmori β-N-acetylglucosaminidase 2 by mass.We speculate that the excessive expression of β-N-acetylglucosaminidase in hemolymph of the resistant silkworm can probably disturb the N-linked glycans of GP64 protein on the cell membrane which is an essential process for initiating second infections, and thus reduce the reproduction of infectious viruses [8,9]. In this paper,the open reading frame (ORF) of BmGlcNAcase2 was cloned and the recombinant enzyme was expressed in E. coli. The amino acid sequence of recombinant BmGlcNAcase2 was verified by mass spectroscopic analysis. Western blotting analysis showed that there is no obvious difference of BmGlcNAcase2 expression in hemolymph, fat body, trachea, ovary, midgut, silk gland and testis of fifth instar larvae. The subcellular localization study through immunofluoresence analysis indicated for the first time that the BmGlcNA case2 was located to cytoplasm or some structure in cytoplasm. The obtained results would facilitate further studies to elucidate molecular mechanisms of Bombyxmoriagainst BmNPV infection.

MATERIALS AND METHODS

Insect, Cell and Virus

B. mori strain C108 (standard strain of silkworm) was maintained in our laboratory. All larvae were reared with fresh mulberry leaves at 27°C under a 12 h light/12 h dark photoperiod.

The BmN cell line was maintained at 27°C in TC-100 insect medium (Gibco, USA) supplemented with 10% (v/v) fetal bovine serum (Gibco, USA) using standard techniques.

The expression vector pET-30a(+) and E. coli strains BL21 (DE3) were obtained from Novagen (CA, USA). All primers, RNase-free DNaseI, EX Taq polymerase, restriction enzymes, T4 DNA ligase and DNA the subcloning vector pMD18-T were purchased from TaKaRa (Dalian, China). Chemicals are all from sigma (MO, USA) or a domestic provider in China if not stated otherwise.

Cloning of the Bmglcnacase2 Gene

The BmGlcNAcase2 specific primers, forward primer (5'-CATGCCATGGCACCGGGACCCGAATAT-3') with a NcoI site (underlined), and reverse primer (5'-ATAAGAATGCGGCCGCCTAAGCGCC TAGGCAGA-3') with an NotI site (underlined) were designed to amplify the ORF of the putative BmGlcNAcase2 gene (GenBank accession no. AB286958). The PCR reaction was carried out with 30 amplification cycles (94°C for 30 s, 55°C for 30 s, and 72°C for 90 s) in a Gene Amp 2400 System thermocycler. The PCR product was ligated into pMD 18-T vector using T4 DNA ligase and then transformed into E. coli TG1. A fragment between NcoI and NotI containing the BmGlcNAcase2 gene was excised from the recombinant plasmid. The purified fragment was subcloned into the pET-30a(+) expression vector and transformed into E. coli BL21 (DE3). DNA sequencing confirmed that the BmGlcNAcase2 gene was correctly fused to the N-terminal 6 × His-tag.

Expression and Purification of Recombinant Protein

To express recombinant protein, a freshly transformed colony was cultured in LB medium supplemented with kanamycin (50 µg/ml) at 37°C overnight. This overnight culture was inoculated into fresh LB medium and cultured at 37°C with vigorous shaking. When OD_{600} reach 0.5, the expression of BmGlcNAcase2 was induced with IPTG (final concentration 0.1-1 mM during optimization) and further cultured at 37°C for another 8 hours and 16°C for another 12 hours respectively. Cells were harvested by centrifugation (4500 g, 4°C, 15 min) and SDSPAGE analysis on 15% gel was performed to estimate the expression level of BmGlcNAcase2. The cell pellet was resuspended in buffer A (50 mM sodium phosphate, 300 mMNaCl, 1 mM EDTA, 0.5 mM PMSF, pH 8.0), then the suspension was lysed by sonication. The lysate was clarified by centrifugation (16,000 g, 4°C, 25 min). The supernatant was loaded onto a Ni-NTA affinity column (Qiagen). Purification conditions were standardized by optimizing pH, the concentration of salt and imidazole. After washing the captured column with 20 mM and 40 mM imidazole, the fusion protein was eluted with 250 mM imidazole. The eluted protein was dialyzed against buffer B (50 mM sodium phosphate, 150 mM NaCl, pH 7.5) at 4°C.

Mass Spectrometry

The specific bands corresponding to BmGlcNAcase2 were excised manually from the gel with a sterile scalpel and digested with trypsin according to Li's [10] method. The digested samples were analyzed by an ultraflex MALDI-TOF-TOF (BRUKER, GERMANY). Peptide mass fingerprinting (PMF) was performed by comparing the masses of peptides to NCBI protein database using the MASCOT search engine (http://www.matrixscience.com).

Antibody Production and Western Blot Analysis

The antibody was prepared by standard techniques [11]. Briefly, purified BmGlcNAcase2 protein (about 2 mg) was injected subcutaneously to immunize New Zealand white rabbits in complete Freund's adjuvant, followed by two booster injections in incomplete Freund's adjuvant within a gap of 2 weeks before exsanguinations. Then, the polyclonal rabbit antibody against 6 × HisBmGlcNAcase2 was obtained and used for immunoassay.

After the SDS-PAGE (Bio-Rad Mini-Protean II, Hercules, CA) was finished, the proteins were transferred to a PVDF membrane with a Bio-Rad liquid transfer apparatus for Western blot. The rabbit anti-BmIDGF polyclonal antibodies (1:1,000 dilution) and horseradish peroxidase (HRP)-conjugated goat anti-rabbit IgG antibodies (1:2,000 dilution) were used, and signals were detected by diaminobenzidine (DAB) (Sigma, USA). To detect if this protein expressed specially in tissues, total protein of each tissues was added to the lines of the SDS-PAGE, and the added protein concentration of each tissues are 50 ug.

Immunofluorescence Microscopy

BmN cells seeded onto coverslips were washed with PBS, and fixed with 2 ml of 4% paraformaldehyde for 15 min. Then cells were washed three times with PBS and permeabilized with 0.1% Triton X-100 in PBS for 15 min. After washing three times with PBS, cells were incubated with anti-BmGlcNAcase2 antibody (1:1000) as primary antibody, fluorescein isothiocyanate (FITC) conjugated goat anti-rabbit IgG antibody as secondary antibody (1:3000) (Qualex, Inc), and nuclei were stained with DAPI (Roche), then examined with a confocal laser scanning microscope (Zeiss lsm 5 live).

RESULTS AND DISCUSSION

Cloning of the BmGlcNAcase2 Gene

The recombinant BmGlcNAcase2 protein was expressed in E. coli BL21 (DE3) harboring the expression vector pET-30a(+)-BmGlcNAcase2. The BmGlcNAcase 2 proteins formed inclusion bodies when host E. coli was cultivated at 37°C and induced with 1 mM IPTG. The expression of 6 His-tagged Bm122 was also confirmed by anti-6xHis monoclonal antibody (**Figure 1(b)**). The soluble BmGlcNAcase2 was purified by a Ni-NTA column (**Figure 1(a), lane 3**).

Mass Spectrometry

To determine whether the amino acid sequence of recombinant BmGlcNAcase2 matches the one predicted from DNA sequencing results, the MALDI-TOF-TOF mass spectra of tryptic digest of recombinant BmGlcN Acase2 was characterized to identify the recombinant protein. 17 peptide fragments were identified. By comparing the masses of indentified peptides to the hypothetical tryptic peptides for proteins in non-redundant NCBI database using the MASCOT search engine, BmGlcNAcase2 was obviously identified with MOWSE score of 86. The identified 17 peptide fragments matched against the deduced amino acid sequence of BmGlcN Acase2 accounted for 41% peptide mass fingerprint sequence coverage (**Figure 2**).

Cellular Localization of BmGlcNAcase2 in BmN Cells

Confocal laser scanning fluorescence microscopy was utilized to determine the cellular localization of BmGlc NAcase2 protein in BmN cells. The result showed that the BmGlcNAcase2 protein was primarily located in the cytoplasm and was scarcely detectable in the nucleus, while no obvious fluorescence signal was observed in cells stained with pre-immune serum of rabbit as primary antibody (**Figure 3**).

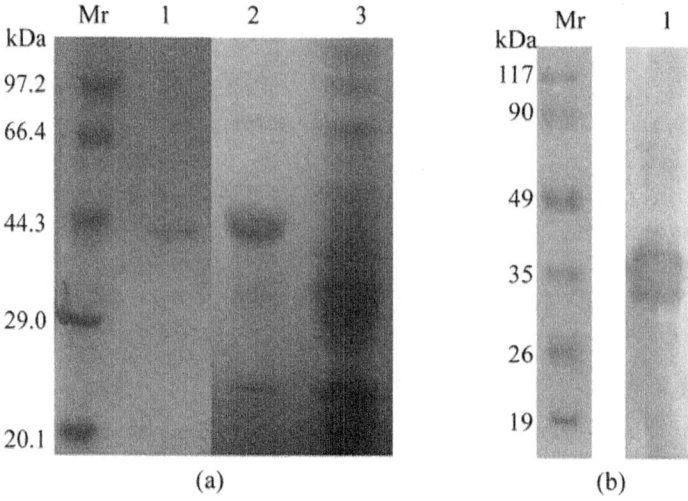

Figure 1. SDS-PAGE stained with commassie blue and western blotting analysis of recombinant Bm GlcNAcase2. (a) Lane1: recombinant BmGlcNAcase2 purified by Ni-NTA resin column; lane2: lysate of BL21 (DE3) harboring pET-30a(+)- -BmGlcNAcase2 induced with 0.5 mM IPTG; lane3: lysate of host cell transformed with empty vector. (b) Lane1: western blotting analysis of recombinant BmGlcNAcase2 with anti-6 his antibody.

Western Blot Analysis

Antibody against BmGlcNAcase2 protein was used to perform western blot analysis Bombyx mori. The result showed that BmGlcNAcase2 was expressed in hemolymph, ovary, testis, fat body, trachea, midgut and silk gland of fifth instar larvae, and there was no obvious difference in expression of these tissues(**Figure 4**).

DISCUSSION

β-N-acetylglucosaminidase of Bombyx mori plays an important role in several physiological process, the enzyme is essential to hydrolyze chitooligosaccharides to their constitutive monomer and contribute to recycling GlcNAc pools for remodeling of the exoskeleton during metamorphosis, as indicated previously [12,13]. It was reported recently that beta-N-acetylglucosaminidase activity in BmNPV resistant strains NB and the near-isogenic line BC_9 was significantly higher than that in the BmNPV susceptible strain 306 [9] so it is interesting to find out the relationship between this enzyme and silkworm NPV disease.

In this report, we have successfully cloned, optimized the expression and purified β-N-acetylglucosaminidase of Bombyx mori in the E.coli strains BL21 (DE3) and we showed that the expression of BmGlcNAcase2 can be detected from almost all tissues(hemolymph, ovary, testis, fat body, trachea, midgut and silk gland) of the fifth instar larvae. Identifying the subcellular localization of protein is particularly helpful in the functional annotation of gene products.

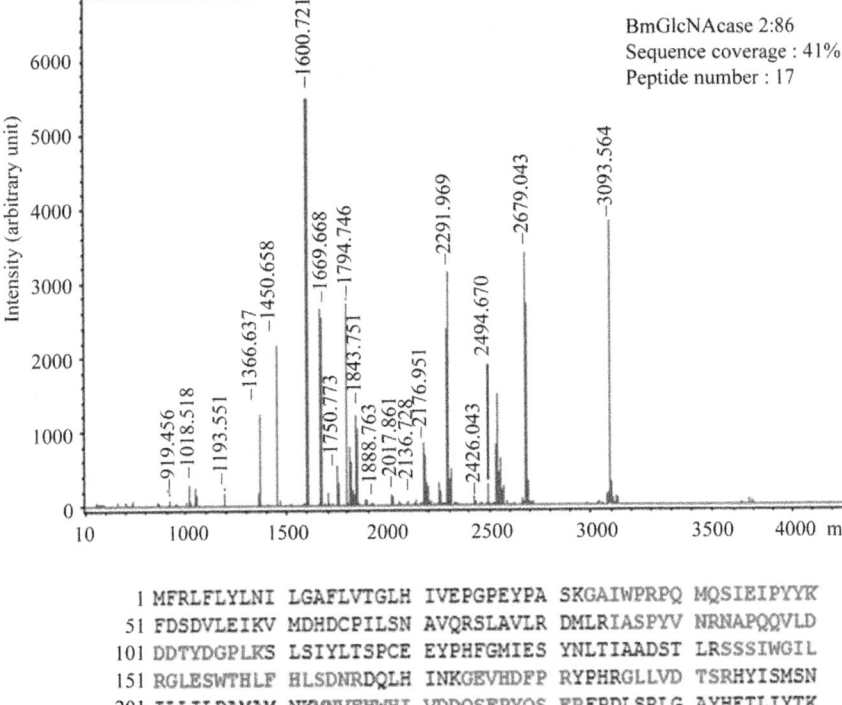

Figure 2. MALDI spectra of tryptic digest of recombinant BmGlcNAcase2. The identified protein, score, amino acid sequence coverage and the number of identified peptides are shown. Matched peptide sequences are showed in red.

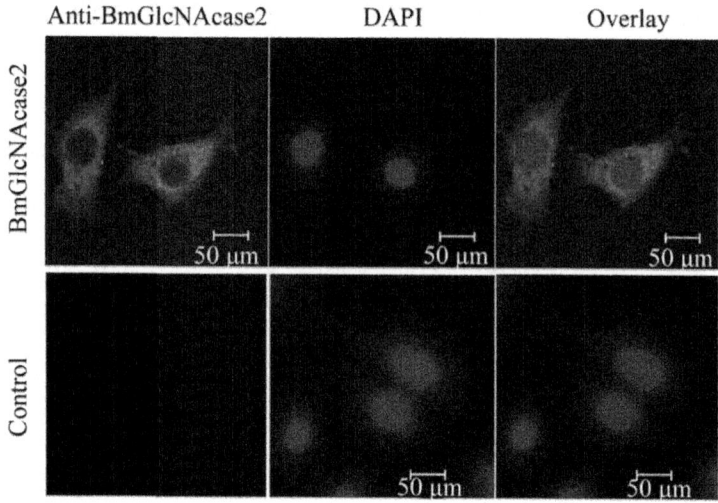

Figure 3. Subcellular localization of BmGlcNAcase2 in BmN cells treated with anti-BmGlcNAcase2 antibody, followed by treatment with FITC-conjugated goat anti-rabbit IgG, and examined in confocal laser microscope. From left to right: green fluorescence for BmGlcNAcase2, DAPI and the overlay images. For the control, pre-immune serum was used as the primary antibody. Nuclei were stained with DAPI (blue). Samples were observed under a confocal laser scanning microscope.

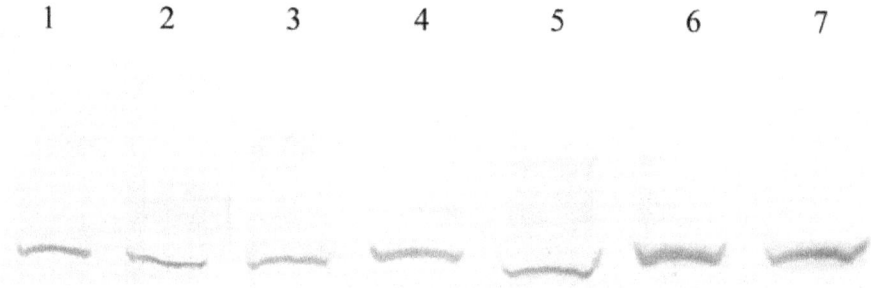

Figure 4. Expression profile of BmGlcNAcase2 in different tissues of Bombyx mori. Lanes 1 to 7 represented fat body, Malpighian tubule, midgut, ovary, hemocytes, testis, and silk gland.

The result showed that BmGlcNAcase2 was a protein distributed mostly in cytoplasm firstly, which is consistent with prediction by computer tool of PLOC (http://www.genome.jp/SIT/plocdir/), whereas the precise biochemical function and the possible role of BmGlcNAcase2 in this process remain to be determined.

ACKNOWLEDGEMENTS

This work was supported by the grants from Jiangsu SciTech Support Project-Agriculture (No. BE2008379).

REFERENCES

1. Chen, H.Q., Chen, K.P., Yao, Q., Guo, Z.J. and Wang, L.L. (2007) Characterization of a late gene, ORF67 from bombyx mori nucleopolyhedrovirus. FEBS Letters, 581, 5836-5842. doi:10.1016/j.febslet.2007.11.059
2. Nakazawa, H., Tsuneishi, E., Ponnuvel, K.M., Furukawa, S., Asaoka, A., et al. (2004) Antiviral activity of a serine protease from the digestive juice of Bombyx mori larvae against nucleopolyhedrovirus. Virology, 321, 154-162. doi:10.1016/j.virol.2003.12.011
3. Ponnuvel, K.M., Nakazawa, H., Furukawa, S., Asaoka, A., Ishibashi, J., et al. (2003) A lipase isolated from the silkworm Bombyx mori shows antiviral activity against nucleopolyhedrovirus. Journal of Virology, 77, 10725- 10729.doi:10.1128/JVI.77.19.10725-10729.2003
4. Aumiller, J.J., Hollister, J.R. and Jarvis, D.L. (2006) Molecular cloning and functional characterization of beta-N-acetylglucosaminidase genes from Sf9 cells. Protein Expression and Purification, 47, 571-590. doi:10.1016/j.pep.2005.11.026
5. Cattaneo, F., Pasini, M.E., Intra, J., Matsumoto, M., Briani, F., et al. (2006) Identification and expression analysis of Drosophila melanogaster genes encoding β- -hexosaminidases of the sperm plasma membrane. Glycobiology, 16, 786-800. doi:10.1093/glycob/cwl007
6. Nagamatsu, Y., Yanagisawa, I., Kimoto, M., Okamoto, E. and Koga, D. (1995) Purification of a chitooligosaccharidolytic beta-N-acetylglucosaminidase from Bombyx mori larvae during metamorphosis and the nucleotide sequence of its cDNA. Bioscience, Biotechnology, and Biochemistry, 59, 219-225. doi:10.1271/bbb.59.219
7. Takahiro, O., Seiji, I., Hideki, S., Akihiro, U., Toshiki, T., et al. (2007) Molecular cloning and expression of two novel β-n-acetylglucosaminidases from silkworm Bombyx mori. Bioscience, Biotechnology, and Biochemistry, 71, 1626-1635. doi:10.1271/bbb.60705
8. Jarvis, D.L., Wills, L., Burow, G. and Bohlmeyer, D.A. (1998) Mutational analysis of the N-linked glycans on autographa californica nucleopolyhedrovirus gp64. Journal of Virology, 72, 9459-9469.
9. Liu, X.Y., et al. (2010) Proteomic analysis of nucleopolyhedrovirus infection resistance in the silkworm, Bombyx mori (Lepidoptera:

Bombycidae). Journal of Invertebrate Pathology, 105, 84-90.

10. Li, X.H., Wu, X.F., Yue, W.F., Liu, J.M., Li, G.L., et al. (2006) Proteomic analysis of the silkworm (Bombyx mori L.) hemolymph during developmental stage. Journal of Proteome Research, 5, 2809-2814. doi:10.1021/pr0603093

11. Kothari, H., Kumar, P. and Singh, N. (2006) Prokaryotic expression, purification, and polyclonal antibody production against a novel drug resistance gene of Leishmania donovani clinical isolate. Protein Expression and Purification, 45, 15-21.doi:10.1016/j.pep.2005.10.002

12. Bade, M.L. and Wyatt, G.R. (1962) Metabolic conversions during putation of the cecropia silkworm. 1. Deposition and utilization of nutrient. Biochemical Journal, 83, 470-478.

13. Zen, K.C., Choi, H.K., Krishnamachary, N., Muthukrishnan, S. and Kramer, K.J. (1996) Cloning, expression and hormonal regulation of an insect β-N-acetylglucosaminidase gene. Insect Biochemistry and Molecular Biology, 26, 435-444. doi:10.1016/0965-1748(95)00111-5

Chapter 10

TERAHERZ VIBRATIONAL SPECTROSCOPY OF E. COLI AND MOLECULAR CONSTITUENTS: COMPUTATIONAL MODELING AND EXPERIMENT

Tatiana Globus[1,2], Igor Sizov[1,2], and Boris Gelmont[1,2]

[1]Department of Electrical and Computer Engineering, University of Virginia, Charlottesville, USA

[2]Vibratess, LLC, Charlottesville, USA

ABSTRACT

In this paper we present the results of our research of E. coli cells and cellular components, DNA and protein thioredoxin, using highly resolved sub-Terahertz (THz) vibrational spectroscopy. In this combined research, the results from experimental spectroscopy are analyzed via molecular dynamics (MD) simulation of vibrational modes and absorption spectra from E. coli cells and constituents in the sub-THz range. Simplified models of DNA macromolecules with a short sequencing have been constructed for several E. coli strains with the goal to predict their absorption spectra. The similarity between spectral characteristics of E. coli cells and cellular components observed in experiments helps us to better understand the mechanism of material interaction with THz radiation and to add genetic information to the characteristic signatures from biological objects. Modeling results supported by experimental characterization using a spectroscopic sensor prototype developed and built by Vibratess confirm that an optical, label and reagent free technique can be used to examine, detect, and identify bacterial cells with high accuracy and selectivity to the level of strains.

INTRODUCTION

In this work, sub-Terahertz (THz) vibrational spectroscopy is explored to characterize Escherichia coli (E. coli) bacterial cells and its cellular components.

E. coli is a diverse bacterial organism that is widely used as a model organism in laboratory studies. However, these bacterial species have strains that can be pathogenic to humans and animals. There are tens of thousands E. coli contamination cases every year in United States. Bacteria can adapt to extreme environments and bacterial pathogens share a common trait that is their ability to live longterm inside the host's cells [1].

Traditionally, to indentify pathogens from different samples one has to collect and isolate species by using culture methods. The identification process is often relied on microscopic examination to determine phenotypic characteristics of bacterial colony. The final identification of microbial species is assisted by molecular biology and biochemistry methods. Cultivation process alone might last more than one week. Therefore, there is an increasing need for alternative methods that makes the identification of microorganisms fast and reliable. Accurate identification of infectious agents (such as pathogenic bacteria) can be critical for the diagnosis and effective treatment of diseases. The monitoring and the detection of pathogens in food are also important for protecting human health.

Terahertz (THz) vibrational spectroscopy is relatively new experimental method that can be more effective than standard methods especially when the quantity of sample material is limited. Emerging highly resolved THz vibrational spectroscopy is an optical, label and reagent free technique that can be used to examine, detect, and identify bacterial cells to the level of strains. This new technology with high spectral and spatial resolution has been recently demonstrated using a spectroscopic sensor prototype developed and built by Vibratess, LLC [2].

Sub-Terahertz (sub-THz) vibrational spectroscopy for bio-sensing is based on specificity of resonance features, fingerprints, observed in absorption (transmission) spectra of large biological molecules and entire bacterial cells/ spores. In our experiments we showed that cellular components contribute to spectroscopic signature of the entire microorganism. As a result, THz vibrational spectroscopy promises to add quantitative genetic information to the characteristic signatures of biological objects, thus increasing the detection accuracy and selectivity.

In addition, it has been shown in our previous work [3, 4] that transmission signatures from sample material in water are well-resolved. As a result, THz spectroscopy takes benefit from working with samples in the natural environment for biological objects with the minimal sample preparation process.

Molecular dynamic (MD) simulations of proteins and nucleic acids, which make the bigest contribution to THz absorption in bacterial cell, can help better understand the mechanism of interaction between radiation and biological cells and their constituents, to further improve indentification of objects and even to predict their signatures.

We have recently simulated spectra of relatively small biological molecules like tyrosine transfer RNA [5] or protein thioredoxin from E. coli [6] using MD simulations. Our approach is based on comparison between measured and simulated spectra using MD of cellular components. We demonstrated a rather good correlation of simulated absorption spectra with experimental data [4]. However a large size of macromolecules (~5 million base pairs for E. coli DNA) prevents direct application of MD simulation at the current level of computational capabilities. Thus, one purpose of this work is to develop a simplified short model of the bacterial genome so that the model would capture the structure and the most important low-frequency vibrational characteristics of the native DNA. MD simulations of the modeled sequences permit us to calculate expected absorption spectra. Another purpose of statistical modeling of short DNA sequences is to compare thier composition and THz spectra obtained for different strains of E. coli. The comparison gives us the possibility to estimate uniqeness of THz DNA signatures of individual pathogenic and non-pathogenic strains.

In addition, we analyzed the results from MD simulations of our new developed models of DNA sequences from E. coli [7]. We demonstrated that the application of molecular dynamics simulations to the 60 base pair DNA models and to relatively small molecules like a protein thioredoxin from E. coli permits us to study directly atomic displacements in molecular dynamics and relaxation processes of intermolecular motions.

TERAHERTZ VIBRATUONAL SPECTROSCOPY OF BIOLOGICAL CELLS AND CELLULAR COMPONENTS

Sub-Terahertz (sub-THz) vibrational spectroscopy for biosensing is based on specific resonance features, vibrational modes or group of modes at close frequencies, in absorption (transmission) spectra of large biological molecules and entire bacterial cells/spores. Significant progress in experimental and computational sub-THz vibrational spectroscopy has been made in the last 2 - 3 years to improve the sensitivity of THz spectroscopic characterization of large biological molecules and microorganisms [4]. Sub-THz spectroscopy was applied to characterize lyophilized and in vitro cultured bacterial cells of non-pathogenic species of E. coli and Bacillus subtilis (BG), spores of BG. Some of cellular components of E. coli, DNA [4,8], transfer RNA [5], and protein thioredoxin [6,9] were characterized as well.

The spectral range below 1 THz is the most attractive for practical applications because of low disturbance from the absorption by water vapors in air and by liquid water or other analytes [3,4]. Although liquid water absorbs and contributes to background in the sub-THz/THz spectral range, the level of water absorption in the low THz range is at least 2.5 orders of magnitude less compared to IR and far-IR. Because of less disturbance from water absorption lines, sensors in sub-THz range do not require evacuation or purging with dry nitrogen. Many synthetic materials are transparent in THz region and can be used as substrates or windows for sample cells.

Till recently, Fourier transform (FT) transmission spectroscopy (Bruker IFS66v) with cooled Si bolometer operating at 1.7 K provided the most detailed information on sub-THz vibrational spectral signatures of biological molecules and microorganisms. Spectral resolution in these studies was 0.25 cm^{-1} [3,4,10]. It was demonstrated that Fourier Transform (FT) spectroscopy in the frequency region of 10 - 25 cm^{-1} is sensitive enough to reveal characteristic spectral features from bio-cells and spores in different environment, to verify the differences between species, and to show the response of spores to vacuum and response of cultured cells to heat [4]. Simultaneously with experimental characterization, computational modeling techniques have been developed using the energy minimization, normal mode analysis and MD approaches to understand and predict low frequency vibrational absorption spectra of short artificial DNA and RNA [11-15] large macromolecules of DNA [5, 8] and proteins [6,9]. Direct comparison of experimental spectra with theoretical prediction for a short chain α-helix RNA fragment with known structure [11], transfer RNA [5] and protein thioredoxin [6] from E. coli showed reasonably good correlation thus validating both, experimental and theoretical results. Vibrational frequencies from simulated spectra for components correlate rather well with the observed features. Thus, multiple resonances due to low frequency vibrational modes within biological macromolecules, components of bacterial organisms, are unambiguously demonstrated experimentally in the sub-THz frequency range in agreement with the theoretical prediction. These results are also in general agreement with analysis of more broad vibrational features observed at higher frequencies using different experimental technologies, mostly on relatively smaller molecules in crystalline form. Organic solid systems and relatively small bio-molecules like protein fragments have been successfully characterized in this range to demonstrate sharp spectral features determined by their individual symmetries and structures [16-19].

Bacteria are very complex biological objects. Because of their small size and relatively low absorption coefficient, the THz radiation propagates through an entire object, allowing the genetic material and proteins all contribute to the THz signature of bacteria or spores. The results of our work confirmed that observed spectroscopic features are caused by fundamental physical mechanism of interaction between THz radiation and biological macro-molecules [4]. Particularly, the analysis of results indicates that the spectroscopic signatures of microorganisms originate from the combination of low frequency vibrational modes or group of modes at close frequencies (vibrational bands) within molecular components of bacterial cells/spores, with the significant contribution from DNA [4]. The obtained results suggest that THz vibrational spectroscopy promises to add quantitative genetic information to the characteristic signatures of biological objects, increasing characterization accuracy and selectivity when appropriate spectral resolution, which is adequate to the widths of spectral lines, is used. The significance of this study is justified by necessity for a fast and effective, label free and reagent free optical technology to protect against environmental biological threats, as well as for general medical research. The ability to discriminate between the different bacterial species quickly and reliably using sub-THz spectroscopy would provide significant benefits. In the medical field it would enable a faster and more tailored treatment once a bacterial organism is identified as the cause of an infection. At the same time, although significant progress in experimental THz spectroscopy was demonstrated and reliable information was received for transmission/absorption spectra from different species, the spectral resolution of Bruker spectrometer (0.25 cm^{-1}) still does not provide a sufficient level of discriminative capability. The shape of the curve and the absorption peak intensities were rather close for different species. It became clear that further improvement of sensitivity and especially of discriminative capability using sub-THz vibrational spectroscopy as an effective method for characterization of bacterial organisms requires even better spectral resolution.

The width of individual spectral lines and the intensity of resonance features observed in sub-THz spectroscopy are sensitive to the relaxation processes of atomic dynamics (displacements) within a macromolecule. It is clear that the decay (relaxation) time, τ, is the factor limiting the spectral width and the intensity of vibrational modes, the required spectral resolution, and eventually the discriminative capability of sub-THz spectroscopy. At the same time, the entire mechanism that determines intra-molecular relaxation dynamics is still not completely understood. The suggested range of molecular dynamics relaxation times for processes without biomolecular conformational change varies from approximately 1.5 ps to 650 ps in different studies [see, for example 20, 21]. The corresponding values for the dissipation factor, γ, and the width of spectral lines, which are reciprocal to τ, are between 0.05 and 20 cm^{-1}. Values of γ above 1 cm^{-1} would result in structure-less sub-THz spectra, since vibrational resonances could not be resolved in this case because of the large density of low intensity vibrational modes. The existence of long-lasting dynamic processes responsible for narrow spectral lines has been confirmed by relaxation dynamics of side chains in macromolecules observed by time-resolved fluorescence experiments [22].

To increase the sensitivity, reliability, spectral and spatial resolution of sub-THz vibrational spectroscopy techniques, Vibratess, LLC, has developed a spectroscopic sensor prototype with imaging capability operating at room temperature, without the need for cryogenic cooling of the detector [2]. This novel CW, frequency-domain instrument is based on a very strong local enhancement of the electro-magnetic field, thus allowing increased coupling of the THz radiation with the sample biomaterials [23,24]. This enhancement was achieved through the use of the discontinuity edge effect and the extraordinary transmission of a sub-wavelength-slit conductive structure [25-28]. Observed multiple intense and specific resonances in transmission/absorption spectra from nano-gram samples with spectral line widths as small as 0.1 cm^{-1} provide conditions for reliable discriminative capability, potentially to the level of the strains of the same bacteria, and for monitoring interactions between biomaterials and reagents in near real-time. Only ~20 ng of biomaterial is required as the sample in our system as compared to the mg sample size required in the previous work done on the Bruker spectrometer (see above). With the complete development of a sealed micro/nanofluidic chip sample holder, liquid samples will be utilized, and the amount of biomaterial required for characterization will be further reduced ~10 to 100 times, thus opening the way for single bio-molecule characterization.

The developed prototype provides spectral resolution better than 0.035 cm^{-1}, and significantly improved the detection sensitivity and reliability in the sub-THz operation range as compared to a commercially available spectrometer with a liquid helium cooled detector. Spatial resolution of the instrument is currently restricted by the opening size of the microdetector waveguide. Highly resolved transmission (absorption) spectra from only 10 - 20 ng of biological macromolecules and bacterial cells/ spores were demonstrated.

The experimental results measured with high spectral resolution reveal very intense and narrow spectral features from biological molecules and bacteria with widths ~0.1 - 0.2 cm^{-1}. This corresponds to much longer scattering time values as compared to those previously evaluated using a spectrometer with a resolution of 0.25 cm^{-1}. The narrow width of the spectral features (or small dissipation factor) in the transmission (absorption) spectra in the THz region makes these lines detectable [29]. Thus, a new sub-THz vibrational spectroscopy technology with high spectral and spatial resolution was developed and experimentally demonstrated in general agreement with modeling results.

MOLECULAR DYNAMICS SIMULATIONS OF SUB-THZ VIBRATIONAL MODES AND ABSORPTION SPECTRA OF DNAS AND PROTEINS

Molecular Dynamics

We were working on MD simulations of sub-THz molecular vibrations and absorption spectra of proteins and DNAs with two goals: 1) to establish theoretical basis for exploring THz region of electro-magnetic (EM) spectrum for the discovery of new spectral signatures from biological materials; and 2) to improve predictive capabilities of MD computational modeling of THz vibrational absorption spectra from biological molecules. The protein thioredoxin from E. coli with known structure is used as a model molecule [pdb ID: 2TRX] to simulate sub-THz vibrational absorption using the software packages Amber 8 [30] and Amber 10 [31]. This small protein contains 108 amino acids, for a total of 1654 atoms. To solvate a sample, an additional 10,500 water atoms are added. Some absorption features predicted by our earlier MD simulations agreed reasonably well with experimental data [9,10,15]. However, the calculated spectra were highly sensitive to the parameter values, and reproducibility was poor. This problem of poor simulation convergence was discussed in work of many authors [32]. Amber was empirically parameterized to correctly represent the structural behavior of nucleic acid and protein as would be needed for predicting non-bond-breaking conformational changes [33]. It was not specifically created to simulate low frequency vibrational modes and THz absorption.

In our recent study [6], MD simulations of sub-terahertz (THz) vibrational modes of the protein thioredoxin was conducted with the goals of finding the conditions needed for simulation convergence, improving the correlation between experimental and simulated spectra, and ultimately enhancing the predictive capabilities of computational modeling. We studied the consistency, accuracy and convergence of MD simulations of the sub-THz vibrational modes by comparing simulations with different initial conditions, protocols and parameters to the experimental results.

Better simulation convergence and improved consistency between simulated vibrational frequencies and experimental data were obtained by using a new procedure for averaging mass-weighted covariance matrices of atomic trajectories in MD simulations. In particular, the open source package ptraj was edited to improve a matrix analyzing function. Averaging of only six matrices gives much more consistent results, with absorption peak intensities exceeding those from the individual spectra and with a rather good correlation between simulated vibrational frequencies and experimental data. We also found that the choice of the production run length considerably influences the obtained absorption spectra. The optimal time for dividing production run into equal subintervals to calculate individual correlation matrices is equal to ~100 ps. This result is in general agreement with relaxation dynamics time scales of the thioredoxin active center, coupled protein-water fluctuations [20,22], and our experimental data on the spectral width of vibrational modes [34,35].

Absorption Coefficient Spectra

Atomic trajectories collected in MD simulations are converted to the covariance matrix of atomic displacements $\langle R_i R_k \rangle$ using a quasi-harmonic approximation. The force-field matrix is found utilizing the relation between the covariance matrix and the inverse of the forceconstant matrix $(\langle R_i R_k \rangle = k_B T [F^{-1}]_{ik})$, where R—displacements, F—force constants [36,37]. Diagonalization of F matrix gives eigenfrequencies (normal mode frequencies) and eigenvectors (displacement vectors-normal modes). The absorption coefficient spectr $a(v)$ as functions of the frequency n are calculated through the relationship between a and the imaginary part of dielectric permittivity [12]:

$$\alpha(v) = \gamma v^2 \sum_k \frac{S_k}{\left(v^2 - v_k^2\right)^2 + \gamma^2 v^2}, \qquad (1)$$

where n_k are normal mode frequencies calculated by diagonalization of the force-constant matrix, and S_k are oscillator strengths computed for all vibrational modes k. Two values of oscillator dissipation for all vibrational modes in sub-THz range were accepted from our experimental works as $\gamma = 0.5$ cm^{-1} (moderate spectral resolution in Bruker spectrometer), and $\gamma = 0.1$ cm^{-1} for a highly resolved spectroscopy using Vibratess spectrometer. **Figure 1** demonstrates correlation between absorption spectrum of thioredoxin simulated with $\gamma = 0.5$ cm^{-1} and experimental results as measured with a moderate spectral resolution of 0.25 cm^{-1}.

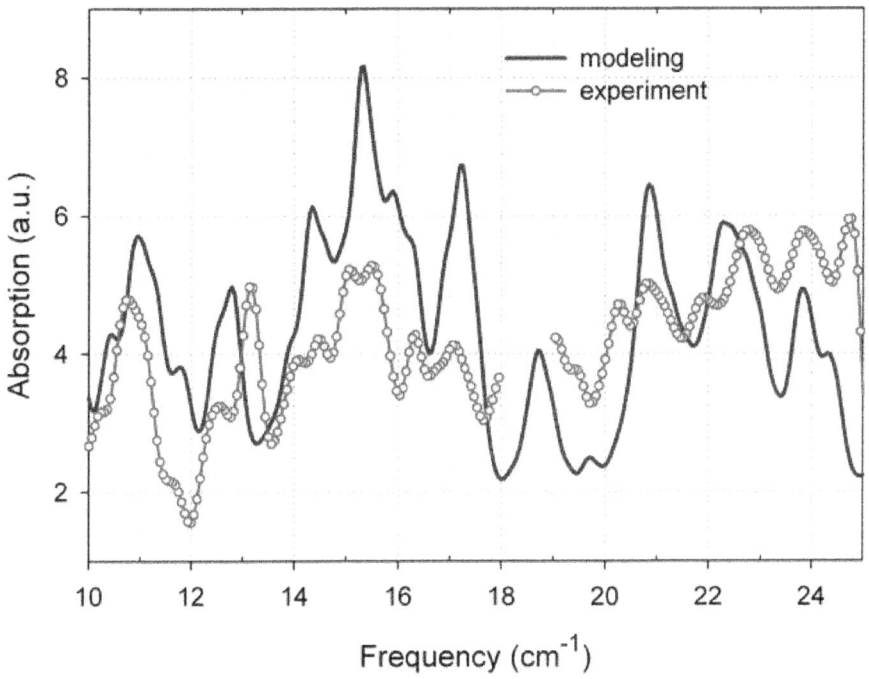

Figure 1. Absorption coefficient spectrum of protein thioredoxin from E. coli measured with spectral resolution of 0.25 cm^{-1} (Bruker spectrometer) is compared with the modeling result (averaging correlation matrix, $\gamma = 0.5$ cm^{-1}. Data are taken from our paper [6]).

STATISTICAL MODEL FOR E. COLI DNA SEQUENCE USING MONTE-CARLO TECHNIQUE FOR MARKOV CHAIN

Terahertz spectroscopy of biological macromolecules reflects low frequency internal molecular vibrations. Relatively new experimental sub-THz spectroscopy has already demonstrated significant achievements in the last decades in regards to sample preparation techniques and enhancing characterization sensitivity and results reproducibility. It was demonstrated that sub-THz radiation can effectively be used to identify various complex biological molecules. However, deeper understanding of interaction mechanism of THz radiation with biological molecules requires development of computational modeling in parallel with experimental studies. We have recently simulated spectra of relatively small biological molecules like transfer RNA or protein thioredoxin from E. coli using molecular dynamic (MD) simulations (sections 2 and 3). We demonstrated a rather good correlation of simulated spectra with experimental data (see, for example, **Figure 1**). However a large size of macromolecules (~5 million base pairs for E. coli DNA) prevents direct application of MD simulation at the current level of computational capabilities. The goal of this work is to develop a simplified model of the DNA macro molecule so that the model would capture the most important low-frequency vibrational characteristics of the native DNA. One way to reach the goal is to build a modeling sequence by using the most frequent repeating fragments (2 - 10 base pairs) occurred in the original DNA. The constructed models and MD simulations of the modeled sequences can permit us to calculate expected absorption spectra, and to better understand the mechanism of interaction of THz radiation with a biological molecule by analyzing dynamics of atoms and correlation of local vibrations in the modeled molecule.

Statistical Model for E. coli DNA Sequence

We developed a new procedure to construct a short DNA sequence much less than the length of the genome to model a whole bacterial genome. We used a second order Markov chain framework combined with a Monte-Carlo technique. The statistical model approach is based on conditional probabilities of occurrence of a single base X_{i+2}, given that two previous bases in the sequence are X_i and X_{i+1} [38].

Using Monte-Carlo technique it is possible to find the most probable sequence of a length L, when random sequences of this length are generated using the conditional probabilities mentioned above [39]. The most probable first two bases have to be found directly from the genome. Then the third base can be found as having the greatest occurrence in random sequences, given that first and second ones have already been determined. Forth base can be found the same way, given that first three bases are known. Applying this algorithm iteratively, each base in the sequence can be specified. An additional condition is applied to every random sequence to be accepted and used in the further analysis:

$$\left(R_g - \Delta < R_s < R_g + \Delta\right), \Delta \ll R_g, \qquad (2)$$

where R_g and R_s are the ratios of the nucleotide of a certain type to the total number of nucleotydes in genome and in the randomly generated sequence, correspondently. Δ is the tolerance parameter which determines how accurately should be the correspondence between R_g and R_s. In our case, $\Delta = 0.007$.

E. coli bacteria include different strain groups with sequences similarity between strains in one group. Statistical models permit us to find strains, which can be discriminated on the basis of their modeled sequencing resulted in specificity of their vibrational spectroscopic signatures. By generating statistical models for different strains, we can predict that if some DNA strains have different modeled sequences they may also have different absorption spectra. E. coli strain BL21 (4534552 bp) derived from E. coli strain B is commonly used as a host strain for protein expression and purification [40-42]. The highly virulent strain CFT073 is one of uropathogenic strains of E. coli— the most common cause of non-hospital-acquired urinary tract infections [43].

The sequences for two E. coli strains, pathogenic CFT073 and non-pathogenic BL21, are quite different: BL21—GCGCGCAGCATTTTTTTCAGCGCAGCGAA AAATTTCGCGCGCAGTTT AACGCGATCAGT, CFT073—GCGCAGC AGCAC ATTTTT TTTCA GC GCAGCAGCAGATT TTCAGCA GATCAGC G ATCAGT, and we can expect a noticeable difference in simulated absorption spectra from these two strains.

We have modeled 20, 40 and 60 base sequences for two E. coli strains, a non-pathogenic strain BL21 and a pathogenic strain CFT073. Calculated sequences for 20, 40 and 60 bases of a CFT073 E. coli strain are compared in the **Table 1**. Since the proposed modeling approach is used to determine the most frequent pattern in the genome, we suggest that with increasing the length, the modeled sequence becomes more representative and able to accurately reflect characteristic features of genomic DNA. E. coli genome contains large numbers of repeated fragments of different length like TTTTT. The presentation of these fragments in a statistical model is improving with increased model length and we expect that the 60 bp sequence is more accurate.

Discriminative Capability

Table 2 lists generated DNA 60 bp sequences from our models for several E. coli strains. By generating statistical models for different strains we can verify that if some DNA strains have different modeled sequences they may also have different absorption spectra. For the first three strains, K-12 DH10B, K-12 BW2952, and K-12 MG- 1655, our statistical models are identical as will be their THz spectra simulated on the bases of these models. As a result, we will not be able to discriminate between these strains in modeling of 60 bases. We also expect that these strains will be more difficult to discriminate experimentally. The sequences for two E. coli strains, pathogenic CFT073 and non-pathogenic BL21, are quite different and we expected a noticeable difference in simulated absorption spectra from these two strains. This is in fact confirmed by the results of MD simulations.

Molecular Dynamic Simulation Results

Structure

Using our statistical models in MD simulations permits us to generate structure of DNA strains. Figures 2(a) and 2(b) compare structure of DNA 60 bp models of two strains in 12 Angstrom water box before and after energy minimization (The pictures are generated using Chimera (http://www.cgl.ucsf.edu/chimera/). The molecules became more compact after minimization. It's clearly seen in upper left corner of the pictures.

Absorption Spectra

We calculated absorption spectra for 20 - 60 base model sequences. MD simulation with explicit water was applied. We investigated the convergence of DNA MD simulation using the approach that was originally developed for E. coli protein thioredoxin to calculate absorption spectra with averaging of atomic displacement correlation matrices as described in [6]. **Figure 3" target="_self"> Figure 3** compares two simulated spectra for 60 bp sequence of CFT073 E. coli strain model presented in Tables 1 and 2. The first spectrum is obtained in a 600 ps production run, and the second one is calculated using averaging correlation matrices procedure for 6 × 100 ps intervals taken from the same run. We consider that the convergence is good (results of MD simulations are stable) if only small differences are observed.

Table 1. Modeled 20, 40 and 60 base sequences for CFT073 E. coli strain.

Model length	Modeled sequence 5'-3'
20 bp	GCAGCATTTCAGCGATCAGT
40 bp	GCGCAGCATTTTTCAGCAGCAGCAGTTTAACGCGATCAGT
60 bp	GCGCAGCAGCAGCATTTTTTTTCAGCGCAGCAGCAGATTTTCAGCAGATCAGCGATCAGT

Table 2. Comparison of different E. coli strains.

Strain	Modeled sequence 5'-3'
K-12 DH10B	GCGCGCAGCATTTTTTTCAGCAGCAGCAGCAGCAGATTTTTAAACGCGCGATTCAGCGAT
K-12 BW2952	GCGCGCAGCATTTTTTTCAGCAGCAGCAGCAGCAGATTTTTAAACGCGCGATTCAGCGAT
K-12 MG1655	GCGCGCAGCATTTTTTTCAGCAGCAGCAGCAGCAGATTTTTAAACGCGCGATTCAGCGAT
K-12 DH1	GCGCGCAGCATTTTTTTCAGCGCAGCAGAAAAATTTCGCGCGCAGTTTAACGCGATCAGT
BL21	GCGCGCAGCATTTTTTTCAGCGCAGCAGAAAAATTTCGCGCGCAGTTTAACGCGATCAGT
B7A	GCGCAGCAGCATTTTTTTCAGCGCAGCAGCAGCAGATTTTTCAGCAGCAGATTCAGCGAT
CFT073	GCGCAGCAGCAGCATTTTTTTTCAGCGCAGCAGCAGCAGATTTTCAGCAGATCAGCGATCAGT
EDL933 (deadly)	GCGCAGCAGCAGCTGATTTTTTTCAGCAGCAGCAGCATTTTAAACGCGCGTTAACGCAGT

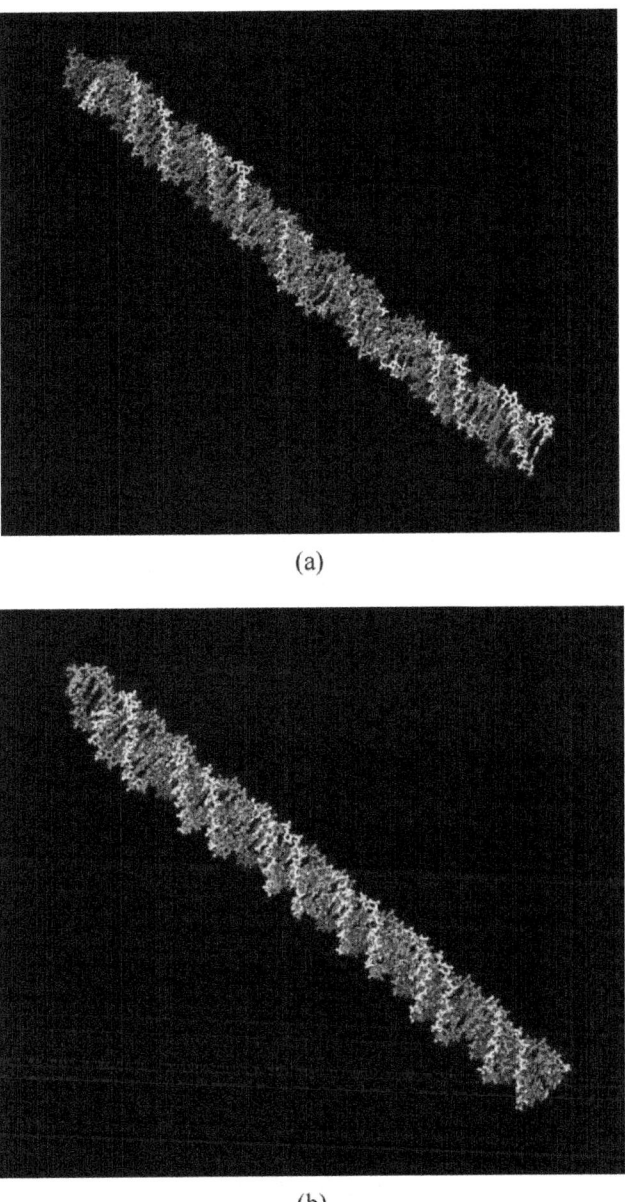

Figure 2. (a) BL21strain: green—before energy minimization; yellow—after minimization; (b) CFT073 (pathogenic) strain: blue—before energy minimization; yellow—after minimization.

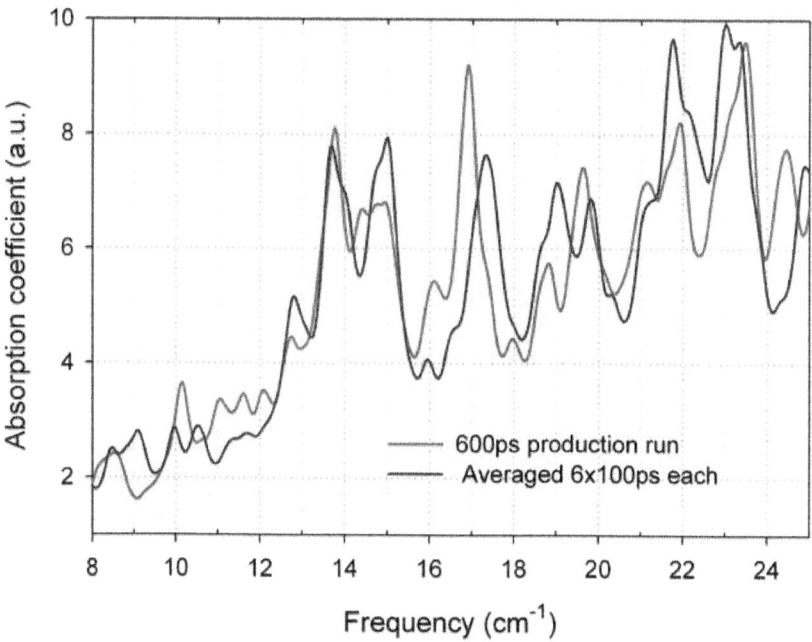

Figure 3. Simulated spectra for 60 bp sequence of CFT073 E. coli strain model. Red—600 ps production run; brown—averaging correlation matrices from six equal intervals of 100 ps each. Dissipation factor is 0.5 cm^{-1} for a moderate spectral resolution of 0.25 cm^{-1} (Bruker FTIR). Water box 12 A.

As it is demonstrated in **Figure 3**, the results of MD simulations are stable thus indicating a good convergence of MD simulation for DNA molecules. This result is consistent with the general opinion that convergence problem in MD is less crucial for DNAs compare to proteins.

As expected, spectra are sensitive to the number of base pairs in the model (see **Figure 4** for 40, and 60 bp). Although models with larger number of base pairs probably give better presentation of absorption spectrum, we are currently limited to 60 bp.

To demonstrate the effect of DNA sequence on their THz vibrational spectra we simulated absorption spectra for CFT073 and BL21 (**Figure 5**). The modeling results predict that we can discriminate between pathogenic and nonpathogenic strains of E. coli 60 bp models using their sub-THz vibrational spectra.

Terahertz Vibrational Spectroscopy of E. coli and Molecular Constituents... 265

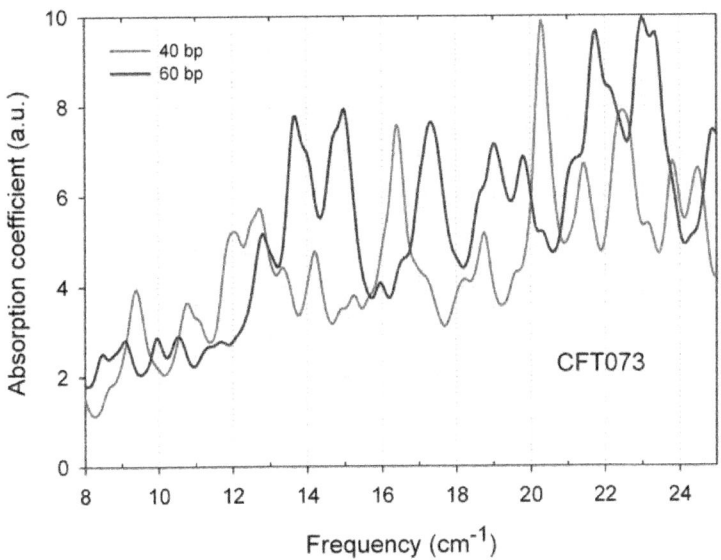

Figure 4. THz absorption spectra of CFT073 E. coli strain models with 40 and 60 bp. Dissipation factor is 0.5 cm^{-1} for a moderate spectral resolution of 0.25 cm^{-1} (Bruker FTIR). Averageing 6 matrices, water box 12A.

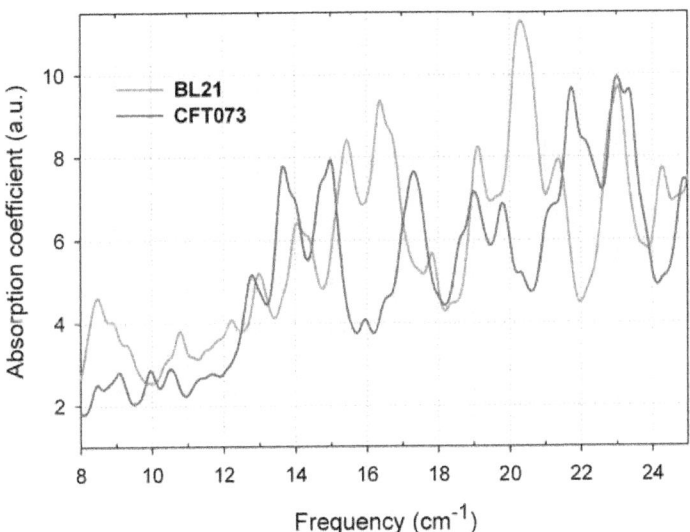

Figure 5. THz absorption spectra of two E. coli strains, pathogenic CFT073 and non-pathogenic BL21, 60 bp models, six matrices, water box 12 A. Dissipation factor is 0.5 cm^{-1} for a moderate spectral resolution of 0.25 cm^{-1} (Bruker FTIR).

Higher spectral resolution gives even better results. Absorption spectra from non-pathogenic BL21strain and deadly strain EDL933 [44] using 60 bp models (water box 12 A, averaged in all three directions, dissipation factor 0.12 cm^{-1}) are shown below in **Figure 6**. Many features for discrimination are available. This spectral resolution is already demonstrated in the range 11 - 16.8 cm^{-1} using Vibratess spectrometer.

Effect of Water

When modeling biological molecules in water solutions, we actually generate and study complexes of biomaterial with water. In our standard simulation we put a biomolecule inside a 12 A water box containing more than 30 thousands of water molecules. It is a known fact that the first several layers of water, the mostly close to a bio-molecule, have different 3D structure and properties compare to water layers far away from a solvate [20]. These "internal" layers are tight bonded and can have almost crystalline structure with large number of hydrogen bonds which might contribute to vibrational spectra.

Figure 7 shows significant changing of E. coli BL21 strain absorption spectrum when water box in simulation was reduced only from 12 to 10 A. By modifying one of MD parameters, the size of water box, we can study variability of THz absorption spectra from bio-molecules depending on material concentration in water solution and even at transition to dry condition. Experimental absorption spectrum for a virtually dry E. coli B strain is also shown, with significantly reduced intensities of peaks.

Highly Resolved Vibrational Spectroscopy (0.03 cm^{-1}): Comparison with Experiment

In this section we demonstrate some results from spectroscopy with the resolution of 0.03 cm^{-1} using Vibratess spectrometer. Transmission spectra were obtained in the sub-THz region between 315 and 480 GHz for both, macromolecules and biological species. Due to high sensitivity, good spectral resolution, and spatial resolution below the diffraction limit, this spectroscopic instrument permits us to observe intense and narrow spectral resonances in transmission/absorption spectra of nano-samples from biological materials. To demonstrate the capabilities of the spectrometer, transmission spectra from bacterial cells and some of their molecular components (DNA, thioredoxin) were measured. From the transmission spectrum of E. coli DNA shown in **Figure 8**, the width of spectral lines can be estimated as ~ 0.1 cm^{-1}.

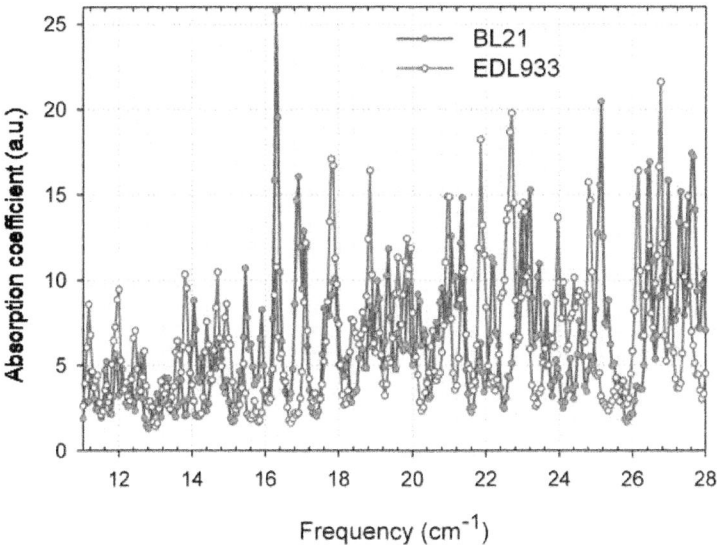

Figure 6. Highly resolved absorption spectra from E. coli nonpathogenic strain BL21, and deadly strain EDL933 (O157:H7), 60 bp models, water box 12 A, averaged in all three directions, dissipation factor 0.12 cm^{-1}.

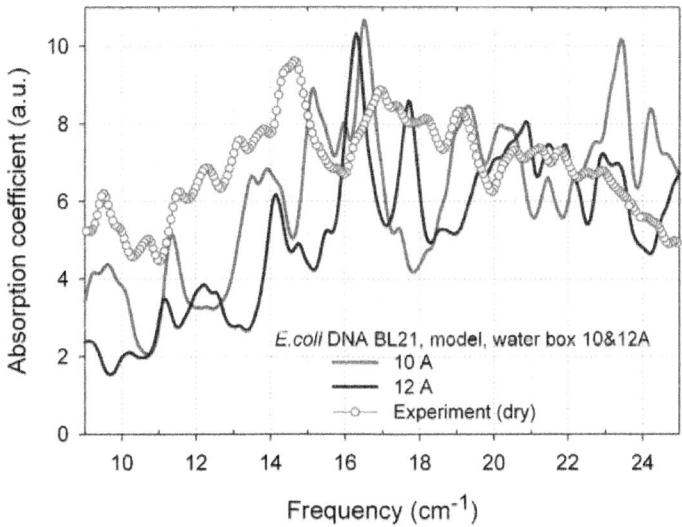

Figure 7. Simulated absorption spectra of E. coli BL21 strain in water box 10 and 12 A, 600 ps, $\gamma = 0.5$ cm^{-1}. Experimental data for virtually dry E. coli B are also shown. Spectral resolution 0.25 cm^{-1}.

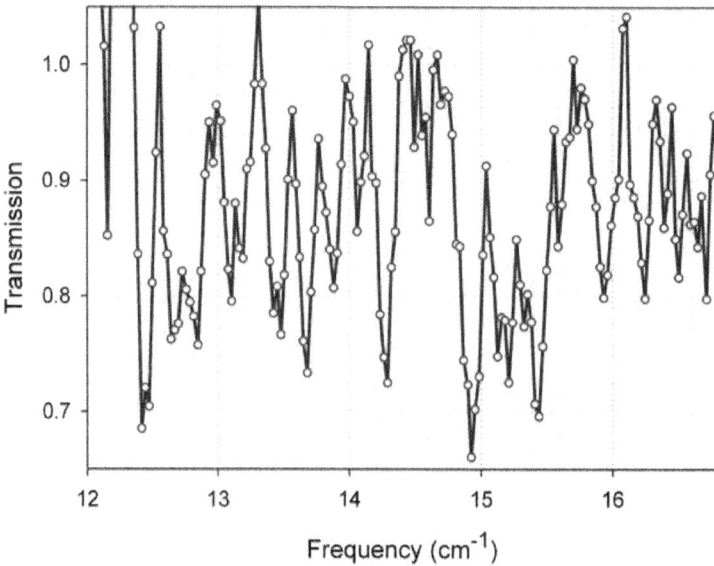

Figure 8. Transmission spectrum of E. coli DNA, (500 ng of material in the drop). Figure is taken from our paper [2].

To further confirm the reality of the observed narrow and intense resonance features in the sub-THz transmission/absorption spectra of biological materials as measured with the new spectroscopic sensor, we compared in **Figure 9** the spectrum from the E. coli protein thioredoxin with computational modeling results using MD simulations with a damping factor of $\gamma = 0.12$ cm^{-1}.

Due to possible contributions from several different modes occurring at close frequencies, the width of spectral lines gives us an upper limit of γ. As seen in **Figure 9**, not all peaks are reproduced in the measured and simulated spectra, since simulation parameters have not yet been optimized. Besides, the same value of γ was used to calculate absorption for all vibrational modes. However, the overall correlation between the theory and experimental data confirms again the existence of intense and narrow absorption lines, which can be used for discrimination between different bacteria and strains.

CONCLUSION

In this work we presented computational and experimental results with the goal to establish a theoretical basis for exploring THz spectrum for the discovery of new spectral signatures from biological materials. We have developed a new statistical model to construct DNA sequences significantly less than the length of the entire genome (20 - 60 base pairs), using the most frequently repeated fragments (2 - 10 base pairs) in the original DNA. We analyzed the results from MD simulations of our new developed models of DNA sequences from E. coli. We demonstrated that the application of molecular dynamics simulations to the 60 base pair DNA models promises high discriminative capability for biosensing in sub-THz regime. MD simulation of the chromosomal DNA of select model organisms revealed that in the case of a good spectral resolution discrimination is possible up to the level of strains of one bacterial species.

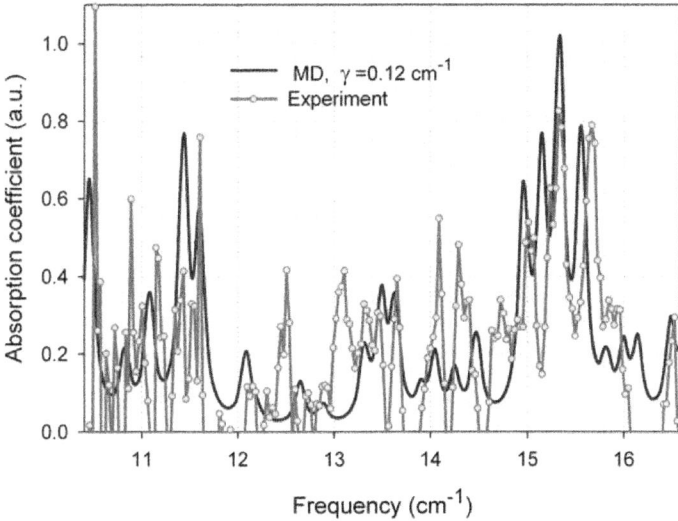

Figure 9. Absorption spectrum of protein thioredoxin from E. coli: MD simulation and experimental results as measured using Vibratess spectroscopic sensor. Figure is taken from our paper [2].

Modeling results are supported by experimental spectroscopic data. The experimental results measured with high spectral resolution demonstrate very intense and narrow spectral features from DNA and a protein thioredoxin from E. coli with the line width ~0.1 cm^{-1}. These results combined with MD simulation confirm that highly resolved sub-THz vibrational spectroscopy can be used for reliable and accurate detection of nanograms of E. coli using optical, highly sensitive biosensors operating at room temperature with significantly improved ability to discriminate between species up to the level of the strains of the same bacteria.

ACKNOWLEDGEMENTS

This work is supported by the Defense Threat Reduction Agency, grant #HDTRA1-08-1-0038 and by the contract from the ARO #W911NF- 08-C-0049.

REFERENCES

1. Finlay, B.B. (2010) The art of bacterial warfare. Scientific American, 302, 56-63.doi:10.1038/scientificamerican0210-56
2. Globus, T., Moyer, A., Gelmont, B., Khromova, T., Lvovska, M., Sizov, I. and Ferrance, J. (2012) Highly resolved sub-Terahertz vibrational spectroscopy of biological macromolecules and cells. IEEE Sensors Journal, 13, 72-79.doi:10.1109/JSEN.2012.2224333
3. Globus, T., Woolard, D., Crowe, T.W., Khromova, T., Gelmont, B. and Hessler J. (2006) Terahertz fourier transform characterization of biological materials in a liquid phase. Journal of Physics D: Applied Physics, 39, 3405- 3413. doi:10.1088/0022-3727/39/15/028
4. Globus, T., Dorofeeva, T., Sizov, I., Gelmont, B., Lvovska, M., Khromova, T., Chertihin, O. and Koryakina Y. (2012) Sub-THz vibrational spectroscopy of bacterial cells and molecular components. American Journal of Biomedical Engineering, 2, 143-154.
5. Bykhovski, A., Globus, T., Khromova, T., Gelmont, B., Woolard, D.L. and Bykhovskaia, M. (2006) An analysis of the THz frequency signatures in the cellular components of biological agents. SPIE Defense and Security Symposium, 6212, 132-141.
6. Alijabbari, N., Chen, Y., Sizovm I., Globus, T. and Gelmont, B. (2012) Molecular dynamics modeling of the subTHz vibrational absorption of thioredoxin from E. coli. Journal of Molecular Modeling, 18, 2209-2218. doi:10.1007/s00894-011-1238-6
7. Sizov, I., Gelmont, B. and Globus, T. (2011) Statistical model for E. coli DNA sequence using Monte-Carlo technique for Markov chain. Study of sub-THz molecular vibrations. DTRA CBD S&T Conference, Las-Vegas, 5 October 2011.
8. Bykhovski, A., Li, X., Globus, T., Khromova, T., Gelmont, B., Woolard, D., Samuels, A. and Jensen, J. (2005) THz absorption signature detection of genetic material of E. coli and B. subtilis. SPIE Proceedings of Chemical and Biological Standoff Detection III, 5995, 2005.
9. Bykhovski, A., Globus, T., Khromova, T., Gelmont, B. and Woolard, D. (2007) Resonant Terahertz spectroscopy of bacterial thioredoxin in water: Simulation and experiment. ISSSR-Proceedings, 888-896.
10. Globus, T. (2010) Low-Terahertz resonance spectroscopy for fingerprinting of biological and organic materials. DTRA CBD S&T Conference, Orlando, 15-19 November 2010.
11. Globus, T., Bykhovskaia, M., Woolard, D. and Gelmont, B. (2003) Sub-

millimeter wave absorption spectra of artificial RNA molecules. Journal of Physics D: Applied Physics, 36, 1314-1322. doi:10.1088/0022-3727/36/11/312

12. Globus, T., Woolard, D., Bykhovskaia, M., Gelmont, B., Werbos, L. and Samuels, A. (2003) THz spectroscopic sensing of DNA and related biological materials. International Journal of High Speed Electronics, 13, 903. doi:10.1142/S0129156403002083

13. Bykhovskaia, M., Gelmont, B., Globus, T., Woolard, D., Samuels, A., Ha-Duong, T. and Zakrzewska, K. (2001) Prediction of DNA far IR absorption spectra basing on normal mode analysis. Theoretical Chemistry Accounts, 106, 22-27.doi:10.1007/s002140100259

14. Globus, T., Bykhovskaia, M., Gelmont, B. and Woolard, D. (2001) Far-infrared phonon modes of selected RNA molecules. Proceedings of SPIE, Instrumentation for Air Pollution and Global Atmospheric Monitoring, 4574, 119. doi:10.1117/12.455149

15. Li, X., Globus, T., Gelmont, B., Salay, L. and Bykhovski, A. (2008) Terahertz absorption of DNA decamer duplex. The Journal of Physical Chemistry A, 112, 12090-12096.doi:10.1021/jp806630w

16. Heilweil, E.J. and Plusquelic, D.F. (2008) Terahertz spectroscopy of biomolecules, In: Terahertz Spectroscopy, CRC Press, Taylor & Francis Group, LLC, London and New York, 269-297.

17. Plusquellic, D.F., Siegrist, K., Heilweil, E.J. and Esenturk, O. (2007) Applications of Terahertz spectroscopy in biosystems. ChemPhysChem, 8, 2412-2431.doi:10.1002/cphc.200700332

18. Zhang, H., Siegrist, K., Douglas, K.O., Gregurick, S.K. and Plusquellic, D.F. (2008) THz investigations of condensed phase biomolecular systems. Methods in Cell Biology, 90, 417-434. doi:10.1016/S0091-679X(08)00818-2

19. Korter, T.M. and Plusquellic, D.F. (2004) Continuouswave Terahertz spectroscopy of biotin: Vibrational anharmonicity in the far-infrared. Chemical Physics Letters, 385, 45-51.doi:10.1016/j.cplett.2003.12.060

20. Li, T., Hassanali, A.A., Kao, Y.T., Zhong, D. and Singer S.J. (2007) Hydration dynamics and time scales of coupled water-protein fluctuations. Journal of the American Chemical Society, 129, 3376-3382. doi:10.1021/ja0685957

21. Furse, K.E. and Corcelli, S.A. (2010) Molecular dynamics simulations of DNA solvation dynamics. The Journal of Physical Chemistry Letters, 1, 1813-1820.doi:10.1021/jz100485e

22. Qiu, W., Wang, L., Lu, W., Boechler, A., Sanders, D.A.R. and Zhong, D. (2007) Dissection of complex protein dynamics in human thioredoxin. Proceedings of the National Academy of Sciences of the United States of America, 104, 5366-5371.doi:10.1073/pnas.0608498104
23. Globus, T., Moyer, A., Gelmont, B., Sizov, I. and Khromova, T. (2011) Dissipation time in molecular dynamics and discriminative capability for sub-Terahertz spectroscopic characterization of bio-simulants. DTRA CBD S&T Conference, Las-Vegas, 14-18 November 2011.
24. Globus, T., Moyer, A., Gelmont, B., Lichtenberger, A., Ferrance, J., Weikle, R. and Lvovska, M. (2011) Sub-Terahertz spectroscopy of bio-materials with high spectral and spatial resolution. Nano DDS, Salt Lake City, 2011.
25. Parthasarathy, R., Bykhovski, A., Gelmont, B., Globus, T., Swami, N. and Woolard, D.L. (2007) Enhanced coupling of sub-Terahertz radiation with semiconductor periodic slot arrays, Physical Review Letters, 98, Article ID: 153906.doi:10.1103/PhysRevLett.98.153906
26. Gelmont, B., Parthasarathy, R., Globus, T., Bykhovski, A. and Swami, N. (2008) Terahertz (THz) electromagnetic field enhancement in periodic subwavelength structures. IEEE Sensors Journal, 8, 791-796. doi:10.1109/JSEN.2008.923222
27. Gelmont, B. and Globus, T. (2011) Edge effect in perfectly conducting periodic subwavelength structures. IEEE Transactions on Nanotechnology, 10, 83-87.doi:10.1109/TNANO.2010.2064785
28. Gelmont, B., Globus, T., Bykhovski, A., Lichtenberger, A., Swami, N., Parthasarathy, R. and Weikle, R. (2012) Method of local electro-magnetic field enhancement of Terahertz (THz) radiation in sub wavelength regions and improved coupling of radiation to materials through the use of the discontinuity edge effect. US Patent No. 8309930.
29. Woolard, D., Brown, E., Pepper, M. and Kemp, M. (2005) Terahertz frequency sensing and imaging: A time of reckoning future applications? Proceedings of the IEEE, 93, 1722.doi:10.1109/JPROC.2005.853539
30. Case, D.A., Pearlman, D.A., Caldwell, J.W., Cheatham III, T.E., Wang, J., Ross, W.S., Simmerling, C.L., Darden, T.A., Merz, K.M., Stanton, R.V., Cheng, A.L., Vincent, J.J., Crowley, M., Tsui, V., Gohlke, H., Radmer, R.J., Duan, Y., Pitera, J., Massova, I., Seibel, G.L., Singh, U.C., Weiner, P.K. and Kollman, P.A. (2004) AMBER 8. University of California, San Francisco.
31. Case, D.A., Darden, T.A., Cheatham III, T.E., Simmerling, C.L., Wang, J., Duke, R.E., Luo, R., Crowley, M., Ross, W.S., Zhang, W., Merz, K.M.,

Wang, B., Hayik, S., Roitberg, A., Seabra, G., Kolossváry, I., Wong, K.F., Paesani, F., Vanicek, J., Wu, X., Brozell, S.R., Steinbrecher, T., Gohlke, H., Yang, L., Tan, C., Mongan, J., Hornak, V., Cui, G., Mathews, D.H., Seetin, M.G., Sagui, C., Babin, V. and Kollman, P.A. (2008) AMBER 10. University of California, San Francisco.

32. Liki'c, V.A., Gooley, P.R., Speed, T.P. and Strehler, E.E. (2005) A statistical approach to the interpretation of molecular dynamics simulations of calmodulin equilibrium dynamics. Protein Science, 14, 2955-2963. doi:10.1110/ps.051681605

33. Lewars, B.G. (2003) Computational chemistry: Introduction to the theory and applications of molecular and quantum mechanics. Springer, Berlin.

34. Bykhovski, A., Globus, T., Khromova, T., Gelmont, B. and Woolard, D. (2008) Resonant Terahertz Spectroscopy of Bacterial Thioredoxin in Water: Simulation and Experiment. In: Selected Topics in Electronics and Systems. Spectral Sensing Research for Water Monitoring Applications and Frontier Science and Technology for Chemical, Biological and Radiological Defense. World Scientific Publishing Co., Singapore City, 48, 367-375.

35. Globus, T., Alijabbary, N., Chen, Y., Sizov, I. and Gelmont, B. (2010) Molecular dynamics modeling of subTHz vibrational absorption of thioredoxin from E. coli. DTRA CBD S&T Conference, Orlando, 15-19 November 2010.

36. Karplus, M. and Kushick, J.N. (1981) Method for estimating the configurational entropy of macromolecules. Macromolecules, 14, 325-332. doi:10.1021/ma50003a019

37. Levy, R.M., Karplus, M., Kushick, J. and Perahia, D. (1984) Evaluation of the configurational entropy for proteins: Application to molecular dynamics simulations of an alfa-helix. Macromolecules, 17, 1370-1374. doi:10.1021/ma00137a013

38. Durbin, R., Eddy, S., Krogh, A. and Mitchison, G. (1998) Biological sequence analysis: Probabilistic models of proteins and nucleic acids. Cambridge University Press, Cambridge, 368.

39. Ching, W.-K., Ng, M.K. and Ching, W. (2005) Markov chains: Models, algorithms and applications. Springer, 208.

40. Studier, F.W. and Moffatt, B.A. (1986) Use of bacteriophage T7 RNA polymerase to direct selective high-level expression of cloned genes. Journal of Molecular Biology, 189, 113-130. doi:10.1016/0022-2836(86)90385-2

41. Grodberg, J. and Dunn, J.J. (1988) ompT encodes the Escherichia coli

outer membrane protease that cleaves T7 RNA polymerase during purification. Journal of Bacteriology, 170, 245-253.

42. Phillips, T.A., VanBogelen, R.A. and Neidhardt, F.C. (1984) Lon gene product of Escherichia coli is a heat-shock protein. Journal of Bacteriology, 159, 283-287.

43. Welch, R.A., Burland, V., Plunkett III, G., Redford, P., Roesch, P., Rasko, D., Buckles, E.L., Liou, S.-R., Boutin, A., Hackett, J., Stroud, D., Mayhew, G.F., Rose, D.J., Zhou, S., Schwartz, D.C., Perna, N.T., Mobley, H.L.T., Donnenberg, M.S. and Blattner, F.R. (2002) Extensive mosaic structure revealed by the complete genome sequence of uropathogenic Escherichia coli. Proceedings of the National Academy of Sciences of the United States of America, 99, 17020-17024. doi:10.1073/pnas.252529799

44. Kendall, M.M., Gruber, C.C., Rasko, D.A., Hughes, D.T. and Sperandio, V. (2011) Hfq virulence regulation in enterohemorrhagic Escherichia coli O157:H7 strain 86-24. Journal of Bacteriology, 193, 6843-6851. doi:10.1128/JB.06141-11

Chapter 11

INFLUENCE OF INITIAL MOLECULAR SUBSTANCE ON THE DIFFUSION FLUX ACROSS CELL MEMBRANES

Bum Joon Jung[1,2] and Dae-Han Ki[2]

[1]Department of Biomedical Engineering, Rensselaer Polytechnic Institute, New York, USA

[2]Proton Therapy Center, National Cancer Center, Goyang, South Korea

ABSTRACT

The influence of initial placement of molecular or ion substance is investigated on the diffusion fluxes across the cell membrane. The diffusion fluxes and recovery curves are obtained by considering both the singlespot and double-spot concentrations inside the cell membrane. The results show that the additional concentration inside the membrane reduces the net fluxes at the cell interior as well as the exterior. In addition, it is found that the change in diffusion flux at the two outer walls of the membrane by the two-spot concentrations in the cell membrane is weaker than that of the single-spot concentration at the center. The variation of the influence of initial locations of the molecular concentrations inside the cell membrane on the diffusion fluxes is also discussed. This result can be applied to the diffusion process in avascular collagenous tissues.
Diffusion Flux; Cell Membrane; Recovery Curves

INTRODUCTION

A diffusion process has been one of the most interesting processes in physical sciences since the diffusion provides physical insights into the flux variation in various scientific areas, such as biology, chemistry, mathematics, and physics. In biological fields, the diffusion process is used in animal dispersal, bacteria motion, cell movement, chemical diffusion such as drug release in membrane and tissues, and plant electrophysiology [1-8]. Especially, the theory of transport into or out of a cell by the intra-cellular diffusion has received considerable attention since this process has wide applications in biological sciences. It has been shown that the diffusion in the human body is noteworthy over small distance scales on the order of $1-100 \mu m$ and the diffusion plays an important role in the metabolic activity [9]. It has been also shown that the diffusion phenomena in cells would be strongly biased by the influence of environments. Hence, it is expected that the diffusion in a cell membrane involving double lipid layers would be significantly influenced by the location of the initial location of a molecular solute. Thus, in this paper we investigate the influence of initial placement of molecular substance on the diffusion fluxes at outer walls of the cell membrane by using the theoretical analysis with the boundary conditions since the diffusion processes are significant for the characterization of cellular and physiological processes and for the operation of biomedical machines [10], such as the heart-lung bypass device, kidney dialysis device, and membrane oxygenator. The diffusion fluxes and recovery curves in the cell membrane are obtained by considering the single-spot and double-spot concentrations, respectively, inside the cell membrane. The influence of initial location of the concentration inside the cell membrane on the net diffusion fluxes at the outer walls of the cell membrane is also discussed.

This paper is composed as follows. In Section 2, the diffusion equation in the cell membrane with the detailed initial and boundary conditions and the Danckwert's method for the diffusion-reaction equation are discussed. In Section 3, we obtain the solutions of the diffusion equation for the single-spot and double-spot concentrations inside the cell membrane and the recovery curves for both single-spot and double-spot concentrations. In Section 4, we obtain the net flues at the wall boundary for the single-spot and double-spot concentrations inside the cell membrane. In Section 5, the influence of initial placement of molecular substance on the diffusion fluxes at outer walls of the cell membrane is discussed. In addition, we discuss the variations of the concentrations and recovery curves inside the cell membrane. Finally, the conclusions are given in Section 6.

DIFFUSION IN THE CELL MEMBRANE

From Fick's first and second equations for the flux $J(x,t)$, the diffusion equation [5] for the concentration $C(x,t)$ in the cell membrane is represented by

$$\frac{\partial}{\partial t} C(x,t) = D \frac{\partial^2}{\partial x^2} C(x,t), \tag{1}$$

where x is the position in one-dimension, t is the time, and D is the diffusion coefficient. It is known that a typical value of the diffusion coefficient for ions such as K^+, Cl^-, and Na^+ is about $2.5 \times 10^{-6} cm^2/s$ [11]. For the sake of the simplicity, we consider the one-dimensional diffusion problem for the investigation of the diffusion fluxes across cell membranes. In Equation (1), we retain only the diffusive flux term, but neglect the electrical drift effect since the influence of initial placement of molecular or ion substance on the diffusion fluxes at outer walls of the cell membrane is the main purpose of this work. Hence, we assume that the cell membrane has been localized in the domain of $0 < x < L$ and the molecular concentrations at the domain boundaries as the boundary conditions are $C(0,t) = C_0$ and $C(L,t) = C_L$. Two cases for the initial intracellular concentration as the initial conditions shall be considered: $C(x,0) = (C_a L)\delta(x-a)$ for the single-spot concentration case and $C(x,0) = (C_a L)\delta(x-a) + (C_b L)\delta(x-b)$ for the double-spot concentration case, where $\delta(x)$ is the Dirac delta function and a and b are initial locations of the molecular concentrations. If the reaction phenomenon is involved in the diffusion process, the righthand side of Equation (1) would have an extra term proportional to the concentration: $-kC(x,t)$, where k is a constant. In this case, the diffusion-reaction equation in the cell membrane is given by

$$\frac{\partial}{\partial t} C(x,t) = D \frac{\partial^2}{\partial x^2} C(x,t) - kC(x,t), \qquad (2)$$

where $kC(x,t)$ represents the rate of removal of diffusing substance. It has been also shown that the diffusion-reaction equation is useful for the mathematical modeling of glioma growth and also for the investigation on the spreading of brain tumors [12] and $kC(x,t)$ term corresponds to the death of cells in the tumor growth model. It has been well known that the general solution of Equation (2) can also be obtained the following transformation based on Danckwert's method [13, 14]:

$$C(x,t) = e^{-kt} C_1(x,t) + k \int_0^t dt' \, e^{-kt'} C_1(x,t'), \qquad (3)$$

where $C_1(x,t)$ is the solution of Equation (1) with the same boundaries. As a result, the standard diffusion equation, Equation (1), is used to examine the change in diffusion fluxes at outer walls of the cell membrane depended the initial molecular substance. In the following Section III, the solutions, $C(x,t)$ for the single-spot and for the double-spot concentrations inside the cell membrane are discussed.

SINGLE-AND DOUBLE-SPOT CONCENTRATIONS

Using the separation of variables method, the solution of the diffusion equation for the domain $0 < x < L$ would be represented by

$$C(x,t) = C_0 + (C_L - C_0)\frac{x}{L} + \sum_{n=1}^{\infty} A_n \sin(\lambda_n x) \exp(-D\lambda_n^2 t), \quad (4)$$

where $\lambda_n (= n\pi/L)$ is the separation constant determined by the boundary condition at $x = L$, A_n are coefficients to be determined by the initial condition at $t = 0$, and the first two terms $C_0 + (C_L - C_0)(x/L)$ represents the solution when the separation constant is zero. Since we are interested in the effect of the initial molecular concentration $C(x,0)$ inside the cell membrane, the coefficients A_n are determined by

$$A_n = \frac{2}{L}\int_0^L dx \left[C(x,0) - C_0 - (C_L - C_0)\frac{x}{L} \right] \times \sin\left(\frac{n\pi}{L}x\right), \quad (5)$$

where $C(x,0)$ would be, respectively, $(C_a L)\delta(x-a)$ for the single-spot concentration and $(C_a L)\delta(x-a) + (C_b L)\delta(x-b)$ for the double-spot concentrations with $0 < a < L$ and $0 < b < L$. The consideration of the initial concentration inside the cell membrane would be also quite useful to understand the behavior of the tumor growth since it has been known that the spatial arrangement effect is crucial for the spatial spread of the cancerous cells and also for the appearance of the tumor in the human body [12]. Hence, the general solutions for the diffusion equations with the single and double-spot concentrations are, respectively, found to be the following forms:

$$C_S(x,t) = C_0 + (C_L - C_0)\frac{x}{L} + 2\sum_{n=1}^{\infty} \left[C_a \sin\left(\frac{n\pi a}{L}\right) + \frac{1}{\pi}\frac{C_L \cos(n\pi) - C_0}{n} \right] \times \sin\left(\frac{n\pi x}{L}\right) \exp\left(-\frac{n^2\pi^2}{L^2}Dt\right), \quad (6)$$

$$C_D(x,t) = C_0 + (C_L - C_0)\frac{x}{L} + 2\sum_{n=1}^{\infty}\left[C_a \sin\left(\frac{n\pi a}{L}\right)\right.$$
$$\left. + C_b \sin\left(\frac{n\pi b}{L}\right) + \frac{1}{\pi}\frac{C_L \cos(n\pi) - C_0}{n}\right]$$
$$\times \sin\left(\frac{n\pi x}{L}\right)\exp\left(-\frac{n^2\pi^2}{L^2}Dt\right). \tag{7}$$

Once the concentration function $C(x,0)$ is obtained, the measurement of the concentration fluctuations would be determined by the signal correlation function $S(0)S(t)$ between the time at $t' = 0$ and $t' = t$, where $S(t) = A\int dx\, I(x)C(x,t)$, A is a factor related to the experimental detection, and $I(x)$ is the intensity of the laser beam [15]. A detailed investigation on the correlation spectroscopy for the influence of initial placement of molecular or ion substance on the concentration fluctuations will be treated elsewhere. The recovery curves for the single-spot $N_S(t)$ and two-spot $N_D(t)$ concentrations are then, respectively, obtained as follows:

$$N_S(t) = \int_0^L dx\, C_S(x,t) = C_0 L + (C_L - C_0)\frac{L}{2}$$
$$-\frac{2L}{\pi}\sum_{n=1}^{\infty}\left[C_a \sin\left(\frac{n\pi a}{L}\right) + \frac{1}{\pi}\frac{C_L \cos(n\pi) - C_0}{n}\right]$$
$$\times \left(\frac{\cos(n\pi) - 1}{n}\right)\exp\left(-\frac{n^2\pi^2}{L^2}Dt\right). \tag{8}$$

$$N_D(t) = \int_0^L dx\, C_D(x,t)$$
$$= C_0 L + (C_L - C_0)\frac{L}{2} - \frac{2L}{\pi}\sum_{n=1}^{\infty}\left[C_a \sin\left(\frac{n\pi a}{L}\right)\right.$$
$$\left. + C_b \sin\left(\frac{n\pi b}{L}\right) + \frac{1}{\pi}\frac{C_L \cos(n\pi) - C_0}{n}\right]$$
$$\times \left(\frac{\cos(n\pi) - 1}{n}\right)\exp\left(-\frac{n^2\pi^2}{L^2}Dt\right). \tag{9}$$

NET DIFFUSION FLUX IN THE CELL MEMBRANE

The diffusion fluxes at the wall boundaries $x=0$, i.e., J_0, and $x=L$, i.e., J_L, for the case of the single-spot concentration are, respectively, given by

$$J_0 = -D\frac{\partial C_S}{\partial x}\bigg|_{x=0}$$

$$= -(C_L - C_0)\frac{D}{L} - 2D\sum_{n=1}^{\infty}\left[C_a \sin\left(\frac{n\pi a}{L}\right)\right.$$

$$\left.+\frac{1}{\pi}\frac{C_L \cos(n\pi) - C_0}{n}\right] \times \left(\frac{n\pi}{L}\right)\exp\left(-\frac{n^2\pi^2}{L^2}Dt\right), \quad (10)$$

$$J_L = -D\frac{\partial C_S}{\partial x}\bigg|_{x=L} = -(C_L - C_0)\frac{D}{L}$$

$$-2D\sum_{n=1}^{\infty}\left[C_a \sin\left(\frac{n\pi a}{L}\right) + \frac{1}{\pi}\frac{C_L \cos(n\pi) - C_0}{n}\right]$$

$$\times\left(\frac{n\pi}{L}\right)\cos(n\pi)\exp\left(-\frac{n^2\pi^2}{L^2}Dt\right). \quad (11)$$

If we retain on the first term in the series, the net flux $|\Delta J_S|(\equiv|J_L - J_0|)$ for the single-spot concentration is then found to be

$$|\Delta J_S(a,t)| = \left|\frac{4D}{L}\left[(C_L - C_0) - C_a \pi \sin\left(\frac{\pi a}{L}\right)\right]\right.$$

$$\left.\times \exp\left(-\frac{\pi^2}{L^2}Dt\right)\right|. \quad (12)$$

As it is seen in Equation (12), the net flux in the cell membrane is completely determined by the location $\sin(\pi a/L)$ and amount C_a of the initial molecular concentration inside the cell membrane. Likewise, the diffusion fluxes at the wall boundaries $x=0$ and $x=L$ for the case of the double-spot concentrations are, respectively, obtained by

$$J_0 = -D\frac{\partial C_D}{\partial x}\bigg|_{x=0} = -(C_L - C_0)\frac{D}{L}$$

$$-2D\sum_{n=1}^{\infty}\left[C_a \sin\left(\frac{n\pi a}{L}\right) + C_b \sin\left(\frac{n\pi b}{L}\right)\right.$$

$$\left. +\frac{1}{\pi}\frac{C_L \cos(n\pi) - C_0}{n}\right] \times \left(\frac{n\pi}{L}\right)\exp\left(-\frac{n^2\pi^2}{L^2}Dt\right), \tag{13}$$

$$J_L = -D\frac{\partial C_D}{\partial x}\bigg|_{x=L} = -(C_L - C_0)\frac{D}{L}$$

$$-2D\sum_{n=1}^{\infty}\left[C_a \sin\left(\frac{n\pi a}{L}\right) + C_b \sin\left(\frac{n\pi b}{L}\right)\right.$$

$$\left. +\frac{1}{\pi}\frac{C_L \cos(n\pi) - C_0}{n}\right]$$

$$\times\left(\frac{n\pi}{L}\right)\cos(n\pi)\exp\left(-\frac{n^2\pi^2}{L^2}Dt\right). \tag{14}$$

If we also retain on the first term as we did in Equation (12), the net flux $|\Delta J_D|(=|J_L - J_0|)$ for the twospot concentrations is then obtained as

$$|\Delta J_D(a,b,t)|$$
$$= \left|\frac{4D}{L}\left[(C_L - C_0) - C_a\pi\sin\left(\frac{\pi a}{L}\right)\right.\right.$$
$$\left.\left. - C_b\pi\sin\left(\frac{\pi b}{L}\right)\right]\exp\left(-\frac{\pi^2}{L^2}Dt\right)\right|. \tag{15}$$

As shown in Equation (15), the net flux in the cell membrane is also completely controlled by the locations $\sin(\pi a/L)$ and $\sin(\pi b/L)$ and amounts C_a and C_b of the initial molecular concentrations inside the cell membrane.

Since the ion diffusion in biological membranes is one of the main transport processes [16], the results of $|\Delta J_S|$ and $|\Delta J_D|$, Equations (12) and (15), in this model analysis would provide useful information on the electric current and the membrane potential. Recently, the diffusional anisotropy has been investigated in avascular collagenous tissues since the diffusion in articular cartilage is the main molecular transport process [17]. Hence, the effects of the initial placement of molecular substance on the two-dimensional anisotropic diffusion process will be treated elsewhere.

INFLUENCE OF INITIAL ION PLACEMENT

In this section, the graph plotting for the 2D and 3D plots are obtained by the technical computing software Mathematica. Figure 1 represents the net flux $|\Delta J_s|$ for the initial single-spot concentration case as a function of the scaled position $\bar{a}(\equiv a/L)$ for various values of the diffusion time t when $C_a = \max\{C_0, C_L\}$. As it is seen, it is found that two non-flux positions for $|\Delta J_s|$ can be existed when the initial molecular concentration inside the cell membrane is equal to the larger one between two molecular concentrations at the domain boundaries. We have also found that that two non-flux positions are placed near the wall boundaries. In addition, it is found that the net flux $|\Delta J_s|$ decreases with an increase of the diffusion time t.

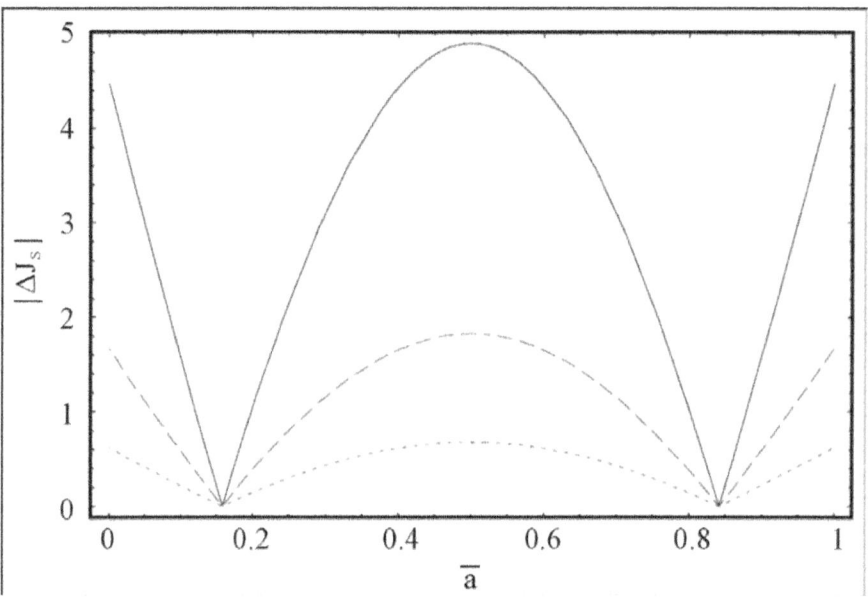

Figure 1. The net flux $|\Delta J_s|$ for the initial single-spot concentration case as a function of the scaled position $\bar{a}(\equiv a/L)$ when $C_0 = 2$, $C_L = 1$, and $C_a = 2$, i.e., $C_a = \max\{C_0, C_L\}$. The solid line represents the case of the scaled diffusion time $\bar{t}(\equiv tD/L^2) = 0.1$. The dashed line represents the case of $\bar{t} = 0.2$. The dotted line represents the case of $\bar{t} = 0.3$.

Figure 2 shows the net flux $|\Delta J_s|$ for the initial single-spot concentration case as a function of the scaled position $\bar{a}$ for various values of the diffusion time t when C_a is smaller than C_0 and C_L. As shown in this figure, it is interesting to note that the non-flux positions for $|\Delta J_s|$ are disappeared inside the cell membrane when the initial molecular concentration inside the cell membrane is smaller than two molecular concentrations at the wall boundaries. In addition, the net flux $|\Delta J_s|$ is found to be quite small at the center of the cell membrane.

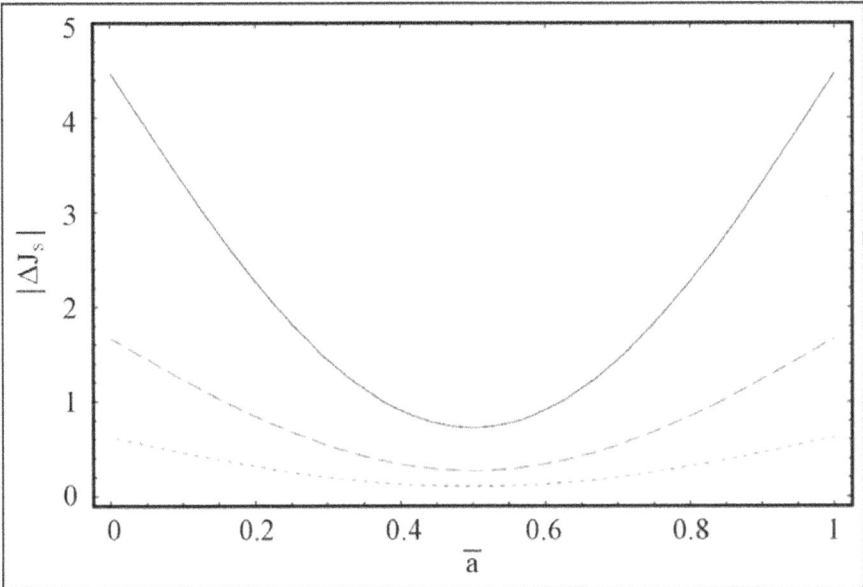

Figure 2. The net flux $|\Delta J_s|$ for the initial single-spot concentration case as a function of the scaled position $\bar{a}$ when $C_0 = 2$, $C_L = 1$, and $C_a = 0.8$, i.e., $C_a < \min\{C_0, C_L\}$. The solid line represents the case of the scaled diffusion time $\bar{t} = 0.1$. The dashed line represents the case of $\bar{t} = 0.2$. The dotted line represents the case of $\bar{t} = 0.3$.

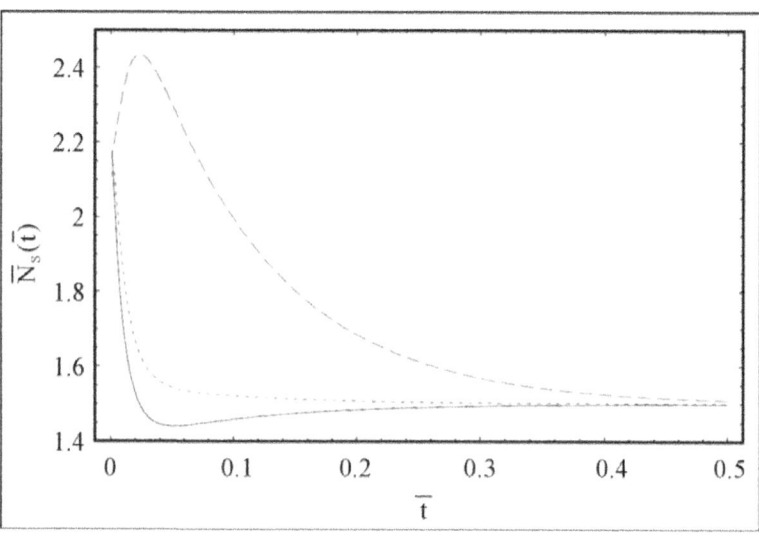

Figure 3. The scaled recovery curve $\bar{N}_s(\equiv N_s/L)$ for the initial single-spot concentration case as a function of the scaled diffusion time $\bar{t}$ when $C_0 = 2$, $C_L = 1$, and $C_a = 2$, i.e., $C_a = \max\{C_0, C_L\}$. The solid line represents the case of $\bar{a} = 1/7$. The dashed line represents the case of $\bar{a} = 1/2$. The dotted line represents the case of $\bar{a} = 5/6$.

Figure 3 represents the scaled recovery curve $\bar{N}_s(\equiv N_s/L)$ for the initial single-spot concentration case as a function of the scaled diffusion time $\bar{t}(\equiv tD/L^2)$ for various values of the initial location $\bar{a}$ when $C_a = \max\{C_0, C_L\}$. From this figure, it is found that the recovery curve $\bar{N}_s$ would have the maximum value when the initial single-spot concentration is placed near the right wall boundary and decreases with an increase of the diffusion time $\bar{t}$. However, we have found that the recovery curve $\bar{N}_s$ would have the minimum value when the initial single-spot concentration is placed near the left wall boundary and increases with an increase of the diffusion time $\bar{t}$. It is also found that that the recovery curve $\bar{N}_s$ decreases monotonically with an increase of the diffusion time when the initial single-spot concentration is placed at the center of the cell membrane.

Figure 4 shows the scaled recovery curve $\bar{N}_s$ for the initial single-spot concentration case as a function of the scaled diffusion time $\bar{t}$ for various values of the initial location $\bar{a}$ when $C_a = \min\{C_0, C_L\}$. As shown in this figure, it is found that the recovery curve $\bar{N}_s$ would have the minimum value when the initial single-spot concentration is placed near the left wall boundary at the center of the membrane and increases with an increase of the diffusion time $\bar{t}$. However, we have found that the recovery curve $\bar{N}_s$ has the maximum value when the initial single-spot concentration is placed near the right wall boundary and decreases with an increase of the diffusion time $\bar{t}$.

Figure 5 represents the surface plot of the net flux $|\Delta J_D|$ for the initial double-spot concentration case as a function of the scaled positions $\bar{a}(\equiv a/L)$ and $\bar{b}(\equiv b/L)$ when $C_a + C_b = \max\{C_0, C_L\}$.

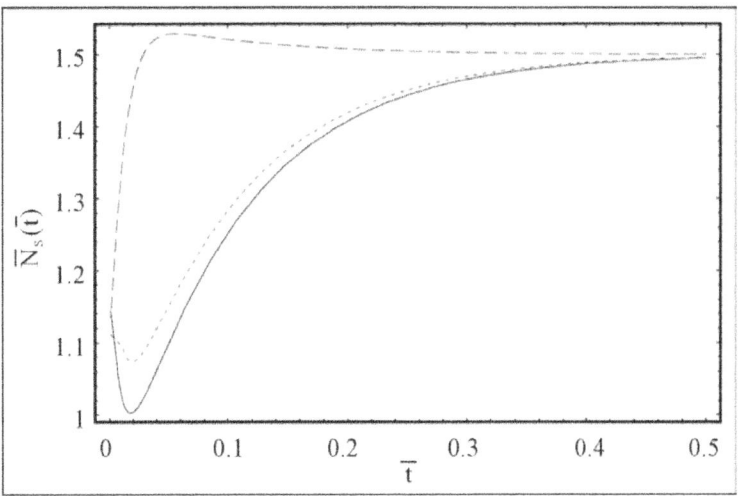

Figure 4. The scaled recovery curve $\bar{N}_s$ for the initial singlespot concentration case as a function of the scaled diffusion time $\bar{t}$ when $C_0 = 2$, $C_L = 1$, and $C_a = 1$, i.e., $C_a = \min\{C_0, C_L\}$. The solid line represents the case of $\bar{a} = 1/7$. The dashed line represents the case of $\bar{a} = 1/2$. The dotted line represents the case of $\bar{a} = 5/6$.

As we can see from this figure, it is found that two non-flux positions for $|\Delta J_D|$ can be existed for the scaled position $\bar{a}$ for a given $\bar{b}$ and also for the scaled position $\bar{b}$ for a given $\bar{a}$. It is also found that the two non-flux positions are located near the wall boundaries. In addition, the net flux $|\Delta J_D|$ has the maximum value at the center of the cell membrane.

Figure 6 represents the surface plot of the net flux $|\Delta J_D|$ for the initial double-spot concentration case as a function of the scaled positions $\bar{a}$ and $\bar{b}$ when the summation of concentrations $C_a + C_b$ is smaller than C_0 and C_L. As it is seen in this figure, it is found that the non-flux positions for $|\Delta J_D|$ are disappeared inside the cell membrane when the initial total molecular concentration $C_a + C_b$ inside the cell membrane is smaller than two molecular concentrations C_0 and C_L at the wall boundaries.

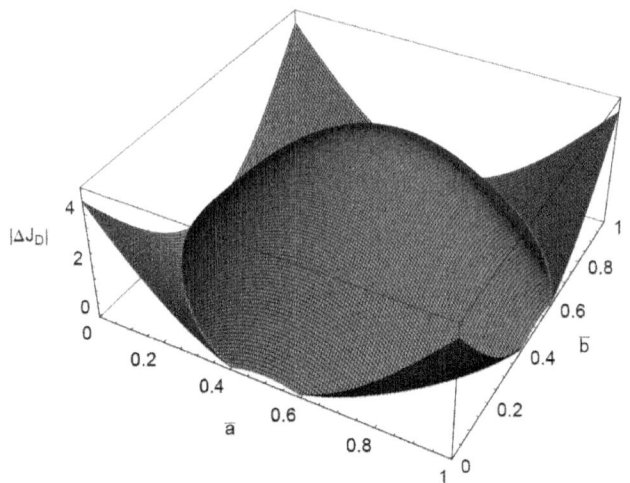

Figure 5. The surface plot of the net flux $|\Delta J_D|$ for the initial double-spot concentration case as a function of the scaled positions $\bar{a}(\equiv a/L)$ and $\bar{b}(\equiv b/L)$ when $\bar{t} = 0.1$, $C_0 = 2$, $C_L = 1$, $C_a = 1$, and $C_b = 1$, i.e., $C_a + C_b = \max\{C_0, C_L\}$.

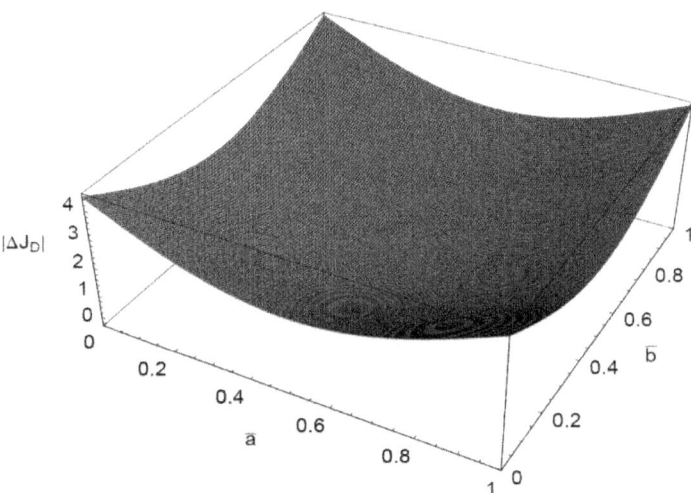

Figure 6. The surface plot of the net flux $|\Delta J_D|$ for the initial double-spot concentration case as a function of the scaled positions $\bar{a}$ and $\bar{b}$ when $\bar{t} = 0.1$, $C_0 = 2$, $C_L = 1$, $C_a = 0.4$, and $C_b = 0.4$, i.e., $C_a + C_b < \min\{C_0, C_L\}$.

The net flux $|\Delta J_D|$ for the initial double-spot concentration case is found to be also quite small at the center of the cell membrane.

Figure 7 shows the scaled recovery curve $\bar{N}_D$ for the initial double-spot concentration case as a function of the scaled diffusion time $\bar{t}$ for various values of the scaled positions $\bar{a}$ and $\bar{b}$ when $C_a + C_b = \max\{C_0, C_L\}$. As it is seen in this figure, the recovery curves $\bar{N}_D$ have maximum values at very short diffusion times $\bar{t}$. It is also found that the maximum time of the recovery curve has been increased as the two positions are getting close to the center of the cell membrane.

Figure 8 represents the scaled recovery curve $\bar{N}_D$ for the initial double-spot concentration case as a function of the scaled diffusion time $\bar{t}$ for various values of the scaled positions $\bar{a}$ and $\bar{b}$ when

$C_a + C_b = \min\{C_0, C_L\}$. From this figure, we have found that the recovery curve $\bar{N}_D$ has no maximum and increases and finally saturates with increasing diffusion time $\bar{t}$ when the two locations $\bar{a}$ and $\bar{b}$ are away from the center of the membrane. However, it is found that the recovery curve $\bar{N}_D$ has the maximum value at the small diffusion time and saturates with an increase of the diffusion time $\bar{t}$.

CONCLUSIONS

In this work, we inspected the influence of initial placement of molecular or ion substance on the diffusion fluxes at outer walls of the cell membrane. The recovery curves in the cell membrane were obtained for the single-spot and double-spot concentrations, respectively.

The net diffusion fluxes were obtained by considering both single-spot and double-spot concentrations across the cell membrane. For the initial single-spot concentration case, it was found that two non-flux positions would exist near the wall boundaries if the initial molecular concentration inside the cell membrane were equal to the larger one between the two molecular concentrations at the domain boundaries. It is also found that the two non-flux positions were placed near the center of the cell membrane when the initial molecular concentration inside the cell membrane was equal to the smaller one between the two molecular concentrations at the wall boundaries.

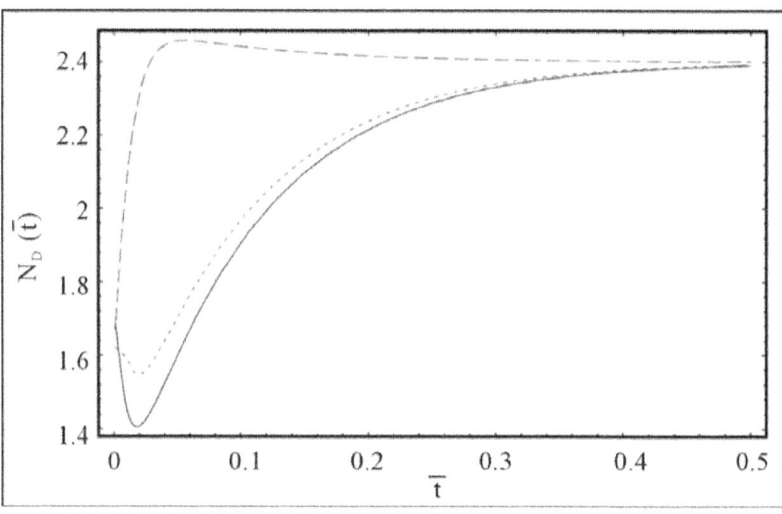

Figure 7. The scaled recovery curve $\bar{N}_D$ for the initial double-spot concentration case as a function of the scaled diffusion time $\bar{t}$ when $\bar{t} = 0.1$, $C_0 = 2$, $C_L = 1$, $C_a = 1$, and $C_b = 1$, i.e., $C_a + C_b < \max\{C_0, C_L\}$. The solid line represents the case of $\bar{a} = 1/4$ and $\bar{b} = 3/4$. The dashed line represents the case of $\bar{a} = 4/9$ and $\bar{b} = 5/9$. The dotted line represents the case of $\bar{a} = 1/3$ and $\bar{b} = 2/3$.

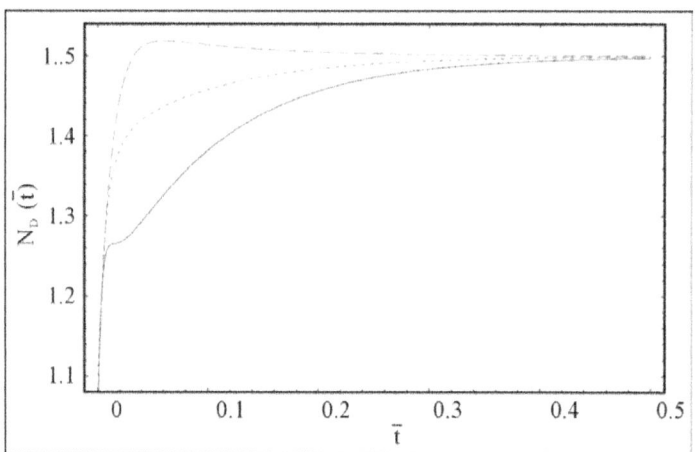

Figure 8. The scaled recovery curve $\bar{N}_D$ for the initial double-spot concentration case as a function of the scaled diffusion time $\bar{t}$ when $\bar{t} = 0.1$, $C_0 = 2$, $C_L = 1$, $C_a = 0.5$, and $C_b = 0.5$, i.e., $C_a + C_b < \min\{C_0, C_L\}$. The solid line represents the case of $\bar{a} = 1/4$ and

$\bar{b} = 3/4$. The dashed line represents the case of $\bar{a} = 4/9$ and $\bar{b} = 5/9$. The dotted line represents the case of $\bar{a} = 1/3$ and $\bar{b} = 2/3$.

In addition, we have found that the non-flux positions disappeared inside the cell membrane when the initial molecular concentration was smaller than two molecular concentrations at the wall boundaries. The molecular concentration C_s could be increased more than C_0 when the initial single-spot concentration was placed near the left wall boundary and $C_a = \max\{C_0, C_L\}$. However, we have found that the molecular concentration monotonically decreases with an increase of the position when the initial single-spot concentration is placed at the center or near the right wall boundary. Moreover, we have found that the molecular concentration can be smaller than C_L when the initial single-spot concentration was placed near the left wall boundary. Thus, the recovery curve would have the maximum vale when $C_a = \max\{C_0, C_L\}$ and the initial single-spot concentration was placed near the right wall boundary and the recovery curve decreases with an increase of the diffusion time. The minimum value would be when $C_a = \min\{C_0, C_L\}$ and the initial single-spot concentration was placed near the left wall boundary at the center of the membrane and increases with an increase of the diffusion time. For the initial double-spot concentration case, it was found that two non-flux positions were located near the wall boundaries when $C_a + C_b = \max\{C_0, C_L\}$. It is also found that the net flux had the maximum value at the center of the cell membrane. We found two non-flux positions could also be existed near the center of the cell membrane when $C_a + C_b = \min\{C_0, C_L\}$ and the net flux was still quite small at the center of the cell membrane. However, we have found that the non-flux positions were disappeared inside the cell membrane when $C_a + C_b < \min\{C_0, C_L\}$ and the net flux for the initial double-spot concentration case was found to be also quite small at the center of the cell membrane. We have found that the molecular concentration C_D had the maximum value when two positions were near the center of the cell membrane when $C_a + C_b = \max\{C_0, C_L\}$. However, we have found that the molecular concentration decreased with an increase of the distance $\bar{x}$ regardless of the locations of the initial double-spot concentrations when

$C_a + C_b = \min\{C_0, C_L\}$. It is found that the recovery curves had maximum values at very short diffusion times and the recovery curve increased as the two positions were getting close to the center of the cell membrane when

$C_a + C_b = \max\{C_0, C_L\}$. It is found that the recovery curve at first had no maximum, but increased and finally saturated with increasing diffusion time when the two initial locations were away from the center of the membrane. It had the maximum value at the small diffusion time and saturated with an increase of the diffusion time when $C_a + C_b = \min\{C_0, C_L\}$. From this work, we have found that the influence of initial locations and amounts of the molecular substance plays a significant role on the diffusion fluxes across the cell membrane. Since the results are obtained in the analytic expressions with the appropriate physical conditions, the equations in this work would be quite reliable for understanding the diffusion process in biological systems. Hence, the results of this work would be useful for understanding the characteristics and properties of the intracellular diffusion process as well as the spread of the cancerous cell and appearance of the tumor. The results in this work can also be applied to the enhancement of the diffusion process in articular cartilage. Moreover, these results would be applied to the biomedical machines related to the biological diffusion processes including the initial ion concentrations inside the wall boundaries.

ACKNOWLEDGEMENTS

The authors gratefully acknowledge Prof. Y.-K. Lim for useful discussions and encouragements while visiting the Proton Therapy Center at National Cancer Center of South Korea. This research was initiated while one of the authors (B. J. Jung) from Rensselaer Polytechnic Institute, USA was affiliated with the Proton Therapy Center at National Cancer Center as a Summer Intern Scholar.

REFERENCES

1. Berg, H.C. and Purcell, E.M. (1967) A method for separating according to mass a mixture of macromolecules or small particles suspended in a fluid, I. Theory. Proceedings of the National Academy of Sciences USA, 58, 862- 869.http://dx.doi.org/10.1073/pnas.58.3.862
2. Berg, H.C. and Purcell, E.M. (1977) Physics of chemoreception. Biophysical Journal, 20, 193-219. http://dx.doi.org/10.1016/S0006-3495(77)85544-6
3. Iwasa, Y. and Teramoto, E. (1984) Branching-diffusion model of the formation of a population's distributional pattern. Journal of Mathematical Biology, 19, 109-124.http://dx.doi.org/10.1007/BF00275934
4. Berg, H.C. (1993) Random walks in biology. Expanded Edition, Princeton University Press, Princeton.

5. Jones, D.S. and Sleeman, B.D. (2000) Differential equations and mathematical biology. Chapman & Hall, London.
6. Murray, J.D. (2001) Mathematical biology, vol. I: An introduction. 3rd Edition, Springer, Berlin.
7. Caputo, M. and Cametti, C. (2007) Diffusion with memory in two cases of biological interest. Journal of Theoretical Biology, 254, 697-703. http://dx.doi.org/10.1016/j.jtbi.2008.06.021
8. Niklas, K.J. and Spatz, H.-C. (2012) Plant physics. The University of Chicago Press, Chicago. http://dx.doi.org/10.7208/chicago/9780226586342.001.0001
9. Herman, I.P. (2008) Physics of human body. Springer, Berlin.
10. Truskey, G.A., Yuan, F. and Katz, D.F. (2004) Transport phenomena in biological systems. Pearson Prentice Hall, Upper Saddle River.
11. Ermentrout, G.B. and Terman, D.H. (2010) Mathematical foundations of neuroscience. Springer, Berlin. http://dx.doi.org/10.1007/978-0-387-87708-2
12. Murray, J.D. (2003) Mathematical biology, vol. II: Spatial models and biomedical applications. 3rd Edition, Springer, Berlin.
13. Danckwerts, P.V. (1951) Absorption by simultaneous diffusion and chemical reaction into particles of various shapes and into falling drops. Transactions of the faraday society, 47, 1014-1023. http://dx.doi.org/10.1039/tf9514701014
14. Crank, J. (1975) The mathematics of diffusion. 2nd Edition. Oxford University Press, Oxford.
15. Jackson, M.B. (2006) Molecular and cellular biophysics. Cambridge University Press, Cambridge. http://dx.doi.org/10.1017/CBO9780511754869
16. Scherer, P.O.J. and Fischer, S.F. (2010) Theoretical molecular biophysics. Springer, Berlin. http://dx.doi.org/10.1007/978-3-540-85610-8
17. Leddy, H.A., Haider, M.A. and Guilak, F. (2006) Diffusional anisotropy in collagenous tissues: Fluorescence imaging of continuous point photobleaching. Biophysical Journal, 91, 311-316. http://dx.doi.org/10.1529/biophysj.105.075283

CITATION

CHAPTER 1
Chauhan, T. and Rajiv, K. (2010) Molecular markers and their applications in fisheries and aquaculture. Advances in Bioscience and Biotechnology, 1, 281-291. doi: 10.4236/abb.2010.14037.

CHAPTER 2
Teich, I. , Verga, A. and Balzarini, M. (2014) Assessing spatial genetic structure from molecular marker data via principal component analyses: A case study in a Prosopis sp. forest. Advances in Bioscience and Biotechnology, 5, 89-99. doi: 10.4236/abb.2014.52013.

CHAPTER 3
Robles, F. , Cano-Roldán, B. , Rejón, C. , Martínez-González, L. , Álvarez-Cubero, M. , Lorente, J. , Cantal, J. , Hoyos, P. , Rus, J. , Sánchez, M. , Vallejo, M. , Rejón, M. and Herrán, R. (2010) Determining the specific status of the Iberian sturgeons by means genetic analyses of old specimens. Advances in Bioscience and Biotechnology, 1, 171-179. doi: 10.4236/abb.2010.13024.

CHAPTER 4
H. Y. Yeap, G. Faruq, H. P. Zakaria, and J. A. Harikrishna, "The Efficacy of Molecular Markers Analysis with Integration of Sensory Methods in Detection of Aroma in Rice," The Scientific World Journal, vol. 2013, Article ID 569268, 6 pages, 2013. doi:10.1155/2013/569268

CHAPTER 5

A'wani Aziz Nurdalila, Hamidun Bunawan, Subbiah Vijay Kumar, Kenneth Francis Rodrigues, and Syarul Nataqain Baharum. Homogeneous Nature of Malaysian Marine Fish Epinephelus fuscoguttatus(Perciformes; Serranidae): Evidence Based on Molecular Markers, Morphology and Fourier Transform Infrared Analysis, Int. J. Mol. Sci. 2015, 16(7), 14884-14900; doi:10.3390/ijms160714884

CHAPTER 6

Yong-An Yang, Wen-Jian Tang, Xin Zhang, Ji-Wen Yuan, Xin-Hua Liu, and Hai-Liang Zhu. Synthesis, Molecular Docking and Biological Evaluation of Glycyrrhizin Analogs as Anticancer Agents Targeting EGFR, Molecules 2014, 19(5), 6368-6381; doi:10.3390/molecules19056368

CHAPTER 7

Matsunaga, G., Karasuda, S., Nishino, R., Fukushima, H. and Matsumiya, M. (2016) Molecular Cloning of a Chitinase Gene from the Ovotestis of Kuroda's Sea Hare Aplysia kurodai. Advances in Bioscience and Biotechnology, 7, 38-46. doi: 10.4236/abb.2016.71005.

CHAPTER 9

Chang, C. , Liu, X. and Chen, K. (2011) Molecular cloning, expression and characterization of a novel geneβ-N-acetylglucosaminidase from Bombyxmori. Advances in Bioscience and Biotechnology, 2, 123-127. doi:10.4236/abb.2011.23019.

CHAPTER 10

Globus, T. , Sizov, I. and Gelmont, B. (2013) Terahertz vibrational spectroscopy of E. coli and molecular constituents: Computational modeling and experiment. Advances in Bioscience and Biotechnology, 4, 493-503. doi: 10.4236/abb.2013.43A065.

CHAPTER 11

Jung, B. and Ki, D. (2014) Influence of initial molecular substance on the diffusion flux across cell membranes.Advances in Bioscience and Biotechnology, 5, 169-176. doi: 10.4236/abb.2014.53021.

INDEX

A

Acinipo scute 58
Acipenser sturio 50, 63, 65, 66
Allele Specific Amplification (ASA) 68, 70, 71, 73
anal fin length (AFL) 85
ariculture 241
Artificial Insemination (AI) 182
Attenuated Total Reflectance-Fourier-transform Infrared (ATR-FTIR) 81
autonomously replicating sequence (ARS) 165

B

Bacterial Artificial Chromosomes (BAC) 165
biochemistry methods 252
body weight (BW) 85

C

caudal fin length (CFL) 85
cell membrane 275, 276, 277, 278, 280, 281, 282, 283, 284, 285, 286, 287, 289
Cell Membrane 275
chitin binding domains (CBDs) 133
Convention on Biological Diversity (CBD) 141
cytochrome oxidase I (COI) 7

D

deoxyribonucleic acid (DNA) 143
diaminobenzidine 244
Diffusion Flux 275
dorsal fin length (DFL) 85

E

electro-magnetic (EM) 257
embryo transfer (ET) 182
enzymes 142, 143, 145, 151, 158, 161, 162, 163, 174, 177, 200, 201, 203, 205, 231
epidermal growth factor receptor (EGFR) 103, 124
Epinephelus 79, 80, 81, 86, 88, 96, 97, 98, 99, 294
Epinephelus fuscoguttatus 79, 81, 86, 88, 97, 98, 99, 294
Epinephelus hexagonatus 79, 81, 86, 88
Expressed Sequence Tags (EST) 6
External Antisense Primer (EAP) 71
External Sence Primer (ESP) 71

F

fluorescein isothiocyanate (FITC) 245
Fork length (FL) 85, 86, 87
Fourier Transform (FT) 254

G

genes 143, 145, 146, 147, 151, 152, 153, 154, 155, 156, 158, 159, 160, 161, 162, 163, 165, 166, 167, 169, 173, 177, 178, 179, 182, 184, 185, 186, 187, 188, 189, 192, 193, 194, 197, 202, 203, 204, 205, 213, 215, 219, 223, 225, 226, 227, 228, 229, 234, 235, 236
genetically modified organisms (GMOs) 144
Genetic variation 1, 2, 9, 16, 20, 22, 25
genotyping technology 27
glycoside hydrolase (GH) 127
glycyrrhetinic acid monoglucuronides 103
Glycyrrhizin (GA) 103
green fluorescent protein (GFP) 177
Green Fluorescent Protein (GFP) 179

H

head length (HL) 85
hepatocellular carcinoma (HCC) 104
heteroduplex analysis 5
horseradish peroxidase (HRP) 244

I

Iberian Peninsula sturgeons 50
Internal Fragrant Antisence Primer (IFAP) 71
Internal Nonfragrant Sence Primer (INSP 71

M

marker-assisted selection (MAS) 181
Maximum Parsimony (MP) 81
microbial species 252

Mitochondrial DNA 3, 14, 15
molecular 275, 276, 277, 278, 279, 280, 281, 282, 283, 286, 287, 290, 291, 294
molecular biology 28, 252
molecular dynamics (MD) 251
Molecular Evolution (ME) 81
mouth length (ML) 85
multiple cloning site (MCS) 163, 168

N

Neighbor-Joining (NJ) 81
nuclear transfer (NT) 191

O

open reading frame (ORF 242
open reading frame (ORF) 131, 151

P

partial hepatectomy (PH) 104
pectoral fin length (PFL) 85
Peptide mass fingerprinting (PMF) 244
phenomena 276, 291
polyethylene glycol (PEG) 174
polymerase chain reaction (PCR) 2, 7
Polymerase chain reaction (PCR) 212
Principal Component analysis (PCA) 27
Principal Components (PC) 31
proteinogenic amino acids 150

Q

quantitative trait loci (QTL) 182

R

Recovery Curves 275
restriction fragment length polymorphism (RFLP) 2
reverse transcription-polymerase chain reaction (RT-PCR) 127
ribonucleic acid (RNA) 150

S

selectable marker gene (SMG) 165, 168

Selectable Marker Genes (SMG) 177
Simple Sequence Repeat (SSR) 68
single nucleotide polymorphism (SNP) 2
Single nucleotide polymorphism (SNP) 5
Single Nucleotide Polymorphism (SNPs) 68
Single Nucleotide Polymorphisms (SNPs) 29
sodium dodecyl sulfate-polyacrylamide gel electrophoresis (SDS-PAGE) 131
Somatic Cell Nuclear Transfer (SCNT) 188
Spatial genetic structure (SGS) 28
spatial Principal Components (sPC) 31
standard length (SL) 85
sturgeon species 50, 51, 54, 57, 58, 61, 62

T

Terahertz 251, 252, 253, 259, 270, 271, 272, 273
Tobacco Mosaic Virus (TMV) 175
total length (TL) 85
tyrosine kinase receptors (TKRs) 103

U

untranslated regions (UTRs) 154

Y

Yeast Artificial Chromosomes (YAC) 165